Springer Tracts in Modern Physics
Volume 168

Springer
Berlin
Heidelberg
New York
Barcelona
Hong Kong
London
Milan
Paris
Singapore
Tokyo

Springer Tracts in Modern Physics

Springer Tracts in Modern Physics provides comprehensive and critical reviews of topics of current interest in physics. The following fields are emphasized: elementary particle physics, solid-state physics, complex systems, and fundamental astrophysics.
Suitable reviews of other fields can also be accepted. The editors encourage prospective authors to correspond with them in advance of submitting an article. For reviews of topics belonging to the above mentioned fields, they should address the responsible editor, otherwise the managing editor.
See also http://www.springer.de/phys/books/stmp.html

Ulrich F. Katz

Deep Inelastic Positron–Proton Scattering in the High-Momentum-Transfer Regime of HERA

With 96 Figures

Springer

Dr. Ulrich F. Katz

University of Bonn
Institute of Physics
Nussallee 12
53115 Bonn, Germany
E-mail: katz@physik.uni-bonn.de

Library of Congress Cataloging-in-Publication Data.

Katz, Ulrich F.: Deep inelastic positron-proton scattering in the high-momentum-transfer regime of HERA/Ulrich F. Katz. - Berlin; Heidelberg; New York; Barcelona; Hong Kong; London; Milan; Paris; Singapore; Tokyo: Springer, 2000. (Springer tracts in modern physics: Vol. 168) (Physics and astronomy online library) ISBN 3-540-67708-9

Physics and Astronomy Classification Scheme (PACS):
01.30.Ee, 12.15.-y, 12.15.Ji, 12.15.Mm, 12.38.-t, 12.38.Bx, 12.38.Qk, 12.60.-i, 12.60.Jv, 13.60.-r, 13.60.Hb, 14.80.-j, 29.40.Gx, 29.40.Vj

ISSN print edition: 0081-3869
ISSN electronic edition: 1615-0430
ISBN 3-540-67708-9 Springer-Verlag Berlin Heidelberg New York

Springer-Verlag Berlin Heidelberg New York
a member of BertelsmannSpringer Science+Business Media GmbH

Printed in Germany

Typesetting: Camera-ready copy by the author using a Springer LaTeX macro package
Cover design: *design & production* GmbH, Heidelberg

Printed on acid-free paper SPIN: 10717104 56/3141/tr 5 4 3 2 1 0

Preface

About three decades after the first experiments on deep inelastic lepton–hadron scattering began to investigate the structure of hadrons, the history of this fruitful field of particle physics continues in the broad spectrum of research performed at the electron– and positron–proton collider HERA at DESY, where the multipurpose detectors ZEUS and H1 access ep scattering at a center-of-mass energy of 300 GeV and explore as yet uncharted kinematic realms of deep inelastic scattering.

After the first years of data taking at HERA, each of the experiments has collected a total of roughly $40\,\mathrm{pb}^{-1}$ of e^+p data, yielding sensitivity to deep inelastic e^+p interactions at high four-momentum transfers, Q^2, where typical cross sections drop into the subpicobarn regime. This kinematic domain is characterized by electroweak unification, manifesting itself most markedly in the neutral- and charged-current cross sections, which approach an equal order of magnitude as Q^2 rises above the square of the W and Z masses. Consequently, HERA allows, for the first time, studies of both types of processes simultaneously with the same initial-state conditions and in the same detector, and thus we can investigate the interplay of electroweak and strong forces governing the respective cross sections.

Positron–proton reactions at HERA involve positron–quark interactions at center-of-mass energies up to more than 200 GeV, providing an ideal scenario in which to search for phenomena beyond the current standard model of particle physics. In 1997, both ZEUS and H1 reported an excess of events at the highest Q^2 over predictions derived from lower-energy measurements which were extrapolated to the HERA regime by using the standard model. These observations have aroused worldwide interest and have triggered intense work both on theoretical aspects and on investigations of the experimental data.

This report summarizes the theoretical concepts related to deep inelastic scattering at high Q^2 and compares the resulting predictions with the latest experimental results on neutral-current and charged-current e^+p cross sections with Q^2 above some $100\,\mathrm{GeV}^2$. Particular emphasis is laid on the description of theoretical scenarios of processes beyond the standard model which might become visible in deep inelastic scattering at HERA, and on the experimental searches for the corresponding signatures. A detailed account is given of the experimental techniques and analysis strategies applied at HERA, and of the evidence provided by the data.

The roots of this work go back to 1995/96, when I had the chance to coordinate the ZEUS physics working group "High Q^2 and Exotics" and met many of those colleagues whose critical judgment and excellence in research are the basis on which many of the results collected in this book have been acquired. I wish to thank H. Abramowicz, D. Acosta, M. Botje, A. Caldwell, R. Cashmore, P. Iacobucci, R. Klanner, D. Krakauer, M. Kuze, R. Nania, A. Quadt, S. Schlenstedt, F. Sciulli, B. Straub, G. Wolf and A. F. Żarnecki for fruitful collaboration and many enlightening discussions, and to apologize to the many others whose names are missing here but to whom I'm no less grateful.

This work was performed in the context of the ZEUS group at the University of Bonn. I thank Prof. E. Hilger, head of the ZEUS group, for his unswerving support, encouragement and confidence, and for his warmhearted and open-minded attitude, which creates an inspiring atmosphere in his group. The individual members of the group have accompanied my struggle to complete the work with sympathy and help. I want to summarily express my gratitude for friendship and good cooperation to all of them.

The research activities of the ZEUS group in Bonn are supported by the German Federal Ministry for Education and Research (BMBF).

I am indebted to J. Crittenden, M. Erdmann, E. Hilger, A. Quadt, S. Schlenstedt and H. Spiesberger for the careful reading of the manuscript or parts of it, for advising valuable improvements and for pointing out mistakes and omissions.

None of the research reported here would have been possible without the dedicated efforts of the HERA machine physicists and technicians, and of all those who contributed to the design, construction, maintenance and operation of the detectors.

Last but not least, a sincere "thank you" to my wife Uschi and my son Jonas, who have given me so much love and backup, and who patiently endured the long times of my absence and occupation with such silly things like quarks. As a little sign of my gratitude I dedicate this book to them.

Bonn, June 2000 *Ulrich F. Katz*

Contents

1. Introduction

The great achievements in particle accelerator science over the past several decades have permitted physicists to study structures and interactions of particles at ever smaller distance scales. The field of deep inelastic scattering (DIS) of leptons on hadrons started out with the experiments at SLAC in the late 1960s, which observed a granular substructure of the nucleon and paved the way for the success of the quark–parton model. Remarkable progress was later achieved by a series of fixed-target experiments at CERN and FNAL with electron, muon and (anti)neutrino beams of energies up to several hundred GeV.[1]

A breakthrough to new frontiers became possible when the first-ever lepton–hadron collider, HERA, at the DESY laboratory came into operation in 1992. HERA provides head-on collisions of electrons or positrons with protons at a center-of-mass energy of approximately 300 GeV, an order of magnitude above previous experiments. Two multipurpose detectors, ZEUS and H1, are installed at HERA to analyze the electron–proton (ep) reactions.[2] By the end of 1997, both detectors had recorded data (mostly e^+p) corresponding to integrated luminosities of $48.9\,\mathrm{pb}^{-1}$ and $42.7\,\mathrm{pb}^{-1}$, respectively.

These data have yielded a wealth of new information about a variety of ep interaction classes, which are interrelated by the common principles governing the dynamics of what we consider to be the elementary building blocks of matter. One intuitive way to categorize ep reactions is by the signatures they leave in the detectors. Important reaction classes of this kind – which are not necessarily mutually exclusive – are, e.g., neutral-current (NC) DIS (the scattered electron is identified in the central detector), charged-current (CC) DIS (the final-state neutrino causes large missing transverse momentum), photoproduction (interaction of the proton with a quasi-real photon emitted by the electron, the latter escaping through the beam hole), diffractive scattering (the proton remains intact, there is no energy flow in a cone around the proton direction) and production of exclusive final states like vector mesons. As may be anticipated, ep reactions probe different aspects of the dynamics and

[1] Throughout this book, we use $\hbar = c = 1$ and multiples of electronvolts (eV) as units for energies, momenta and masses. In particular, $1\,\mathrm{GeV} = 10^9\,\mathrm{eV}$.

[2] The term *electron* will generically denote both e^- and e^+, unless explicitly stated otherwise. Similarly, *neutrino* will be used both for ν and for $\bar{\nu}$.

structure of the proton, photon and electron and therefore allow us to study both the electroweak and the strong sectors of the standard model (SM).

From analyzing differential cross sections and final-state topologies of various *ep* reaction classes, an impressive spectrum of new results has been derived and published. Prominent amongst these are measurements of the proton structure functions [1–11], of multijet production in both photoproduction and deep inelastic reactions [12–26], of inclusive cross sections and final-state properties of diffractive processes [27–35], of exclusive vector-meson production [36–39] and of total photoproduction cross sections [40, 41]. The first two categories open unique new testing grounds for quantum chromodynamics (QCD, the SM gauge theory of strong interactions) in its perturbative regime. However, one can also study these two categories and the other reaction classes in kinematic regimes where they receive contributions from processes without "hard" scales where the QCD perturbation series is not expected to converge, and thus investigate the transition to phenomenological models like the Regge formalism. Examples of such border areas, into which the HERA experiments have provided new insights, are DIS at very low four-momentum transfers between electron and proton, the partonic aspects of diffraction and the dynamics of exclusive vector-meson production. The physics potential of HERA has been comprehensively explored in the workshops *Physics at HERA* held in 1991 [42] and *Future Physics at HERA* held in 1995/96 [43]. Recent reviews cover the investigation at HERA of *ep* collider physics [44], DIS [45], structure functions [46], the photon structure [47], QCD [48] and exclusive vector-meson production [49].

Both HERA collaborations are also investigating *ep* DIS at four-momentum transfers far beyond the reach of previous fixed-target experiments. The distinctive feature of these reactions is that they are mediated by the t-channel exchange of a space-like gauge boson, with virtuality scales reaching beyond the masses of the heavy gauge bosons W and Z. NC and CC interactions at such four-momentum transfers occur at comparable rates, and the study of DIS in this kinematic regime hence provides access to manifestations of the unification of electromagnetic and weak forces. The first ZEUS and H1 measurements of the NC and CC cross sections at high four-momentum transfers [50–54] still suffered from large statistical uncertainties, but agreed within errors with the predictions derived from lower-energy structure function results which were extrapolated to the HERA domain using the SM prescription. Also, no signs of new phenomena were found in searches for nonstandard electron–quark interactions [55–65] and for other hypothesized non-SM processes [66–70]. However, the increased data samples and refined analysis methods available after the 1996 running period revealed an indication of possible deviations from the SM: both experiments observed an excess of events at very high four-momentum transfers and at electron–quark invariant masses of roughly 200 GeV [71, 72]. Meanwhile, results from analyses of the complete 1994–97 data sets are available (albeit partly preliminary) on high-Q^2 DIS cross sections [73–75] and on searches for phenomena beyond the SM [76–88]. These measurements do not corroborate the previously observed excesses, but on the other hand they also cannot exclude the existence of anomalies. The situation has hence arrived at a point where conclusive statements about the existence and origin of non-SM signals in ep scattering appear to be possible only in the distant future. In addition, an immense spectrum of phenomenological work points into different directions which have to be illuminated by experimental facts.

In this situation, it seems appropriate to take stock of the status of experimental investigation of DIS reactions in the regime of high momentum transfer which has been made accessible by HERA (Q^2 above some $100\,\mathrm{GeV}^2$), of theoretical suggestions, and of the many bits and pieces which were used to construct the "established knowledge" against which the experimental data is tested. In Chap. 2 we first review the ingredients of the DIS cross section predictions in the SM framework and discuss a selection of those hypothesized scenarios beyond the SM which would manifest themselves in reactions with DIS-like event topologies. In Chap. 3 we then summarize the relevant experimental techniques and describe the data samples and, finally, present in Chap. 4 the experimental results and their comparisons with predictions and models. We cover e^+p measurements publicized up to the beginning of 2000, and only in a few places allude to preliminary e^-p results derived from the 1998 and 1999 data. An outlook on the future experimental investigation of DIS and related processes (Chap. 5) and a summary (Chap. 6) conclude the book. Before embarking on this program, a few more introductory pages will provide a synopsis of the historical achievements of DIS experiments, a summary of the HERA parameters and performance since 1993, and an introduction to DIS kinematics and typical event topologies, defining the notions used in the following sections.

1.1 History of Deep Inelastic Scattering

Deep inelastic scattering in the contemporary sense implies accelerator-based experiments with center-of-mass energies sufficiently large to produce multiparticle final states. Even so, the first successful application of the basic principle of DIS was the famous experiment by E. Rutherford et al. [89] which dates from 1911, when the highest particle energies available for experimentation were those released in radioactive decays. Bombarding a target of unknown structure (a gold foil) with a beam of α-particles, Rutherford deduced from the distribution of scattering angles the existence of "pointlike" charged structures inside the gold which carry almost all of the bulk mass – the discovery of the atomic nucleus.

Five decades later, the fundamental tenets of relativity and quantum mechanics, and their development to relativistic quantum field theory, had established the theoretical basis on which particle reactions were described. In the late 1950s, the experimental verification of parity violation in beta decay [90] and the postulation of the $V-A$ theory as an improved extension of Fermi's description of four-fermion interactions [91] had laid the foundation of the theory of weak interactions. At that time, the spectrum of known particles comprised the electron and the muon and their antiparticles, the neutrino, the nucleons and a variety of further hadrons, including some carrying strangeness. Gell-Mann and Ne'eman [92] suggested in 1961 the "Eightfold Way" as an ordering scheme for the hadrons, and 1964 it was postulated[3] that the underlying fundamental triplet representation is actually realized in nature in the guise of three fractionally charged *quarks*, which are the building blocks of hadrons. However, the failure of searches for free quarks jeopardized this ansatz and impeded further progress. It was in this situation in 1969 that the SLAC DIS experiments with multi-GeV electron beams verified the substructure of the nucleon [94] in much the same way as Rutherford demonstrated the existence of nuclei: they observed a scattering-angle distribution compatible with the existence of pointlike constituents (*partons*) inside the nucleon, rather than a continuous charge distribution in a sphere of the nucleon radius (which was known to be of the order of 10^{-15} m at that time). Furthermore, initial measurements of the nucleon structure functions revealed the scaling behavior predicted by Bjorken [95]. These results stimulated Feynman to postulate the quark–parton model (QPM) [96], which identified the partons with the quarks and provided a natural explanation for the Bjorken scaling.

In the following years, DIS experiments with electrons at SLAC and with neutrinos at CERN confirmed that the properties of the nucleon constituents, i.e. the partons, are actually consistent with those assigned to the quarks. A further historical step towards constructing the SM was the discovery of weak neutral currents [97] in neutrino DIS in 1973, supporting the Glashow–Salam–Weinberg (GSW) model [98] of spontaneously broken $\mathrm{SU}(2)_\mathrm{L} \times \mathrm{U}(1)_\mathrm{Y}$ electroweak local gauge symmetry. The period between 1974 and 1981 was then marked by the discoveries of the charm quark, the τ lepton and the bottom quark. A consistent picture of properties and dynamics of the now three families of fundamental particles began to consolidate when the next generation of DIS experiments with muon and neutrino beams at CERN and FNAL became operational in the late 1970s and early 1980s, using beams of energies up to several hundred GeV and increased intensities. The study of DIS in these experiments (EMC, BCDMS, NMC, E665 with muons, BEBC, CDHS(W), CHARM and CCFR with neutrinos) contributed two essential blocks of information: the confirmation of the GSW predictions for the weak mixing angle and coupling constants, and the high-precision assessment of the nucleon structure functions and parton distributions.

The GSW theory not only successfully passed the tests of the DIS experiments but also predicted the existence and masses of the heavy gauge bosons

[3] A collection of original papers can be found in [93].

$W^{\pm}$ and Z (which were discovered in 1983 in $p\bar{p}$ collisions at CERN [99]), and has meanwhile been verified to the radiative-correction level of precision by the LEP experiments at CERN. The other sector of the SM, the QCD, received two crucial pieces of support from the structure function measurements in DIS: (i) the evidence for gluons, deduced first indirectly from the observation that the quarks carry only about half of the nucleon momentum (early 1970s [100]) and later directly from the observation of reactions of the type $e^+e^- \to 3$ jets at the e^+e^- accelerator PETRA at DESY (1979 [101]); and (ii) the discovery (late 1970s) and precise determination (1980s) of violations of Bjorken scaling, which are quantitatively predicted by the QCD evolution equations [102]. An impressive spectrum of further corroborating experimental results in conjunction with theoretical progress finally established QCD as the standard theory of strong interactions. DIS remains today one of the most important sources of experimental investigations of QCD and its parameters.

The 1980s saw an ongoing effort of the fixed-target experiments to study DIS with growing precision and also with polarized beams and targets. However, it was obvious that the kinematic reach of these experiments was limited and could only be overcome by an electron–hadron collider. Corresponding ideas were first summarized in 1979 in a feasibility study of the European Committee for Future Accelerators [103] and finally led to the HERA proposal [104], which was approved in 1984. The first ep collisions in HERA were achieved in 1991, and data taking has been in progress since 1992.

1.2 The HERA Collider: Parameters and Performance

The HERA accelerator facility at Hamburg, Germany, consists of two separate synchrotron storage rings of 6.336 km circumference which are equipped with superconducting and normal-conducting magnets for the protons and electrons, respectively. Four interaction regions are available, two of which accommodate the multipurpose detectors ZEUS and H1 recording ep reactions. A third experiment, HERMES, studies polarized lepton–nucleus scattering on an internal target. In the fourth interaction region the HERA-B experiment is being set up, which will use the proton beam halo in fixed-target mode to investigate decays of bottom mesons.

During the past years of operation, HERA has collided longitudinally unpolarized positrons and protons with beam energies of $E_e = 27.5$ GeV and $E_p = 820$ GeV, respectively. Since 1996 beam currents of typically 40 mA (e^+) and 70 mA (p) have been achieved, distributed in about 180 (190) bunches in the p (e) beam, some of which were left unpaired for background studies. The bunch colliding frequency was 10.4 MHz; the length of the interaction region (about 12 cm, except in 1993, when it was about 20 cm) was determined by the proton bunch length.

The history of the HERA parameters and of the integrated luminosities delivered to ZEUS and H1 are summarized in Table 1.1 for the period relevant

Table 1.1. HERA beam energies and integrated luminosities in the period 1993–97. Note that the delivered luminosity includes all *ep* collisions provided by HERA and is slightly different at the two interaction regions (quoted for ZEUS here). The integrated luminosities actually usable for physics analyses are summarized in the last four columns

Year	Beam energies (GeV)		Integrated luminosity (pb^{-1})					
			Delivered		ZEUS		H1	
	E_e	E_p	e^-p	e^+p	e^-p	e^+p	e^-p	e^+p
1993	26.7	820.0	1.1	—	0.5	—	0.5	—
1994	27.5	820.0	1.1	5.1	0.3	3.0	0.5	3.2
1995	27.5	820.0	—	12.3	—	6.6	—	5.5
1996	27.5	820.0	—	17.2	—	10.7	—	9.6
1997	27.5	820.0	—	36.4	—	27.7	—	24.5

to this book. Note the changeover from electron to positron running in 1994, which avoided electron-beam lifetime limitations due to space-charge effects and resulted in an immediate increase of operation efficiency.

During the 1997/1998 winter shutdown an improved electron vacuum system was installed, and from 1998 onwards HERA was operated at an increased proton beam energy of 920 GeV, corresponding to an *ep* center-of-mass energy of 318 GeV. Both ZEUS and H1 have collected roughly 15 pb^{-1} of useful e^-p data up to April 1999, and subsequently more than 50 pb^{-1} of e^+p data. Currently, e^+p data taking is in progress. In a nine-month shutdown to be completed in the first half of 2001, a major upgrade of the HERA machine and the ZEUS and H1 interaction regions is foreseen, yielding higher specific luminosities and the possibility to provide *ep* collisions with longitudinally polarized electrons and positrons. It is hoped that after this upgrade, substantially increased e^+p data samples will become available which may help to answer some of the questions left open by the available data.

1.3 Deep Inelastic Scattering at HERA

Deep inelastic *ep* scattering in its two major modes, NC and CC, proceeds in leading order (LO) via the two Feynman graphs shown in Fig. 1.1. Accordingly, the final state of these reactions consists of the scattered lepton (which is a neutrino and hence unobservable for CC reactions) and a partonic system (the scattered quark and the proton remnant) which evolves into observable hadrons by QCD radiation and fragmentation processes.

HERA offers for the first time the possibility to observe the full spectrum of gauge boson exchanges in DIS simultaneously for a given initial state. In previous fixed-target experiments, processes involving the heavy gauge bosons could only be studied with neutrino beams, whereas electron and muon experiments were only sensitive to the photon exchange. A few characteristics

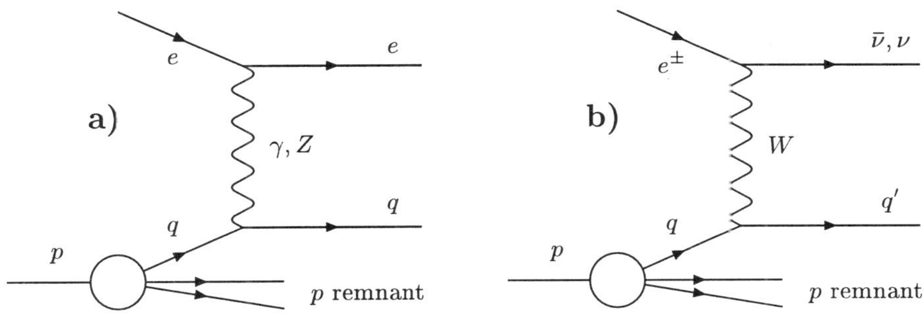

Fig. 1.1. Leading-order Feynman graphs for deep inelastic *ep* scattering, (**a**) for NC and (**b**) for CC reactions. Note that scattering on antiquarks, produced in the proton by virtual processes, would also be possible. The backward-pointing arrows for the antilepton lines (e^+, $\bar{\nu}$) have been omitted for simplicity

of NC and CC DIS events at HERA will be summarized in Sect. 1.3.2, after introducing the relevant kinematic variables in Sect. 1.3.1.

1.3.1 Kinematic Variables

The kinematics of the DIS reactions in Fig. 1.1, and of the their most prominent counterparts with photon radiation from one of the electron lines, are described using the following four-momenta:

k	incoming electron,
P	incoming proton,
k'	scattered electron or neutrino,
k_γ	photon radiated from the initial- or final-state electron,
$q = k - k' - k_\gamma$	exchanged gauge boson,
$q_\ell = k - k'$	lepton four-momentum transfer.

Of course $q = q_\ell$ in the absence of photon radiation. In addition to the particle masses, there are five relevant Lorentz invariants which can be formed from these four-momenta:

$$s = (k + P)^2 \quad \text{square of } ep \text{ center-of-mass energy,} \tag{1.1}$$

$$Q^2 = -q^2 \quad \text{negative square of four-momentum transfer,} \tag{1.2}$$

$$x = Q^2/(2\, q \cdot P) \quad \text{Bjorken scaling variable } x, \tag{1.3}$$

$$y = (q \cdot P)/(k \cdot P) \quad \text{Bjorken scaling variable } y, \tag{1.4}$$

$$W = \sqrt{(q + P)^2} \quad \text{invariant mass of the hadronic system.} \tag{1.5}$$

Whereas s is given by the beam energies, Q^2, x, y and W characterize the kinematics of individual DIS reactions. In the QPM, where *ep* DIS is interpreted as the scattering of the electron on a quasi-free quark with zero mass and transverse momentum, x is the fraction of the proton momentum carried by the quark, and y is related to the scattering angle, θ^*, in the electron–quark center-of-mass system:

$$y = \frac{1}{2}(1 - \cos\theta^*) \tag{1.6}$$

($\theta^* = 0$ corresponds to no deflection). Note that in some places *lepton variables* will be used, which are marked by an index ℓ and are derived from (1.2)–(1.5) by replacing q by q_ℓ.

In the following, terms of the order $\mathcal{O}(M_p^2/s)$ (M_p being the proton mass) and lepton mass terms will be neglected (even for $\nu_\mu p \to \mu X$ scattering, where the lepton mass corrections are largest, they turn out to be negligible in the relevant kinematic domain). In this approximation, the following relations hold:

$$Q^2 = x\,y\,s\,, \tag{1.7}$$

$$W^2 = Q^2\left(\frac{1}{x} - 1\right) = (1-x)\,y\,s\,, \tag{1.8}$$

demonstrating that for given s only two of the four variables defined in (1.2)–(1.5) are independent. Note, however, that for radiative events the photon four-momentum, k_γ, has to be known to evaluate these variables. The frequent and experimentally relevant case of an emission of a photon which is approximately collinear with the incoming electron can be described by an effective reduction of the electron beam energy, i.e. the photon energy, E_γ, alone suffices to quantify $k - k_\gamma$. Note that this case corresponds to a reduced center-of-mass energy, $\tilde{s} = s(1 - E_\gamma/E_e)$.

A detailed account of the reconstruction of kinematic variables from experimentally measurable quantities is given in Sect. 3.3. Here we only want to quote the relations between x, y, Q^2 and the energy, E', and scattering angle, θ_ℓ, of the final-state lepton in the laboratory system[4] which hold for $k_\gamma = 0$ ("electron method", see Sect. 3.3.3); note that here, in contrast to the center-of-mass case in (1.6), $\theta_\ell = 180°$ corresponds to no deflection:

$$Q^2 = 2\,E_e\,E'(1 + \cos\theta_\ell)\,, \tag{1.9}$$

$$x = \frac{E_e}{E_p}\,\frac{E'(1+\cos\theta_\ell)}{2E_e - E'(1-\cos\theta_\ell)}\,, \tag{1.10}$$

$$y = 1 - \frac{E'}{2E_e}(1 - \cos\theta_\ell)\,. \tag{1.11}$$

[4] Both ZEUS and H1 use a right-handed coordinate system with the Z axis pointing into the proton beam direction (referred to as "forward") and the X axis horizontal, pointing towards the center of HERA. Polar and azimuthal angles are defined with respect to this system in the usual way.

Evidently, reactions at $Q^2 \gtrsim 10^3\,\mathrm{GeV}^2$ require leptons scattered at angles significantly less than 180° and with energies exceeding about 10 GeV. In general, this also implies large lepton transverse momenta, $p'_\mathrm{t} = E' \sin\theta_\ell$.

Although the variables defined in (1.1)–(1.5) are sufficient for inclusive studies, it will also be useful to consider single particles with momenta $\boldsymbol{p}$, which are characterized by the polar and azimuthal angles in the laboratory system, θ and ϕ, and by $|\boldsymbol{p}|$. Two derived single-particle quantities are

$$\eta = -\ln\left(\tan\frac{\theta}{2}\right) \quad \text{pseudorapidity}\,, \tag{1.12}$$

$$p_\mathrm{t} = |\boldsymbol{p}|\,\sin\theta \quad \text{transverse momentum}\,. \tag{1.13}$$

1.3.2 Characteristics of High-Q^2 NC and CC Reactions

For the kinematic domain under study here, the final-state lepton typically has a transverse momentum of at least $\mathcal{O}(10\,\mathrm{GeV})$, which is balanced by the hadronic system, implying that the lepton and the hadrons are well separated in space in most cases. The topology of the hadronic final state reflects the pattern of partons emerging from the electron–proton scattering process: high-momentum quarks or gluons give rise to *jets*, i.e. bundles of hadrons and their decay products collimated along the direction of flight of the original parton. High-Q^2 DIS events at HERA usually contain one jet in the proton direction (*proton remnant jet*) and one or more jets with high transverse momentum (*current jet* plus possibly further jets caused by hard QCD radiation processes). Event topologies with n jets in addition to the proton remnant jet will in the following be denoted by the shorthand "$(n+1)$-jet".

In order to relate the kinematic variables of DIS to each other and to typical event topologies, Fig. 1.2 depicts the scattering angles and energies of the electron and quark as a function of x and y. The QPM picture of quasi-free, massless quarks, as well as the absence of QED radiation (i.e. $k_\gamma = 0$), has been assumed for relating x and y to the angles and energies. For simple (1+1)-jet topologies, the displayed quark angle approximately corresponds to the polar angle of the current jet.

Event displays of two archetypal (1+1)-jet high-Q^2 ep DIS reactions are shown in Fig. 1.3: an NC event in the H1 detector and a CC event in ZEUS (both detectors and their relevant components are presented in detail in Sect. 3.1).

In the NC event, the scattered positron is identified by its characteristic topology: one high-p_t track pointing to a well-localized and isolated energy deposit in the innermost section of the calorimeter. In contrast to the positron, the hadronic system is a multiparticle state containing (1+1) jets, characterized by several tracks and by hadronic showers in the calorimeter which penetrate into its outer sections and are laterally much more extended than

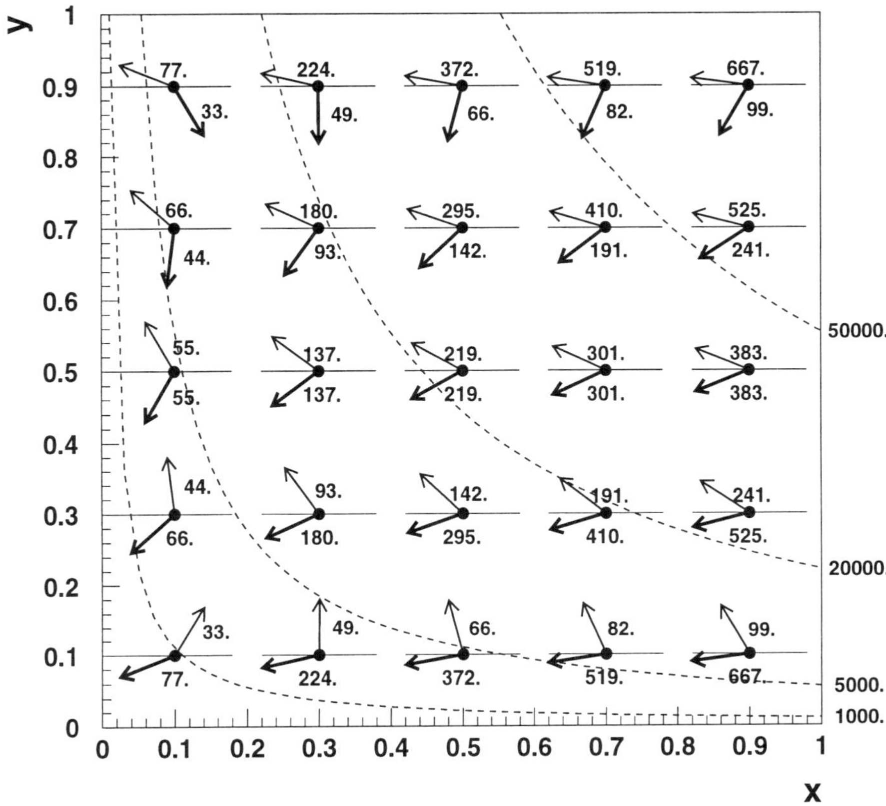

Fig. 1.2. Event topologies of DIS reactions at HERA for beam energies $E_e = 27.5\,\text{GeV}$ and $E_p = 820\,\text{GeV}$. Each event pictogram shows the directions in which the lepton (*upper arrow*) and quark (*lower arrow*) are emitted, where x and y are indicated by the position of the *central dot*. The electron and quark energies in GeV are denoted by the numbers next to the arrows. The QPM assumption of quasi-free, zero-mass quarks has been used to calculate the angles and energies. The *dashed lines* indicate constant values of Q^2 (given in GeV^2 on the *right margin*)

the electron shower. The CC event has similar hadronic activity, but there is no sign of a scattered electron: the final-state neutrino escapes detection and leaves the typical "missing transverse momentum" signature.

The experimental detection and reconstruction of DIS events crucially depends on identification and measurement of the scattered electron (in NC events) and on a good survey of the hadronic system. Correspondingly, particular problems are posed by NC events with the electron outside the geometrical acceptance of the central tracking detectors, or by events with the current jet close to the beam hole, at the edge of calorimetric coverage. From Fig. 1.2 it becomes obvious that very forward-going electrons appear at high values of y, whereas low y implies forward-going hadronic jets.

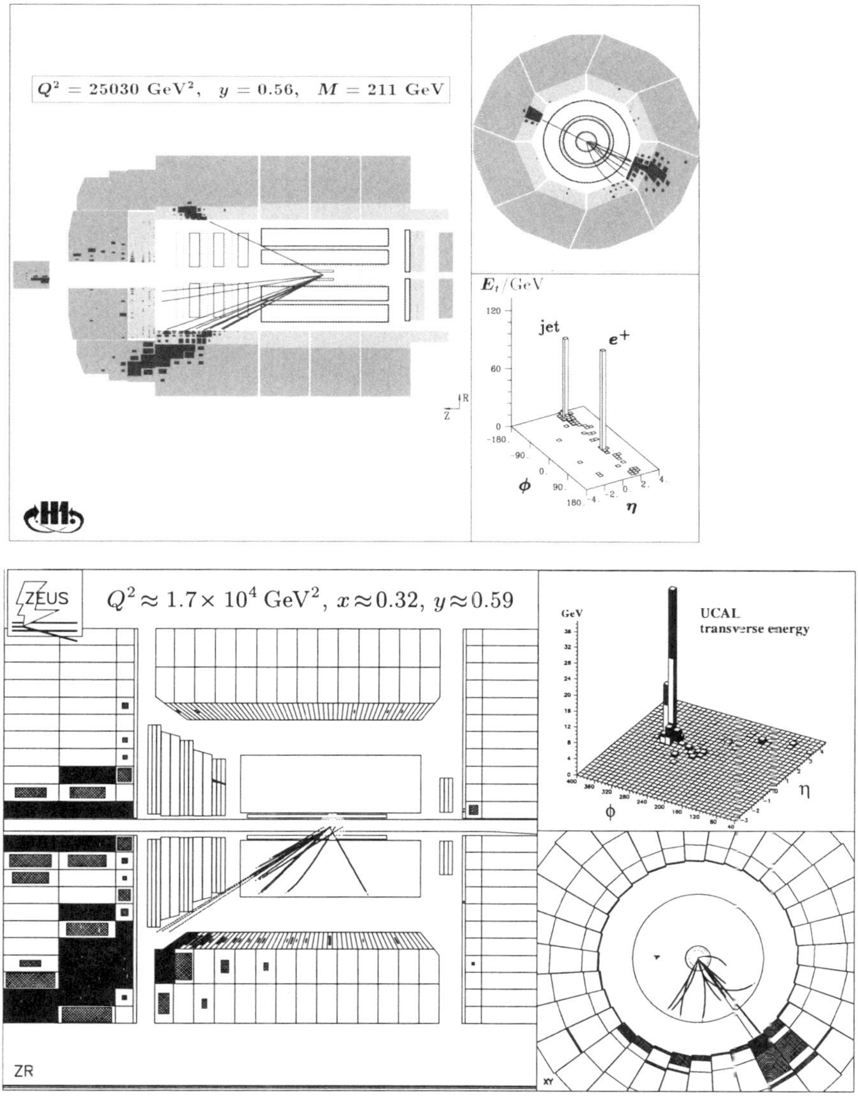

Fig. 1.3. Event displays of high-Q^2 e^+p reactions: an NC event in the H1 detector (*top*) and a CC event in ZEUS (*bottom*). The *left* parts of both pictures show vertical sections through the detectors along the beam direction, with the positron entering from the left and the proton from the right. The tracks reconstructed from the hits in the inner tracking chambers and the energy deposits in the surrounding calorimeters are indicated. The two displays at the *right* show transverse sections through the central detectors and the flow of transverse energy in the pseudorapidity–azimuth plane, respectively. The reconstructed kinematic variables of each event are indicated on top of the displays ($M = \sqrt{xs}$). Note that the topologies of both events are similar, except that the scattered lepton (identified as a track and energy deposit in the top half of the NC display) is missing in the CC event

2. Deep Inelastic *ep* Cross Sections

Confronting theoretical predictions for DIS cross sections with experimental measurements serves both to test the underlying theories and to extract information characterizing the reacting particles, such as structure functions or parton distribution functions (PDFs). For both purposes, an exact knowledge of the theoretical expectation and of its uncertainty are crucial.

The first few of the following sections review the theoretical and experimental ingredients of the standard-model (SM) prediction for deep inelastic *ep* cross sections and consider the different sources of uncertainty. Thereafter, selected phenomenological models are described which apply to DIS or involve reactions with DIS-like event topologies and potentially could modify the measured DIS cross sections at high Q^2. Many of these models – including modifications in the PDF sector and theoretical concepts beyond the SM – have been suggested as possible explanations of the HERA excess of events in this kinematic region. The different scenarios are discussed with respect to their compatibility with the experimental results, and also regarding distinctive signatures which might serve to differentiate between them.

Although the kinematic region of interest for this book is defined by Q^2 above some 100 GeV^2, it is important to note that the predicted cross sections depend strongly on experimental results obtained at lower energy scales, in particular on structure-function measurements in fixed-target deep inelastic scattering experiments at much smaller values of Q^2. Hence, the Q^2 range in which methods of perturbative QCD (pQCD) are used and have to be quantitatively understood extends down to values of a few GeV^2, reaching the domain where their applicability becomes questionable (since the strong coupling constant, α_s, approaches unity), and where higher-twist terms of $\mathcal{O}(1/Q^{2n})$ cannot be neglected.

The use of the term *deep inelastic* is not unambiguous or well-defined. In the following, "deep inelastic" is understood to characterize a kinematic domain where Q^2 is large enough to serve as the hard scale for pQCD.

2.1 Deep Inelastic Scattering in the Standard Model

One of the striking successes of the first-generation fixed-target DIS experiments was to establish the concept of pointlike, quasi-free constituents of hadrons, the partons, with which the leptons react (quark–parton model, QPM). By relating the cross sections of different reactions to the parton content of the respective initial-state hadrons and to the quantum numbers of the partons, this approach provided predictive power and helped us to understand the properties of the partons. Even today the fundamental QPM structure of DIS cross sections (i.e. products of PDFs and lepton–parton cross sections) still holds, though a much more rigorous theoretical treatment than thirty years ago is employed to reach this conclusion. The two basic ingredients of DIS cross sections are the matrix elements for lepton–parton interactions (the so-called *hard subprocesses*), which are governed by the electroweak sector of the SM, and the PDFs, which include the effects of the strong interaction and are, at sufficiently high Q^2, subject to pQCD calculus.

After the leading-order differential cross sections are introduced in Sect. 2.1.1, theoretical aspects and uncertainties related to QCD and electroweak radiative corrections are summarized in Sects. 2.1.2 and 2.1.3, respectively. The experimental data constraining the PDFs and the parameters governing the lepton–parton matrix elements, and the corresponding associated sources of uncertainty, are reviewed in Sect. 2.1.4. In Sect. 2.1.5, the fit procedures applied to derive PDF parameterizations are discussed. Finally, some figures illustrating the most relevant PDF and cross section results and the corresponding error estimates are collected in Sect. 2.1.6.

The calculation of the inclusive DIS cross sections involves implicit summation over the degrees of freedom of the hadronic final state, which are therefore irrelevant for the above discussion. However, their accurate modeling is a necessary prerequisite for applying simulation techniques in the analysis of experimental data to infer acceptance and migration corrections as well as background estimates. It is hence mandatory to test the models employed and to be aware of their imperfections. A synopsis of the standard approaches to assigning hadronic degrees of freedom to given partonic final-state configurations of lepton–hadron reactions is given in Sect. 2.1.7.

2.1.1 The Leading-Order Differential Cross Sections

In the QPM, the PDF $p_{i/h}(x)$ is defined to be the probability density to find a parton of type i carrying a fraction x of the momentum of a hadron h. In order to simplify the notation, the proton PDFs will be abbreviated using the customary quark flavor symbols, i.e. $u(x) \equiv p_{u/p}(x)$ for the u quark density etc., or $q(x)$ in sums where q runs over multiple flavors and/or over quarks and antiquarks. Likewise, the gluon density will be denoted by $g(x)$.

By supposing that a charged lepton (ℓ) or a neutrino ($\overset{(-)}{\nu}$) scatters on a parton without initial transverse momentum with respect to the hadron direction of flight, and that all partons involved in the process are massless, the relevant x for a given DIS reaction is identified to be the Bjorken scaling variable defined in (1.3). In the following we already use the fact that the PDFs acquire a Q^2 dependence owing to QCD corrections to the simple QPM picture (see Sect. 2.1.2). The leading-order (LO) cross section for lepton–proton scattering is then given by [96, 105, 106]

$$\frac{\mathrm{d}^2\sigma^{\ell p,\overset{(-)}{\nu} p}}{\mathrm{d}x\,\mathrm{d}Q^2}(x,Q^2) = \sum_q q(x,Q^2)\,\frac{\mathrm{d}\,\hat{\sigma}^{\ell q,\overset{(-)}{\nu} q}}{\mathrm{d}Q^2}(x,Q^2)\ , \tag{2.1}$$

where $\hat{\sigma}^{\ell q}$ is the cross section of the hard lepton–parton subprocess and q runs over all quark and antiquark flavors. The gluon does not contribute to this sum since it does not couple to leptons in leading order. Higher-order correction factors are absorbed into the quark PDFs.

The neutral-current (NC) cross sections for electron–quark scattering, or equivalently for the scattering of other charged leptons on quarks, are calculated from the corresponding electroweak matrix elements:

$$\frac{\mathrm{d}\,\hat{\sigma}^{eq\to eq}}{\mathrm{d}Q^2} \propto$$

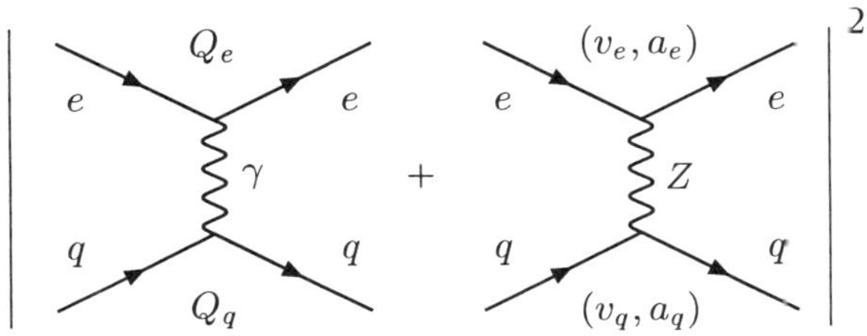

(2.2)

The couplings indicated in (2.2) are fixed by the $\mathrm{SU}(2)_\mathrm{L} \times \mathrm{U}(1)_\mathrm{Y}$ gauge structure of the electroweak theory and are given in units of the elementary charge, e. Q_e and Q_q denote the charges of the electron and the quark, and (v_e, a_e) and (v_q, a_q) are their vector and axial-vector couplings to the weak neutral current. For a fermion f, the latter are related to its charge, Q_f, and the z component of its weak isospin, I_f^3:

$$v_f = \frac{-I_f^3 + 2\sin^2\theta_\mathrm{W}\, Q_f}{2\sin\theta_\mathrm{W}\cos\theta_\mathrm{W}} \quad \text{and} \quad a_f = \frac{-I_f^3}{2\sin\theta_\mathrm{W}\cos\theta_\mathrm{W}}\ , \tag{2.3}$$

where θ_W is the weak mixing angle (*Weinberg angle*). Calculating $\hat{\sigma}^{eq\to eq}$ from (2.2) and inserting it into (2.1) finally yields the LO cross section [107]

$$\frac{\mathrm{d}^2\sigma^{e^\pm p\to e^\pm X}}{\mathrm{d}x\,\mathrm{d}Q^2} = \frac{2\pi\alpha^2}{Q^4}\left\{\left[1+(1-y)^2\right]\mathcal{F}_2^\mathrm{NC} \mp \left[1-(1-y)^2\right]\mathcal{F}_3^\mathrm{NC}\right\}\ , \tag{2.4}$$

$$\mathcal{F}_2^\mathrm{NC}(x,Q^2) = \sum_{q=d,u,s,c,b} A_q(Q^2)\ \left[q(x,Q^2)+\bar{q}(x,Q^2)\right]\ , \tag{2.5}$$

$$\mathcal{F}_3^\mathrm{NC}(x,Q^2) = \sum_{q=d,u,s,c,b} B_q(Q^2)\ \left[q(x,Q^2)-\bar{q}(x,Q^2)\right]\ . \tag{2.6}$$

The $\mathcal{F}_i^\mathrm{NC}$ are the NC structure functions.[1] They contain the PDFs, the coupling constants and the coefficient functions $A_q(Q^2)$ and $B_q(Q^2)$, given by

[1] Different conventions for defining and symbolizing structure functions are in widespread use. The $\mathcal{F}_i$ defined in (2.4)–(2.6) are related to the commonly used structure functions F_2 and F_3 by $\mathcal{F}_2 \equiv F_2/x$ and $\mathcal{F}_3 \equiv F_3$.

$$A_q(Q^2) = Q_q^2 - 2Q_q v_e v_q \, P_Z + (v_e^2 + a_e^2)(v_q^2 + a_q^2) \, P_Z^2 \, , \tag{2.7}$$

$$B_q(Q^2) = \quad - 2Q_q a_e a_q \, P_Z + 4 v_e a_e v_q a_q \, P_Z^2 \, , \tag{2.8}$$

where

$$P_Z = \frac{Q^2}{Q^2 + M_Z^2} \tag{2.9}$$

denotes the ratio of the Z and γ propagators and $\alpha = e^2/(4\pi)$ is the fine-structure constant. Note that the Z exchange and γ–Z interference contributions to the cross section (2.4) are suppressed by P_Z at low Q^2, whence the high $Q^2 = \mathcal{O}(M_Z^2)$ accessible at HERA allows us for the first time to study DIS in a kinematic domain where these contributions have a similar size to that of γ exchange. The $\mathcal{F}_3$ term in (2.4) is parity-violating and is absent if Z exchange is neglected. Scattering on top (t) quarks is excluded from the sums in (2.5) and (2.6) since, in the absence of flavor-changing neutral currents, it would imply final states containing $t\bar{t}$ pairs and hence exceeding the HERA center-of-mass energy. In contrast, the phase space reduction due to the masses of final-state charm or bottom particles is assumed to be negligible in high-Q^2 NC reactions at HERA. The suppression of heavy quarks in the initial state is taken into account by the corresponding PDFs (see Sect. 2.1.2).

For CC reactions, the equivalent to (2.2) is given by

$$\frac{\mathrm{d}\,\hat{\sigma}^{e^\pm q \to \overset{(-)}{\nu}_e q'}}{\mathrm{d}Q^2} \propto \sum_{q'} |V_{qq'}|^2 \left| \; \text{[diagram]} \; \right|^2 \, , \tag{2.10}$$

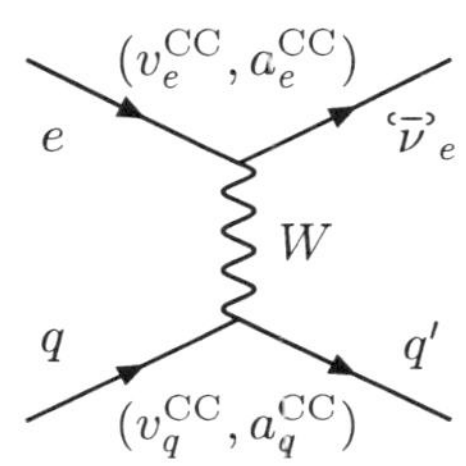

where the $V_{qq'}$ denote the elements of the Cabibbo–Kobayashi–Maskawa (CKM) matrix and, in units of e, the fermion–W couplings are given by

$$v_f^{\mathrm{CC}} = a_f^{\mathrm{CC}} = \frac{1}{2\sqrt{2}\sin\theta_{\mathrm{W}}} \, . \tag{2.11}$$

As in the NC case, the CC *ep* cross section and the CC structure functions are obtained by inserting $\hat{\sigma}^{e^\pm q \to \overset{(-)}{\nu}_e q'}$ into (2.1) [107]:

$$\frac{\mathrm{d}^2\sigma^{e^\pm p \to \overset{(-)}{\nu} X}}{\mathrm{d}x\,\mathrm{d}Q^2} = \frac{\pi\alpha^2}{8\sin^4\theta_{\mathrm{W}}} \frac{1}{\left(Q^2 + M_W^2\right)^2} \times \left\{ \left[1 + (1-y)^2\right] \mathcal{F}_2^{\mathrm{CC}\pm} \mp \left[1 - (1-y)^2\right] \mathcal{F}_3^{\mathrm{CC}\pm} \right\} \, , \tag{2.12}$$

$$\mathcal{F}_{2,3}^{\mathrm{CC}+} = \sum_{q=d,s,b} \zeta_q q(x, Q^2) \pm \sum_{q=u,c,t} \zeta_q \bar{q}(x, Q^2) \, , \tag{2.13}$$

$$\mathcal{F}_{2,3}^{\mathrm{CC}-} = \sum_{q=u,c,t} \zeta_q q(x, Q^2) \pm \sum_{q=d,s,b} \zeta_q \bar{q}(x, Q^2) \, . \tag{2.14}$$

The sign of the $\mathcal{F}_3$ term in (2.12) depends on the charge of the initial-state electron, whereas the signs in the right-hand sides of (2.13) and (2.14) are positive for $\mathcal{F}_2$ and negative for $\mathcal{F}_3$. Note that electrons and positrons couple to quarks and antiquarks of different flavor, as required by charge conservation: $Q_{q'} = Q_e + Q_q$.

As a consequence of lepton flavor universality, the above discussion is valid for muon CC scattering ($\mu^\pm p \to \overset{(-)}{\nu}_\mu X$) as well. The cross section formulae (2.12)–(2.14) can also be applied to (anti)neutrino CC scattering ($\overset{(-)}{\nu}_\ell q \to \ell q'$) by replacing $\sigma^{e^+p\to\bar{\nu}X}$ with $\sigma^{\nu_\ell p\to\ell^- X}$ and $\sigma^{e^-p\to\nu X}$ with $\sigma^{\bar{\nu}_\ell p\to\ell^+ X}$, and by reversing the sign of the $\mathcal{F}_3$ term in (2.12).

The ζ factors in (2.13)–(2.14) are given by

$$\zeta_q = \sum_{q'} |V_{qq'}|^2 \, \kappa_{q'} \ , \tag{2.15}$$

where the $\kappa_{q'}$ quantify the phase space suppression of heavy final-state quarks q' and are, in general, functions of the kinematic variables (x, Q^2) assuming values between zero and unity. In high-Q^2 reactions at HERA, we have $Q^2, W^2 \gg m^2_{d,u,s,c}$, whence the production of the four lightest quark flavors is unrestrained to a good approximation, i.e. we use $\kappa_{d,u,s,c} \equiv 1$ (however, this approximation is not valid for fixed-target experiments with neutrino beams which probe the threshold region of charm production; specific assumptions on κ_c have to be made in order to include their results in PDF parameterizations). As in the NC case, the suppression of heavy quarks in the initial state is taken into account by the corresponding PDFs (see Sect. 2.1.2).

CC processes with b or t quarks in the initial or final state are severely suppressed if they involve off-diagonal elements of the CKM matrix. Employing the unitarity of the CKM matrix and the 90% C.L. upper limits on the third-family $V_{qq'}$ values from [108], we arrive at

$$\begin{aligned} 1 - \zeta_d &= (1-\kappa_t)\,|V_{td}|^2 < |V_{td}|^2 < 1.7\times 10^{-4} \ , \\ 1 - \zeta_u &= (1-\kappa_b)\,|V_{ub}|^2 < |V_{ub}|^2 < 2.0\times 10^{-5} \ , \\ 1 - \zeta_s &= (1-\kappa_t)\,|V_{ts}|^2 < |V_{ts}|^2 < 1.8\times 10^{-3} \ , \\ 1 - \zeta_c &= (1-\kappa_b)\,|V_{cb}|^2 < |V_{cb}|^2 < 1.8\times 10^{-3} \ . \end{aligned} \tag{2.16}$$

Finally, the family-diagonal subprocesses $e^-t \to \nu b$ and $e^+b \to \bar{\nu} t$ and their charge-conjugate counterparts are all associated with ep final states containing a top quark or antiquark (either produced in the hard subprocess, or the "surviving" one of the initial virtual $t\bar{t}$ pair) and will be considered to be completely suppressed owing to the high top mass, $m_t = (174.3 \pm 5.1)$ GeV [109]. Using this and again $\kappa_{d,u,s,c} \equiv 1$, one obtains

$$\begin{aligned} \zeta_b &= |V_{ub}|^2 + |V_{cb}|^2 < 1.8\times 10^{-3} \ , \\ \zeta_t &= |V_{td}|^2 + |V_{ts}|^2 < 1.9\times 10^{-3} \ , \end{aligned} \tag{2.17}$$

which demonstrates together with (2.16) that third-family quarks can be safely ignored both in the initial and in the final state of CC lepton–nucleon scattering at HERA and also in lower-energy experiments.

Explicitly making use of the approximations implied by (2.16) and (2.17) and inserting the Fermi constant,

$$G_{\mathrm{F}} = \frac{\pi\alpha}{\sqrt{2}\sin^2\theta_{\mathrm{W}} M_W^2} = \frac{\sqrt{2}}{M_W^2}\left(e\, v_f^{\mathrm{CC}}\right)^2 = \frac{\sqrt{2}}{M_W^2}\left(e\, a_f^{\mathrm{CC}}\right)^2 , \tag{2.18}$$

the cross sections in (2.12) can be written as

$$\frac{\mathrm{d}^2\sigma^{e^+p\to\nu X}}{\mathrm{d}x\,\mathrm{d}Q^2} = \frac{G_{\mathrm{F}}^2}{2\pi}\left(\frac{M_W^2}{Q^2+M_W^2}\right)^2 \times\left[\sum_{q=u,c}\bar{q}(x,Q^2) + (1-y)^2\sum_{q=d,s} q(x,Q^2)\right] , \tag{2.19}$$

$$\frac{\mathrm{d}^2\sigma^{e^-p\to\bar{\nu} X}}{\mathrm{d}x\,\mathrm{d}Q^2} = \frac{G_{\mathrm{F}}^2}{2\pi}\left(\frac{M_W^2}{Q^2+M_W^2}\right)^2 \times\left[\sum_{q=u,c} q(x,Q^2) + (1-y)^2\sum_{q=d,s} \bar{q}(x,Q^2)\right] . \tag{2.20}$$

The use of (2.18) to eliminate α and $\sin\theta_{\mathrm{W}}$ from the cross section formula has the advantage of reducing the virtual electroweak radiative corrections to the percent level (see [107] and Sect. 2.1.3).

The y terms in (2.19) and (2.20) reflect the angular dependence of $\hat{\sigma}^{e^\pm q\to \overset{(-)}{\nu} q'}$ caused by the $V-A$ structure of weak interactions, which enforces definite helicities of the fermions participating in the hard subprocess (left-handed fermions, right-handed antifermions). An initial state with fermion number $|F|=2$ carries spin 0 and produces an isotropic final state ($\mathrm{d}\hat{\sigma}/\mathrm{d}y$ is flat in y), whereas $F=0$ implies spin 1 and a strongly forward-peaked cross section ($\mathrm{d}\hat{\sigma}/\mathrm{d}y \propto (1-y)^2 = \cos^4(\theta^*/2)$, where (1.6) has been used).

2.1.2 Including QCD Effects

This section introduces some aspects of the application of pQCD to deep inelastic scattering by collecting the relevant basic arguments and formulae. The intention is less to present a formally correct and complete derivation of the central results, than to illustrate which theoretical concepts are applied and where the theoretical uncertainties originate from. A profound theoretical coverage can be found in [110–115], and recent reviews [44–46] cover theoretical and experimental aspects of hadron structure in the light of the HERA results at lower x and Q^2 than discussed here.

Let us assume for the moment that the PDFs $p_{i/h}(\xi)$ of a hadron h are defined as in the QPM, i.e. $p_{i/h}(\xi)$ denotes the probability density that a parton of type i carries the momentum fraction ξ of h. In this case the lepton–hadron cross section $\sigma^{\ell h}$ can be written as

$$\sigma^{\ell h}(x, Q^2) = \sum_i \int_x^1 \frac{\mathrm{d}\xi}{\xi}\, p_{i/h}(\xi)\, \hat{\sigma}^{\ell i}\left(\frac{\xi}{x}, Q^2\right) , \tag{2.21}$$

where $\hat{\sigma}^{\ell i}$ is the cross section of the hard lepton–parton subprocess, x is the Bjorken scaling variable (see (1.3)), the sum runs over all partons in h and $\mathrm{d}\xi/\xi$ is a relic of the Lorentz-invariant integration over the internal quark line (see Fig. 1.1). Note that ξ is a priori unrelated to x. Equation (2.21) is only valid in the limit of massless, collinear partons (i.e. for ${k^i}^2/Q^2, {\boldsymbol{k}^i}_\perp^2/Q^2 \ll 1$, where k^i is the four-momentum of parton i and ${\boldsymbol{k}^i}_\perp$ is its momentum component transverse to the hadron direction of flight). Actually, this condition, together with the assumption that the hadron mass is negligible, simultaneously ensures that ξ can be consistently interpreted as the four-momentum fraction of the parton in the hadron. The requirement of vanishing hadron masses is formally met by a description in the infinite-momentum frame, where the hadron (and with it all partons) travels at the speed of light and hence the partons have no masses. However, one has to keep in mind that the experimental situation corresponds to finite momenta, so there are validity limits of (2.21) for low x or high parton masses.

In the simplest case of a $2 \to 2$ scattering $\ell\, i \to \ell'\, j$ with massless partons i and j, energy and momentum conservation require $\xi = x$, in accordance with the QPM interpretation of x as the momentum fraction carried by the struck quark. In this particular case, $\hat{\sigma}^{\ell i} = \sigma_0^{\ell i}\delta(x/\xi - 1)$ and the convolution integral reduces to the QPM cross section $\sigma^{\ell h}_{\mathrm{QPM}}(x, Q^2) = \sum_i p_{i/h}(x)\sigma_0^{\ell i}$. However, once the partonic final state consists of more than one parton (e.g. because of gluon radiation) and hence in general has finite invariant mass, $\hat{\sigma}^{\ell i}$ is nonzero for $\xi > x$. Taking into account the leading-order QCD processes contributing to two-parton final states in lepton–parton reactions as shown in Fig. 2.1, one obtains[2]

$$\hat{\sigma}^{\ell i \to \ell' kj}(z, Q^2) = \frac{\alpha_{\mathrm{s}}}{2\pi}\sigma_0^{\ell i}\left[P(z)\ln\frac{Q^2}{m^2} + f(z) + \mathcal{O}\left(\frac{1}{Q^2}\right)\right] . \tag{2.22}$$

In (2.22), $z = x/\xi$, α_{s} is the strong coupling constant and m^2 is a cutoff mass or virtuality parameter regularizing the infrared divergences corresponding to collinear gluon emission or splitting. The m^2 singularity can be avoided by introducing an arbitrary but finite *factorization scale* μ_{f} and subsequently absorbing terms of $\mathcal{O}\left[\ln(\mu_{\mathrm{f}}^2/m^2)\right]$ into $p_{i/h}(\xi)$, which becomes $p_{i/h}(\xi, \mu_{\mathrm{f}}^2)$.

The *QCD factorization theorem* states that this procedure is process-independent and justified for all orders in α_{s} and in $t = \ln(Q^2/\mu_{\mathrm{f}}^2)$ and hence allows the use of the $p_{i/h}(\xi, \mu_{\mathrm{f}}^2)$ in a consistent way in the context of

---

[2] The following discussion mostly adopts the line of argument of [110] and [112].

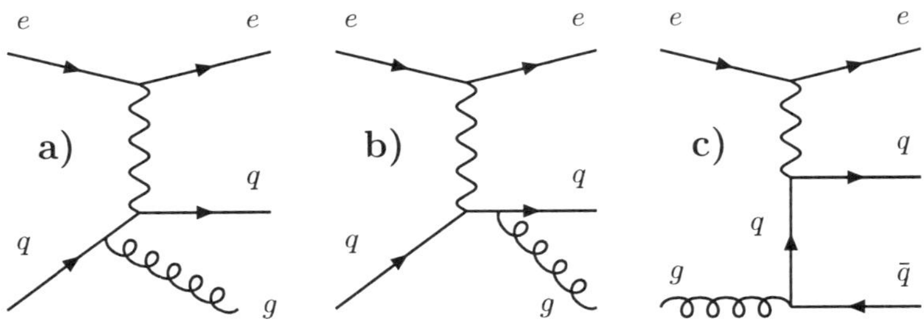

Fig. 2.1. Leading-order Feynman graphs for lepton–parton reactions with two-parton final states: (**a**,**b**) gluon radiation, (**c**) gluon splitting

pQCD applications to inclusive cross sections of reactions involving initial-state hadrons. Actually, the $p_{i/h}(\xi, \mu_f^2)$ and not the original "bare" functions $p_{i/h}(\xi)$ enter the cross section formulae and are thus experimentally observable. The $p_{i/h}(\xi, \mu_f^2)$ – or rather the functions derived from them by applying the pQCD calculus described below – are identified as the QCD equivalent of the QPM parton distribution functions and are the objects which we have already used as PDFs in Sect. 2.1.1.

Clearly, for high Q^2 the remaining $\ln(Q^2/\mu_f^2)$ term in (2.22) yields the dominant correction to σ_{QPM}, which again can be absorbed into the PDFs by defining an additive correction term

$$\delta p_{i/h}(x, Q^2) = \frac{\alpha_s}{2\pi} \ln\left(\frac{Q^2}{\mu_f^2}\right) \int_x^1 \frac{d\xi}{\xi} p_{i/h}(\xi, \mu_f^2) P\left(\frac{x}{\xi}\right) + \mathcal{O}(\alpha_s^2 t) , \qquad (2.23)$$

inducing an effective Q^2 dependence of the PDFs. Since $(\alpha_s \ln Q^2)$ is not small compared with unity, the contributions to $\sigma^{\ell h}$ have to be taken into account to all orders in $\alpha_s t$. This is achieved by turning (2.23) into a differential equation:

$$\frac{dp_{i/h}(\xi, Q^2)}{dt} = \frac{\alpha_s}{2\pi} \int_x^1 \frac{d\xi}{\xi} p_{i/h}\left(\xi, Q^2\right) P\left(\frac{x}{\xi}\right) + \mathcal{O}(\alpha_s^2) . \qquad (2.24)$$

Note that the PDF in the integral kernel is evaluated at scale Q^2, indicating that the factorization scale μ_f^2 is taken to be Q^2. This choice is not enforced by theory, and using alternatives like $\mu_f^2/Q^2 = \text{const.} \neq 1$ yields modifications of the $f(z)$ term in (2.22). Equation (2.24) represents a simplified form of the DGLAP evolution equations [102] which govern the Q^2 dependence of the PDFs in a given hadron type once these PDFs are known as a function of x for a fixed $Q^2 = Q_0^2$. It is important to observe that the integral in (2.24) only depends on the PDFs for $\xi > x$, indicating that the DGLAP evolution can be performed in a limited x range $x > x_{\min}$.

Let us look a little closer at the right-hand side of (2.24). The kernel function $P(z)$ originates from the QCD splitting processes shown in Fig. 2.1 and quantifies the probability that an incoming parton i undergoes one of the LO splitting processes ($q \to qg$, $g \to q\bar{q}$, $g \to gg$) and thereby produces a parton j with momentum x. As a consequence, there are four independent such *splitting functions* $P_{ji}(z)$ for $i, j = q, g$, and the full set of DGLAP equations includes integrals over the gluon densities which account for *sea quarks* produced by gluon splitting. For the proton, the DGLAP evolution equations can be written as

$$\frac{\mathrm{d}}{\mathrm{d}t}\begin{pmatrix} q^{\mathrm{S}}(x,Q^2) \\ q^{\mathrm{NS}}(x,Q^2) \\ g(x,Q^2) \end{pmatrix} = \frac{\alpha_{\mathrm{s}}}{2\pi}\begin{pmatrix} P_{qq} & 0 & P_{qg} \\ 0 & P_{qq} & 0 \\ P_{gq} & 0 & P_{gg} \end{pmatrix} \otimes \begin{pmatrix} q^{\mathrm{S}}(x,Q^2) \\ q^{\mathrm{NS}}(x,Q^2) \\ g(x,Q^2) \end{pmatrix} \tag{2.25}$$

with

$$q^{\mathrm{S}}(x,Q^2) = \sum_q \left[q(x,Q^2) + \bar{q}(x,Q^2)\right] , \tag{2.26}$$

$$\begin{aligned} q^{\mathrm{NS}}(x,Q^2) = \, & q_i(x,Q^2) - \bar{q}_i(x,Q^2) \quad \text{or} \\ & q_i(x,Q^2) - q_j(x,Q^2) \; (i \neq j) , \end{aligned} \tag{2.27}$$

$$P_{ji} \otimes q(x,Q^2) \equiv \int_x^1 \frac{\mathrm{d}\xi}{\xi}\, q(\xi,Q^2)\, P\left(\frac{x}{\xi}\right) . \tag{2.28}$$

The singlet distribution, $q^{\mathrm{S}}(x,Q^2)$, includes all quark flavors which can be produced by gluon splitting at a given Q^2 (see below for a discussion of heavy-quark PDFs). In the nonsinglet PDFs, $q^{\mathrm{NS}}(x,Q^2)$, the contributions of P_{qg} cancel by construction since gluon splitting is independent of flavor and charge. Prominent examples for nonsinglet PDFs are the *valence quark* distributions, $q_{\mathrm{v}}(x,Q^2) \equiv q(x,Q^2) - \bar{q}(x,Q^2)$.

Inserting (2.22) into (2.21) and using PDFs which are solutions of (2.25) finally results in

$$\begin{aligned} \sigma^{\ell h}(x,Q^2) = & \sum_i \int_x^1 \frac{\mathrm{d}\xi}{\xi}\, p_{i/h}(\xi,Q^2)\, \sigma_0^{\ell i}\left[\delta\left(\frac{x}{\xi} - 1\right) + \frac{\alpha_{\mathrm{s}}}{2\pi} f\left(\frac{x}{\xi}\right)\right] \\ & + \mathcal{O}(\alpha_{\mathrm{s}}^2) \\ = & \sum_i \sigma_0^{\ell i}\, p_{i/h}(x,Q^2) + \mathcal{O}(\alpha_{\mathrm{s}}) . \end{aligned} \tag{2.29}$$

The leading $(\alpha_{\mathrm{s}} t)$ terms are included in the effective PDFs, leaving only $\mathcal{O}(\alpha_{\mathrm{s}})$ corrections to the approximation which is obtained by inserting these PDFs into the QPM formula.

Some important remarks about PDFs and the DGLAP evolution are appropriate at this point.

Q^2 Dependence of PDFs and Spatial Resolution. The Q^2 dependence of the parton distributions can intuitively be understood as a consequence of the fact that increasing the virtuality Q^2 of a probe (like the gauge boson exchanged between lepton and hadron in DIS) corresponds to improving the spatial resolution, so that more virtual QCD subprocesses can be resolved and the probe "sees" e.g. a quark and a gluon instead of a quark carrying the sum of the momenta of both.

This argument can be put on a more quantitative footing. The maximum transverse momentum kinematically allowed in one of the QCD splittings in Fig. 2.1 is given by [110]

$$(k_\perp^2)_{\max} = Q^2 \, \frac{1-z}{4z} \qquad \text{with} \qquad z = \frac{x}{\xi} < 1 \, . \tag{2.30}$$

Consequently, the integration over $k_\perp$ yields a probability to observe the result of a splitting which increases with growing Q^2. Also, since the splitting is a virtual process, Heisenberg's uncertainty principle applies and yields a transverse spatial resolution $\propto 1/\sqrt{Q^2}$.

Universality of PDFs. Although an explicit probing process, i.e. *eh* scattering, was assumed in the above synopsis, the evolution equations can actually be derived using QCD only (see e.g. [112]). The PDFs are hence process-independent global quantities characterizing the respective hadron. However, corrections of $\mathcal{O}(\alpha_s)$ or higher orders to the leading-order cross section in (2.29) are process-dependent and have to be taken into account explicitly in the respective calculations (see the paragraph on higher-order corrections below).

Running α_s. So far, virtual QCD corrections due to Feynman graphs like those in Fig. 2.2 have been neglected. It can be shown that they are correctly accounted for by using in (2.25) the running coupling constant $\alpha_s = \alpha_s(Q^2)$, which is a solution of the renormalization group equation (RGE):

$$\begin{aligned}
\frac{\partial \alpha_s(\mu_r^2)}{\partial \ln \mu_r^2} &= -\beta_0 \frac{\alpha_s^2(\mu_r^2)}{4\pi} - \beta_1 \frac{\alpha_s^3(\mu_r^2)}{(4\pi)^2} + \mathcal{O}(\alpha_s^4) \, , \\
\beta_0 &= 11 - \frac{2N_f}{3} \, , \\
\beta_1 &= 102 - \frac{38N_f}{3} \, ,
\end{aligned} \tag{2.31}$$

where N_f is the number of quark flavors the gluon can split into (*active flavors*; see below for a discussion of the Q^2 dependence of N_f). The α_s^3 term in (2.31) is the result of a next-to-leading-order calculation, i.e. it corresponds to the $\mathcal{O}(\alpha_s)$ term in (2.29). Similarly to the evolution (2.25), solutions of the RGE involve implicitly the resummation of all terms of $\mathcal{O}\left[(\alpha_s \ln \mu_r^2)^n\right]$.

The purpose of the *renormalization scale* μ_r in (2.31) is, similarly to that of μ_f, to absorb infrared singularities into observable quantities. Usually $\mu_r^2 = Q^2$ is considered a natural choice, which, however, is not fixed by theory.

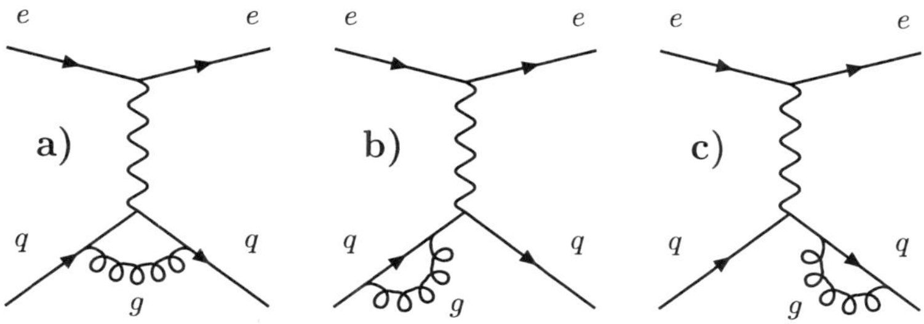

Fig. 2.2. Leading-order Feynman graphs giving rise to virtual QCD corrections to lepton–parton scattering: (**a**) vertex correction, (**b**,**c**) self-energy corrections

Making use of the convention of evaluating experimental measurements of α_s at $\mu_r = M_Z$, the solution of (2.31) can be written as

$$\frac{1}{\alpha_s(\mu_r^2)} = \frac{1}{\alpha_s(M_Z^2)} + \beta_0 \ln\left(\frac{\mu_r^2}{M_Z^2}\right) - \frac{\beta_1}{\beta_0} \ln\left\{\frac{\alpha_s(\mu_r^2)[\beta_0 + \beta_1 \alpha_s(M_Z^2)]}{\alpha_s(M_Z^2)[\beta_0 + \beta_1 \alpha_s(\mu_r^2)]}\right\} + \dots . \tag{2.32}$$

According to (2.32), the coupling constant α_s vanishes logarithmically in the limit $\mu_r \to \infty$ (unless $N_f > 16$, which is, however, avoided by nature). This result, which is a consequence of the RGE and is true to all orders of pQCD, is called *asymptotic freedom* and resolves the seeming contradiction that partons can be considered as quasi-free in DIS while they are tightly bound in hadrons at the same time: the existence of a *hard scale* (Q^2 in the case of DIS), which can be identified with μ_r^2, allows us to "decouple" the hard subprocess from the nonperturbative *soft physics* which happens at energy scales of the order of hadronic masses.

Following the above arguments, one would expect that the validity of the DGLAP equation breaks down for Q^2 values for which $\alpha_s(\mu_r^2 = Q^2)$ approaches unity, i.e. for Q^2 less than a few GeV^2. Actually, the latest HERA structure function measurements at low Q^2 [6, 11, 116] (see [46] for a detailed review of experimental data and theoretical aspects) have demonstrated that the Q^2 range of validity extends down to 0.5–$1.0\,\mathrm{GeV}^2$, at least at small $x \lesssim \mathcal{O}(10^{-4})$. This observation may indicate that actually α_s/π is the relevant quantity which has to be smaller than unity, as argued by Glück, Reya and Vogt [117]. For the objectives of this book, it is sufficient to state that the DGLAP evolution is expected to hold in the full Q^2 range from the fixed-target data to the HERA high-Q^2 domain.

Higher-order Corrections. So far, terms of $\mathcal{O}(\alpha_s)$ in the cross section (2.29) and terms of $\mathcal{O}(\alpha_s^2)$ in the evolution (2.25) have been neglected. This approximation consistently takes into account terms of $\mathcal{O}[(\alpha_s t)^n]$, resummed to all orders of n, and is usually named the *leading-log approximation* (LLA).

Actually, the leading-order (LO) results presented in Sect. 2.1.1 are of LLA type if the PDFs are solutions of the DGLAP equations. The terms LLA and LO are often (but not always) used synonymously.

Adding higher-order diagrams to the hard-subprocess cross section $\hat{\sigma}^{\ell i}$ results in contributions of $\mathcal{O}(\alpha_s^n t^{n-1})$, yielding corrections to the LLA formulae of the same order as those due to the $f(x/\xi)$ term in (2.29). This not only implies that both types of terms have to be included at the next level of accuracy (the *next-to-leading-log approximation*, NLLA), but also that it becomes ambiguous which of these terms are assigned to the PDFs. In particular, in the NLLA and higher orders it is no longer possible to assign all corrections to the PDFs in a process-independent way. Consequently, the structure functions introduced in (2.5), (2.6) and (2.13), (2.14) acquire corrections, which, in the representative case of $\mathcal{F}_2^{\rm NC}$, are given by

$$\mathcal{F}_2^{\rm NC}(x,Q^2) = \sum_q (q+\bar{q}) \otimes C_2^q + g \otimes C_2^g \ , \tag{2.33}$$

$$C_2^q(z) = Q_q^2 \left[\delta(1-z) + \frac{\alpha_s(Q^2)}{2\pi} f_2(z) + \mathcal{O}(\alpha_s)^2\right] \ , \tag{2.34}$$

$$C_2^g(z) = \frac{\alpha_s(Q^2)}{2\pi} f_g(z) + \mathcal{O}(\alpha_s)^2 \ , \tag{2.35}$$

where the folding operator $\otimes$ is defined in (2.28). The *coefficient functions* $C_2^{q,g}$ differ for different structure functions and different processes and are furthermore related to the splitting functions in (2.25) by the *factorization scheme*, which fixes how the coefficient and splitting functions share same-order contributions. Note that any PDF set always has to be used together with coefficient functions calculated in the same factorization scheme and to the same order; inconsistent combinations will yield wrong results. Two factorization schemes are in widespread use: the DIS scheme (defined such that $\mathcal{F}_2^{\rm NC}$ retains the simple form from (2.5) also in the NLLA), and the $\overline{\rm MS}$ scheme [118], which has other advantages like preserving the momentum sum rule (see Sect. 2.1.4) to all orders in α_s.

It can be shown that the complete perturbation series of corrections of $\mathcal{O}(\alpha_s^n)$, including those beyond the NLLA, can be written as a sum over terms of $\mathcal{O}[\alpha_s^n (Q^2)^{n-k} \ln^{n-l}(1/x)]$ according to the following scheme (see [119] and references therein):

$$\begin{array}{rlcl}
 & \underbrace{\begin{array}{l} \alpha_s^n \ln^n Q^2 \ \times \left[\ln^n(1/x) + \ln^{n-1}(1/x) + \ldots\right] \\ \alpha_s^n \ln^{n-1} Q^2 \times \left[\ln^n(1/x) + \ln^{n-1}(1/x) + \ldots\right] \\ \alpha_s^n \ln^{n-2} Q^2 \times \left[\ln^n(1/x) + \ln^{n-1}(1/x) + \ldots\right] \\ \ldots \end{array}}_{\text{BFKL}} & & \begin{array}{l}\text{LLA} \\ \text{NLLA} \\ \text{NNLLA} \\ \end{array}
\end{array} \tag{2.36}$$

The LLA (NLLA) scheme discussed above corresponds to using PDFs which are solutions of the LO (NLO) DGLAP evolution equations and hence include the first (first two) rows of (2.36), resummed to all orders of n. The abbreviation NLO (*next-to-leading order*), although often used synonymously with NLLA, refers to the next-to-leading order of the underlying hard subprocess but does not imply resummation.

The summation over the $\ln(1/x)$ terms is implicitly performed by assuming a certain x dependence of the PDFs at a fixed $Q^2 = Q_0^2$. It is expected that below some very small x, far outside the region of interest for this book, the leading logarithms of $1/x$ will dominate the perturbation series. Alternative evolution equations which are based on the summation of these logarithms have been derived by Balitskii, Fadin, Kuraev and Lipatov (BFKL) [120], as indicated in (2.36). This approach will not be pursued further here.

Splitting and coefficient functions have been calculated in the LLA and NLLA (see [121, 122] and references therein), and progress is being made on the *next-to-next-to-leading-log approximation* (NNLLA) [121, 123]. For the calculation of cross section predictions for the HERA high-Q^2 analyses, as well as for the Monte Carlo simulations, the NLLA prescription is used.

The Longitudinal Cross Section. Both for NC and for CC scattering, the NLLA (and higher-order) cross sections receive a nonzero contribution from the respective longitudinal structure function, $\mathcal{F}_{\rm L}$, which is of the type of the $\mathcal{O}(\alpha_{\rm s})$ contribution to (2.33) [110]:

$$\frac{{\rm d}^2\sigma_{\rm L}^{ep\to eX}}{{\rm d}x\,{\rm d}Q^2} = -\frac{2\pi\alpha^2}{Q^4}\,y^2\mathcal{F}_{\rm L}^{\rm NC}(x,Q^2)\,, \tag{2.37}$$

$$\frac{{\rm d}^2\sigma_{\rm L}^{e^\pm p\to \overset{(-)}{\nu} X}}{{\rm d}x\,{\rm d}Q^2} = -\frac{G_{\rm F}^2}{2\pi}\left(\frac{M_W^2}{Q^2+M_W^2}\right)^2 y^2\mathcal{F}_{\rm L}^{{\rm CC}\pm}(x,Q^2)\,, \tag{2.38}$$

with

$$\mathcal{F}_{\rm L}^{\tau}(x,Q^2) = \mathcal{F}_2^{\tau}\otimes C_{\rm L}^{\mathcal{F}} + g\otimes C_{\rm L}^{g}\,. \tag{2.39}$$

The coefficient functions $C_{\rm L}^{\mathcal{F}}$ and $C_{\rm L}^{g}$ do not depend on the type τ of the DIS reaction. The longitudinal cross section is sizable and dominated by the gluon distribution at low x, but for the kinematic regime considered here it only induces a small correction to the LO cross section. It is taken into account by the standard NLLA calculations.

Higher-Twist Terms. With decreasing Q^2, corrections of $\mathcal{O}(1/Q^{2n})$ to the perturbative results described above become important. They originate from the corresponding terms in (2.22), which are related to reinteractions of the struck quark with the hadron remnant and from target-mass effects. A set of leading terms of the higher-twist contributions related to finite target mass can be calculated using the operator product expansion [124]. However, currently there is no complete theoretical description of higher-twist terms (see [46] for more details and [125] for some recent calculations).

Lately, quantitative investigations of the size of higher-twist contributions to DIS structure functions [126, 127] have corroborated the expectation that higher-twist contributions are largest[3] at high x and hence are concentrated at low W, indicating that they are irrelevant for the HERA data but play a rôle when deriving PDFs from fixed-target data at high x. Typically, a lower cutoff in W is applied in such analyses, and the results are checked with respect to their stability against variations of this cutoff. For the HERA high-Q^2 analyses, higher-twist terms have been neglected.

Scale Dependences. Two mass scales have been introduced in this section so far: the factorization scale μ_f^2 and the renormalization scale μ_r^2. Both are arbitrary scales, and any result obtained by taking into account the full theory must not depend on them. It is customary to identify these scales with Q^2, and we have followed this convention without having mandatory reasons to do so. It can be shown that multiplying the scales μ_f^2 or μ_r^2 by constant factors corresponds to a modification of the next-higher-order terms in the RGE or the DGLAP evolution, i.e. of the lowest-order terms which have been neglected (and hence this has no effect in the limit of infinite order, i.e. if the full perturbation series is summed).

Usually the sensitivity of cross section calculations to the scales μ_f^2 and μ_r^2 is evaluated by varying the scales in the range $\mu^2 = Q^2/f_\mu$ to $f_\mu\, Q^2$, with typical values $f_\mu = 4$, and using the results as estimates both of the scale-related cross section uncertainties and of the impact of neglected higher-order QCD corrections. On the other hand, one can also argue that the relative difference between results obtained in two successive orders of accuracy in the pQCD expansion (often referred to as the *K factor*) may serve as an upper estimate of the contributions of higher-order corrections and hence also of the uncertainties related to the choice of factorization and renormalization scales. Some illustrations of the cross section modifications in the HERA high-Q^2 domain corresponding to μ_f and μ_r variations by a factor $f_\mu = 2$ can be found in Sect. 2.1.6 (see in particular Fig. 2.12).

Heavy Quarks. The charm (c), bottom (b) and top (t) quarks carry masses ($m_c \approx 1.3\,\mathrm{GeV}$, $m_b \approx 4.3\,\mathrm{GeV}$, $m_t \approx 174\,\mathrm{GeV}$ [108]) which are not necessarily negligible compared with the typical scales of HERA high-Q^2 kinematics. As a consequence, the dynamical behavior of these *heavy quarks* differs from their lighter counterparts, both in their virtual interactions inside a hadron and in the hard subprocess of lepton–quark scattering.

Virtual processes involving heavy quarks affect the RGE, (2.31), describing the Q^2 dependence of α_s. The number of active flavors in the RGE, N_f, grows by one unit once Q^2 increases beyond the threshold below which virtual splitting processes involving charm, bottom or top quarks are suppressed. The standard procedure [128] to approximate these thresholds in DIS is to assume

[3] Recently it has also been speculated that there may be significant higher-twist contributions in the BFKL regime at very low x, which, however, do not affect the high-Q^2 domain at HERA.

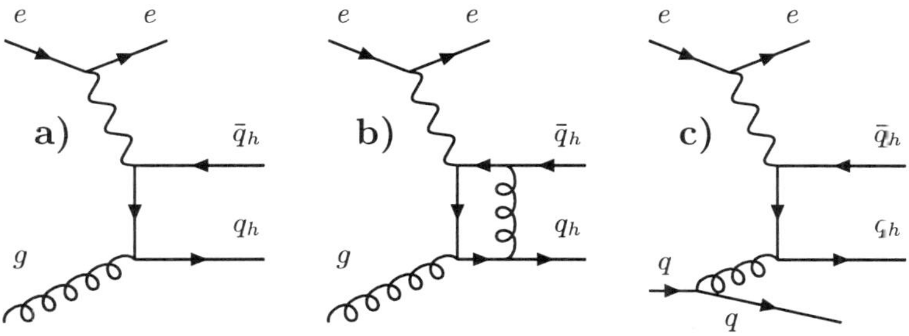

Fig. 2.3. Examples of mechanisms for heavy-quark production: (**a**) boson–gluon fusion (BGF), (**b**) gluon one-loop correction to the BGF graph, (**c**) leading-order example of heavy-quark production in a scattering of the electron on a light quark

a steady Q^2 dependence of α_s and a step-function-like increase $N_f = 3 \to 4$ at $Q^2 = m_c^2$, $N_f = 4 \to 5$ at $Q^2 = m_b^2$ and $N_f = 5 \to 6$ at $Q^2 = m_t^2$. Since N_f enters into the RGE dominantly via $\beta_0 = 11 - 2N_f/3$, each flavor threshold corresponds to a change of the $\ln(Q^2)$ slope of α_s by $\mathcal{O}(8\%)$.

There are two extreme approaches to calculating the cross section of deep inelastic reactions with a heavy quark in the initial state of the hard subprocess. In the *variable-flavor-number scheme* (VFNS, sometimes referred to as *massless evolution*), these cross sections are entirely assigned to heavy-quark PDFs, which are required to vanish for Q^2 below some scale $\mu_{q_h}^2 = \mathcal{O}(m_{q_h}^2)$ ($q_h = c, b, t$) and are generated by the DGLAP evolution equation above this threshold. This ansatz was used e.g. for the earlier PDF parameterizations by Martin, Roberts and Stirling [129, 130] (see Sect. 2.1.4). Alternatively, the *fixed-flavor-number scheme* (FFNS) assumes that heavy quarks are exclusively produced via *boson–gluon fusion* (BGF) (Fig. 2.3a) or similar higher-order graphs like those shown in Figs. 2.3b,c. In this scheme, which is used by Glück, Reya and Vogt [117, 131, 132], there is no concept of heavy-quark PDFs. Instead, the cross section contributions from heavy quarks are calculated using suitable BGF coefficient functions, which yield the heavy-quark structure functions $\mathcal{F}_i^{q_h}$ according to (2.33) and have been calculated to leading order ($\mathcal{F}_i^{q_h} \propto \alpha_s$) [133] and also to next-to-leading order [134].

In the threshold region, the FFNS is expected to yield a realistic description of heavy-quark production, whereas the VFNS is obviously not capable of accommodating such processes for $Q^2 < \mu_{q_h}^2$ (where, at sufficiently low x, the invariant hadronic mass W is still large enough to contain heavy-quark pairs). On the other hand, the VFNS is expected to hold in the limit $Q^2 \gg m_{q_h}^2$, where the quark masses become negligible and the heavy quarks effectively behave like light quarks. In this limit, the FFNS is inappropriate since it does not contain the resummation of higher-order contributions implicit in the DGLAP evolution.

Triggered by recent HERA measurements of the proton charm structure function $\mathcal{F}_2^c$ [4, 10], different approaches to connect the FFNS and VFNS have been intensively discussed [135, 136], amongst these in particular the *on-mass-shell scheme* suggested by Aivazis, Collins, Olness and Tung (ACOT) [137], which contains the FFNS and VFNS as limiting cases for $Q^2 \lesssim m_{q_h}^2$ and $Q^2 \gg m_{q_h}^2$, respectively. Recently, a generalized factorization theorem to all orders of pQCD was proven for this scheme [138]. The CTEQ collaboration has meanwhile provided PDF parameterizations [139] using the ACOT scheme, and Martin, Roberts, Stirling and Thorne (MRST) used a variant of this approach [140] for their latest PDF set [141] (see also Sect. 2.1.4). One of the results of these studies is that the VFNS-like heavy-quark PDFs rapidly dominate the $\mathcal{F}_i^{q_h}$ once Q^2 rises above the thresholds.

For the HERA high-Q^2 studies, obviously the VFNS is appropriate for the charm and bottom contributions, whereas the top-quark contributions can be neglected (see Sect. 2.1.1). However, the correct treatment of the heavy-quark thresholds is an essential ingredient for the determination of PDFs in analyses using the lower-Q^2 fixed-target data (see Sects. 2.1.4 and 2.1.6).

2.1.3 Electroweak Radiative Corrections

In contrast to the QCD corrections illustrated in the previous section, the effects of higher-order terms in the electroweak perturbation series do not significantly alter the hadronic structure, i.e. the PDFs or evolution equations, but modify the hard-subprocess cross section. A comprehensive discussion of radiative corrections at HERA can be found in [42].

Electroweak radiative corrections can be categorized according to whether they are caused by (i) graphs involving loops, like the examples shown in Fig. 2.4 (*virtual corrections*), or (ii) the radiation of real photons from one of the charged lines in the LO graphs in Fig. 1.1, as in the examples of Fig. 2.5.

The dominant effect for NC reactions is due to the radiation of real photons from the electron lines, $ep \to e\gamma X$ (see Fig. 2.5). The fivefold differential cross section for this process, $\mathrm{d}^5\sigma/(\mathrm{d}x\,\mathrm{d}Q^2\,\mathrm{d}^3\boldsymbol{k}_\gamma)$, exhibits three poles, corresponding to zero virtuality of either the electron or the photon propagator, i.e. for $z_1 = k_\gamma \cdot k = 0$, $z_2 = k_\gamma \cdot k' = 0$ (electron) or for $z_3 = (k - k' - k_\gamma)^2 = 0$ (photon), where k_γ denotes the photon four-momentum and the electron mass has been neglected. The cross section for radiative events is largest close to these poles and falls off with $1/z_i$ in their vicinity. The three poles correspond to photons which (i) are emitted collinearly with the initial-state electron (*initial-state radiation*, ISR), (ii) are emitted collinearly with the final-state electron (*final-state radiation*, FSR) or (iii) balance the transverse momentum of the final-state electron (*QED-Compton* process; the electron scatters on a quasi-real photon emitted from the proton). The latter pole is absent for Z exchange. The NC radiative cross section can be calculated [142] as a function of an arbitrary but small lower cutoff on the photon energy, and

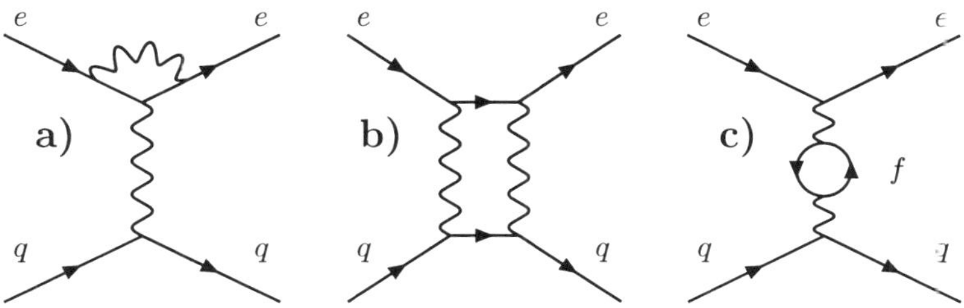

Fig. 2.4. Examples of virtual electroweak corrections to NC electron–quark scattering: (**a**) vertex correction, (**b**) box graph, (**c**) self-energy correction of the γ/Z propagator due to a fermion loop. In all graphs the *photon-type lines* denote one of the neutral electroweak gauge bosons, γ or Z

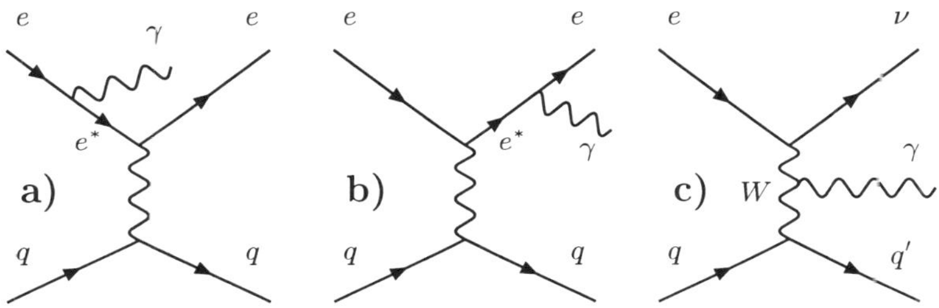

Fig. 2.5. Radiation of a real photon from (**a**) the incoming and (**b**) the scattered electron line in an NC eq reaction, and (**c**) from the W propagator in a CC reaction. Lines representing the electron propagators are denoted by e^*

the remaining soft-photon corrections are added to the virtual radiative corrections of the leading-order cross section. A similar procedure is applied to CC reactions [143], $ep \to \nu X$, where the FSR and QED-Compton poles are absent and photon radiation from the W propagator (see Fig. 2.5c) yields an additional contribution, which, however, does not exhibit poles.

In addition to the cross section modification, ISR processes have a particular experimental significance since the photon can escape detection. In this case, the reconstruction of kinematic variables is degraded, irrespective of which reconstruction method is used (see Sect. 3.3.3 and [144]). The relative change of $\mathrm{d}^2\sigma/(\mathrm{d}x_\ell\,\mathrm{d}Q_\ell^2)$ due to electroweak radiative corrections may serve as an example of the size of the resulting effects:[4] for $y_\ell > 0.8$ and $x_\ell < 0.01$ the corrections exceed 50%, whereas the cross section is reduced by more than 30% for $y_\ell < 0.2$ and $x_\ell > 0.5$ [145].

For photon radiation from the initial- or final-state quark, the distortion of kinematic variables is absent in the dominant cases where the photon is emitted in the proton or current-quark direction. Furthermore, since quarks have larger masses and smaller charges than electrons, these processes are suppressed and therefore contribute little to the overall radiative corrections.

[4] See Sect. 1.3.1 for the definition of the leptonic variables Q_ℓ^2 and x_ℓ^2.

The leading terms of the complete set of electroweak soft and virtual corrections mentioned above have been calculated [142, 143], and the dominant contributions can be taken into account by introducing various gauge-invariant form factors which modify the amplitudes related to the exchange of the different gauge bosons. For the photon exchange this corresponds to using a Q^2-dependent "running" coupling constant $\alpha = \alpha(Q^2)$. Likewise, the weak mixing angle is effectively modified by a Q^2-dependent form factor. In general, the virtual corrections to the leading-order cross section are of the order of a few percent and vary slowly and smoothly with Q^2 and x (see [107] for a detailed survey). They depend on SM parameters like the top and Higgs masses to which they thus yield some sensitivity.

The algorithms and assumptions involved in calculations of virtual electroweak corrections have been proven to be highly accurate and to provide impressive predictive power. The most prominent example is the top mass, m_t, which was found at Tevatron [146] to lie in the narrow window which had been predicted [147] from the LEP and SLD electroweak precision measurements by exploiting the radiative-correction-induced m_t dependences of the SM parameters. For lepton–hadron scattering, the precision with which radiative corrections are known is limited by higher-order contributions with multiphoton emission and by the choice of the input used for their calculation. The corresponding uncertainty is largest at the edges of phase space, in particular at high y, where it amounts to a few percent [145]. The errors on virtual one-loop corrections are negligible in the context of this book.

For almost all aspects of data analysis, a detailed Monte Carlo simulation of *ep* reactions including all relevant radiative corrections is mandatory. This simulation is provided by the HERACLES program package which includes photon radiation to $\mathcal{O}(\alpha)$ and the full set of one-loop electroweak corrections [148]. All comparisons of experimental results with theoretical predictions covered in this book have been performed using HERACLES. See Sect. 3.2 for a summary of Monte Carlo generators used for the analyses.

2.1.4 Experimental Input Data from Previous Experiments

In the previous sections the various ingredients for the calculation of *ep* DIS cross sections have been summarized. We will now collect the experimental information needed as input for this calculation. An overview of the procedures and assumptions used to derive the commonly used PDF parameterizations from the data discussed in this section will be given in Sect. 2.1.5.

The parameters governing the electroweak sector of the SM (referred to as *electroweak parameters*) comprise α, $\sin\theta_{\mathrm{W}}$, the gauge boson masses M_W and M_Z and, in addition, the elements of the CKM matrix, the fermion masses and the Higgs mass. The first four of these parameters are interrelated by the underlying $\mathrm{SU(2)_L \times U(1)_Y}$ gauge symmetry, with additional mass dependences (mainly on the top and Higgs masses) induced by radiative corrections. The values of those parameters entering into (2.2)–(2.12), as

Table 2.1. Latest world average values and combined statistical and systematic uncertainties of the electroweak parameters entering the *ep* DIS cross sections

Parameter	Value	Relative error	Cited from
$\alpha(0)$	1/137.0359895	0.045 ppm	[108]
$\alpha(M_Z^2)$	1/128.878	0.095%	[149, 150]
$\sin^2\theta_{\mathrm{W}}(M_Z^2)$	0.23157	0.08%	[149]
G_{F} (GeV^{-2})	1.16639×10^{-5}	9 ppm	[108]
M_Z (GeV)	91.1867	30 ppm	[149]
M_W (GeV)	80.371	0.032%	[149]

summarized in Table 2.1, are taken from the Particle Data Group [108] and from a combined SM analysis [149] of the precision electroweak measurements at LEP, SLD and the Tevatron. The increase of the error on α with the mass scale reflects theoretical uncertainties on the radiative corrections inducing this scale dependence. The M_Z value has been obtained assuming lepton universality; the value of M_W is the result of a fit presupposing the validity of the SM. An estimate of the uncertainty of the SM *ep* DIS cross section prediction due the errors quoted in Table 2.1 can be found in Sect. 2.1.6.

In contrast to the electroweak sector, the QCD input to the DIS cross section calculation requires knowledge about a sector of the theory which cannot be assessed perturbatively, i.e. the low-Q^2 QCD phenomenology of hadron structure which is absorbed into the PDFs. In the following we will summarize the experimental data which are used to constrain the PDFs. The PDF fit procedures are discussed in Sect. 2.1.5, and some selected results and investigations of the related uncertainties are collected in Sect. 2.1.6.

In principle, data for any reaction involving hard subprocesses can be used to gain information on the PDFs of the participating hadrons. This opens a wide field of experimental data that can be used as input for nucleon PDF parameterizations, the most important of which are the following.

NC Structure Functions. Several fixed-target experiments (BCDMS [151–153] and NMC [154] at CERN, E665 [155] at Fermilab, SLAC [156])[5] have performed high-precision measurements of the differential cross sections for eN or μN DIS (the nuclear targets, N, being protons, deuterons or heavier nuclei) and have extracted the structure functions $\mathcal{F}_2^{\mathrm{NC}}$. Corresponding measurements have also been published based on the data taken by H1 [8, 9, 11] and ZEUS [2, 3, 5, 6] until 1995, filling up a large portion of those parts of the HERA (x, Q^2) plane which were left empty by the fixed-target experiments (see Fig. 2.6). Note that structure function results from the more recent HERA data discussed in this book have not yet been included in global PDF analyses. The combined $\mathcal{F}_2^{\mathrm{NC}}$ data are the major input for PDF deter-

[5] This list is restricted to data which are actually used in contemporary PDF analyses. A summary of past experiments can be found in e.g. [157].

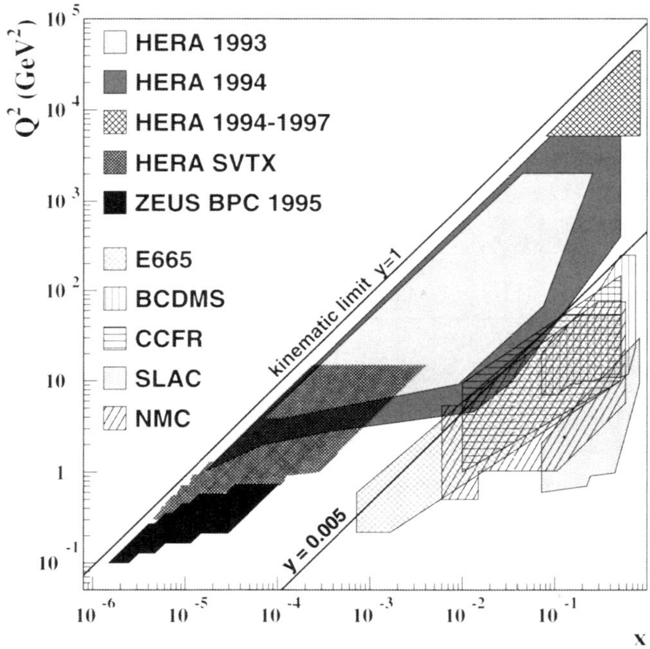

Fig. 2.6. Regions in the HERA kinematic (x, Q^2) plane covered by the structure function measurements of ZEUS and H1 and of the fixed-target experiments with charged lepton beams (BCDMS, NMC, E665, SLAC) and with neutrino beams (CCFR). The kinematic regions covered by the HERA analyses of the 1993 [2, 8] and the 1994 [3, 9] data are marked as "HERA 1993" and "HERA 1994", respectively; "HERA SVTX" and "ZEUS BPC 1995" indicate results obtained from HERA runs with longitudinally displaced interaction points [6, 11] and from a ZEUS measurement using a dedicated calorimeter close to the rear beam pipe [5]. The *cross-hatched triangle* at the highest Q^2 marks the region which has become accessible to precision measurements only with the data taken after 1995. The cross section results for this region, which are one of the major subjects of this book, are not yet used for PDF analyses

minations; as an illustration, the $\mathcal{F}_2^{\mathrm{NC}}$ results for proton targets are shown in Fig. 2.7 as functions of Q^2 for different fixed x values (the results from [11] are not included in this plot). The lines indicate the result of a QCD fit and reproduce well the scaling-violating tendencies of the data. In particular, the logarithmic rise (fall-off) of $\mathcal{F}_2^{\mathrm{NC}}$ with increasing Q^2 at low (high) x is clearly evident.

Given the fact that the QCD evolution of a parton distribution to a given point (x, Q^2) depends only on the PDFs at lower Q^2 and higher x (see (2.24)), it becomes obvious that the main input for the PDFs in the HERA high-Q^2 domain comes from the fixed-target experiments, whereas the high-x HERA data are still statistics-limited and are roughly an order of magnitude less precise than the BCDMS, NMC and E665 data at the same x. It is important

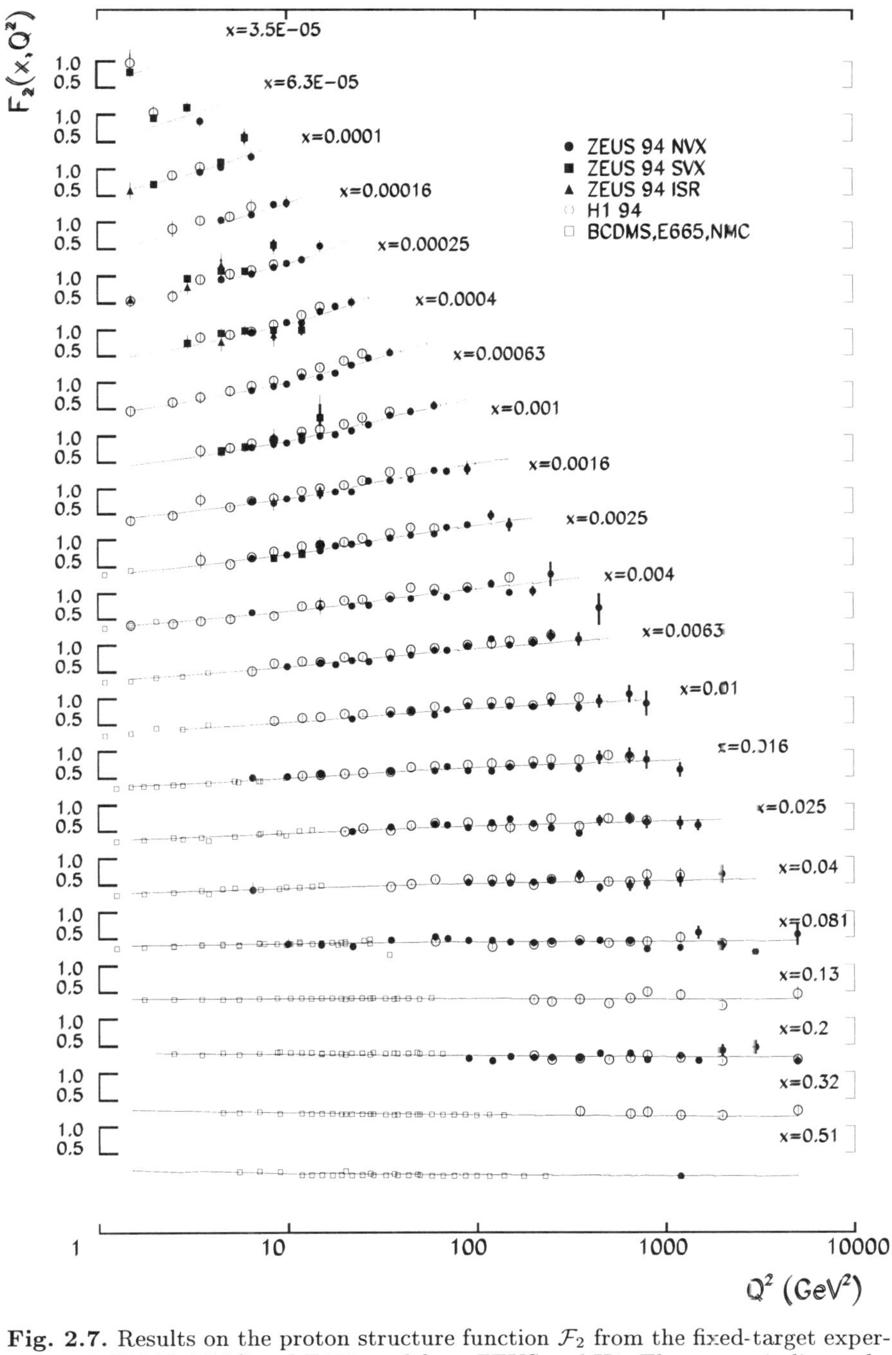

Fig. 2.7. Results on the proton structure function $\mathcal{F}_2$ from the fixed-target experiments BCDMS, NMC and E665, and from ZEUS and H1. The *curves* indicate the result of an NLLA QCD fit taken from the ZEUS analysis

to note that the NC proton data, in particular in the fixed-target regime, are dominated by the photon exchange graph and hence mainly carry information on the PDF combination $(Q_u^2 u + Q_d^2 d) \propto (4u + d)$. In order to disentangle the u and d distributions, one has either to include CC data (see below) or to use NC DIS data obtained with deuterons or heavier nuclear targets [154]. The analysis of the latter is commonly based on the assumption of isospin symmetry between the quark distributions in the proton and neutron, and requires corrections for nuclear effects (cf. Sect. 2.1.6).

On the basis of the identification of charm particles in the hadronic final state, H1 [10] and ZEUS [4], as well as previously EMC [158], have measured the differential cross section for NC DIS reactions in which charm quarks participate in the hard subprocess, albeit as yet with limited precision (typically more than 15% uncertainty). The structure function $\mathcal{F}_2^c$ derived from these results is uniquely related to $c(x, Q^2)$ (see Sect. 2.1.2) and provides constraints on this PDF and on the treatment of heavy quarks in general. In the latest PDF fits from CTEQ [139] and MRST [141] it has been demonstrated that a consistent theoretical approach combining the advantages of the FFNS and VFNS is able to reproduce the $\mathcal{F}_2^c$ data.

In addition to measurements of structure functions at particular values of (x, Q^2), with ranges determined by the experimental setup, there are also constraints on integrals over structure functions or PDFs, usually referred to as *sum rules.* One example is the Gottfried sum rule, which is derived using isospin symmetry between the proton and neutron, whereby in LO

$$I_{\mathrm{G}} = \int_0^1 \mathrm{d}x \left(\mathcal{F}_2^{\ell p \to \ell X} - \mathcal{F}_2^{\ell n \to \ell X} \right) = \frac{1}{3} + \frac{2}{3} \int_0^1 \mathrm{d}x \, (\bar{u} - \bar{d}) \, . \tag{2.40}$$

The NMC collaboration has reported a measurement [159] of $I_{\mathrm{G}} = 0.235 \pm 0.026$, indicating $\langle \bar{d} \rangle > \langle \bar{u} \rangle$.

CC Structure Functions (Neutrino Scattering). Structure function results from deep inelastic $\overset{(-)}{\nu} N$ scattering have been reported by the experiments CHARM [160, 161], CDHS(W) [162] and BEBC [163] at CERN and by CCFR [164–166] at Fermilab. Only the latter data have been included in recent PDF analyses. The relevant measurements are those of the CC structure functions $\mathcal{F}_{2,3}^{\mathrm{CC}}$ for neutrino and for antineutrino beams, which allow a distinction between u-type and d-type PDFs. In addition, unlike-sign dimuon production, $\overset{(-)}{\nu} N \to \mu^- \mu^+ X$, tags charm production via $\nu s \to \mu c$ or the charge-conjugate process and hence constrains the strange-quark PDF, $s(x, Q^2)$. The cross section for this process has been measured by CCFR [167].

W Production in $p\bar{p}$ Scattering. Exploiting the paradigm that the antiquark PDFs in the antiproton are identical to the quark PDFs in the proton, W production in $p\bar{p}$ scattering provides another possibility to separate the u and d PDFs. The dominant reaction mechanisms are $u\bar{d} \to W^+$ and the charge-conjugate process, and the charge and rapidity of the muon from $W \to \mu\nu$

decays hence carry information on the type and x values of the initial-state quarks. The total cross sections for W production (measured by UA2 [168] at CERN and by the Tevatron experiment CDF [169]) and the rapidity distributions of the muons (CDF [170]) are used as input to the PDF fits.

Drell–Yan Data. Drell–Yan production of lepton pairs $\ell^+\ell^-$ in hadron–hadron collisions proceeds via the hard subprocess $q\bar{q} \to \ell^+\ell^-$, with a virtual photon or Z in the s channel. Experimentally, the signature of this reaction is cleanest if muon pairs are produced, $\ell = \mu$. Drell–Yan cross section measurements have been reported by the E605 [171] and E772 [172] collaborations for pN and pp initial states, respectively. These data and in addition the ratio of the $pp \to \mu^+\mu^- X$ and $pd \to \mu^+\mu^- X$ cross sections, as measured by the NA51 [173] and recently also the E866 [174] collaborations, are used for PDF analyses. The p/d Drell–Yan cross section ratio significantly constrains $\bar{d}/\bar{u}$, and hence adds to the information provided by the Gottfried sum rule measurement.

Prompt-Photon Data. Isolated photons with high transverse momentum produced in hadron–hadron collisions via $qg \to q\gamma$ yield information on the gluon density. Some of the standard PDF analyses include corresponding data from the WA70 experiment [175] and, most recently, also from precision measurements by the E706 collaboration [176]. The use of these data for constraining $g(x, Q^2)$ is, however, limited by the observation that their interpretation depends strongly on assumptions on the initial-state partonic transverse-momentum spectrum [139, 141, 177].

Inclusive Jet Production in $p\bar{p}$ Collisions. Jets produced at high transverse momentum, $p_{\rm t}^{\rm jet}$, in hadron–hadron reactions signal the transverse momentum of the final-state partons in the underlying hard subprocesses ($q\bar{q} \to gg$, $qg \to qg$ and $gg \to gg$) and therefore allow us to constrain the PDFs of the initial-state partons, in particular that of the gluon at $x = \mathcal{O}(0.1)$. Inclusive jet cross sections $\mathrm{d}\sigma/\mathrm{d}p_{\rm t}^{\rm jet}$ have recently been reported by the CDF [178, 179] and the DØ [180] collaborations at Tevatron and are used by the CTEQ collaboration for their latest PDF analysis [139], although both experiments are not in perfect agreement and the CDF results show a marked cross section excess over NLO QCD expectations derived from earlier PDF parameterizations.

$\alpha_{\rm s}$ Measurements. Although $\alpha_{\rm s}$ can in principle be inferred from the Q^2 dependence of the structure function data (cf. e.g. Fig. 2.7), the value of $\alpha_{\rm s}(M_Z^2)$ is usually imposed as an additional constraint. This serves both to preserve consistency with data not included in the analyses and to reduce the uncertainty of the gluon distribution, which, in particular at low x, is closely related to $\partial\mathcal{F}/\partial\ln(Q^2)$, which in turn is proportional to $\alpha_{\rm s}$ (see (2.25)). The current world average of $\alpha_{\rm s}(M_Z^2)$ in next-to-leading order is [181]

$$\alpha_{\rm s}(M_Z^2) = 0.119 \pm 0.004 \,. \tag{2.41}$$

Note that the uncertainty in (2.41) exceeds the value quoted in [108], reflecting correlations between measurements which have previously been neglected.

2.1.5 Parton Distribution Fits

Several theory groups have specialized in PDF analyses which exploit the universality of PDFs to determine PDF parameterizations obeying the QCD evolution equations and fitting all the aforementioned experimental data (*QCD fits*). The "conventional" approach chosen by Martin, Roberts, Stirling and Thorne (MRS/MRST) [129, 130, 135, 141, 182, 183] and by the CTEQ collaboration [139, 184, 185] is to parameterize the PDFs as functions of x at a given Q_0^2, to perform the QCD evolution and to calculate the predictions for the various experimental quantities. The free parameters are finally determined by iteratively minimizing the χ^2 calculated from the differences of predictions and measurements and from the experimental errors. Special care is taken to account for correlated errors like normalization uncertainties.

The initial PDF parameterizations at Q_0^2 are typically of the type

$$x\, p(x, Q_0^2) = A_p\, x^{a_p}\, (1-x)^{b_p}\, P(x; c_p, \dots)\;, \tag{2.42}$$

where $P(x; c_p, \dots)$ is a slowly varying function of x. Independent parameter sets $A_p, a_p, b_p, c_p, \dots$ are defined for the valence distributions $p = u_\mathrm{v}$ and $p = d_\mathrm{v}$, for the singlet distribution $p = q^\mathrm{S}$ (see (2.26)) and for the gluon distribution $p = g$. In addition, in recent analyses allowing for a flavor-asymmetric sea, $\bar{u}+\bar{d}$ and $\bar{u}-\bar{d}$ are parameterized separately. CTEQ employs a further parameter set for the strange sea, $s = \bar{s}$, whereas MRS(T) use $s = \bar{s} = (\bar{u}+\bar{d})/4$ [141, 183]. The parameters a_p govern the low-x behavior of the PDFs and are, at $Q_0^2 \approx 5\,\mathrm{GeV}$, typically found to be of $\mathcal{O}(0.25-0.6)$ for the valence quarks, and in the ranges -0.3 to 0.1 for the sea quarks and -0.4 to 0.2 for the gluon. The exponents b_p determine the high-x fall-off of the PDFs. They assume values between 3 and 5 for valence quarks and between 6 and 10 for sea quarks and gluons, where the exact results depend on the functional form of $P(x; c_p, \dots)$ and on Q_0^2.

In addition to the experimental data, the sum rule constraints

$$\int_0^1 \mathrm{d}x\, [u(x, Q_0^2) - \bar{u}(x, Q_0^2)] = 2\;, \tag{2.43}$$

$$\int_0^1 \mathrm{d}x\, [d(x, Q_0^2) - \bar{d}(x, Q_0^2)] = 1\;, \tag{2.44}$$

$$\int_0^1 \mathrm{d}x\, x \left\{ g(x, Q_0^2) + \sum_q [q(x, Q_0^2) + \bar{q}(x, Q_0^2)] \right\} = 1 \tag{2.45}$$

are applied, where the first two fix the number of valence quarks (and hence the baryon number) of the proton, and the third reflects the fact that (anti-) quarks and gluons carry the entire proton momentum (*momentum sum rule*).

Equations (2.43)–(2.45) determine three of the normalization parameters A_p in (2.42). Furthermore, imposing sum rule constraints generally introduces sensitivity of the PDFs at a given $x = x_0$ to the complete x spectrum of input data and not only to the region $x > x_0$ as relevant for the DGLAP evolution.

For the analysis of the HERA high-Q^2 data discussed in this book, PDF parameterizations by the MRS [183] and CTEQ [185] groups have been used which take into consideration the ZEUS and H1 structure function results based on data taken up to and including 1994. These PDF sets have been tested against, but not fitted to, the inclusive jet cross sections at high transverse momenta, p_t^{jet}, which have been reported by the Tevatron experiments CDF [178] and DØ [180]. The excess of jets with very high p_t^{jet} seen by CDF could be accommodated by a reasonable modification of the CTEQ gluon distribution [185], which, however, only weakly affects the HERA cross sections. This CTEQ variant, and also the latest PDF sets from MRST [141] and CTEQ [139], are not used in the HERA high-Q^2 analyses.

A slightly different approach from the one by MRS and CTEQ is pursued by Glück, Reya and Vogt (GRV) [117, 131, 132], who start the QCD evolution at a very low scale, $Q_0^2 = \mu_{\mathrm{GRV}}^2 \lesssim 0.5\,\mathrm{GeV}^2$, where they use a valence-like input for the u_{v}, d_{v}, sea quark and gluon distributions (i.e. parameterizations of the type (2.42), with all $a_p > 0$). The free parameters are fixed by a fit to DIS and other relevant data after DGLAP-evolving the input parameterizations to the appropriate experimental scales and thereby radiatively producing the typical rise of sea quark and gluon distributions towards low x. This procedure leaves fewer degrees of freedom than the conventional fits, in particular for describing the low-x part of the PDFs. Nevertheless the latest higher-order (NLLA) GRV set [132], which uses $\mu_{\mathrm{GRV}}^2 = 0.40\,\mathrm{GeV}^2$, yields good agreement with the various experimental measurements and in particular with the NC structure function data for $Q^2 \gtrsim 1.5\,\mathrm{GeV}$.

To summarize, three groups provide modern PDF parameterizations which obey the QCD evolution equations by construction and describe well the more than one thousand experimental data points used to constrain the PDF parameters in χ^2 fits. Typical χ^2 values per degree of freedom range between unity and 1.25. A considerable problem for the interpretation of the HERA data is posed by the fact that none of the groups derive PDF uncertainties from propagating the errors of the input data and their correlations through the PDF fit procedures. However, estimates of these uncertainties, and their impact on extrapolations of the PDFs to the HERA high-Q^2 region, have been inferred by the ZEUS collaboration and are discussed in the next section.

2.1.6 Selected Parton Distribution and Cross Section Results

The relevant DIS NC and CC cross sections are calculated using the electroweak parameter values and the PDF parameterizations discussed in the previous section. In the following, we will collect a few plots illustrating the

x and Q^2 dependence of PDFs and differential cross sections and then discuss the approaches used to determine the cross section uncertainties. Unless explicitly stated otherwise, the PDFs from the set MRS-R1 (NLLA, $\alpha_s(M_Z^2) = 0.113$) [183] are used in these sections for calculating the cross sections shown in the figures. Neither photon radiation nor virtual electroweak radiative corrections (see Sect. 2.1.3) are included in these calculations.

Parton Distribution Functions and Cross Sections

In order to set the scales and to quantify the objects we have been discussing quite extensively in the preceding sections, we now present some plots of PDFs and differential cross sections.

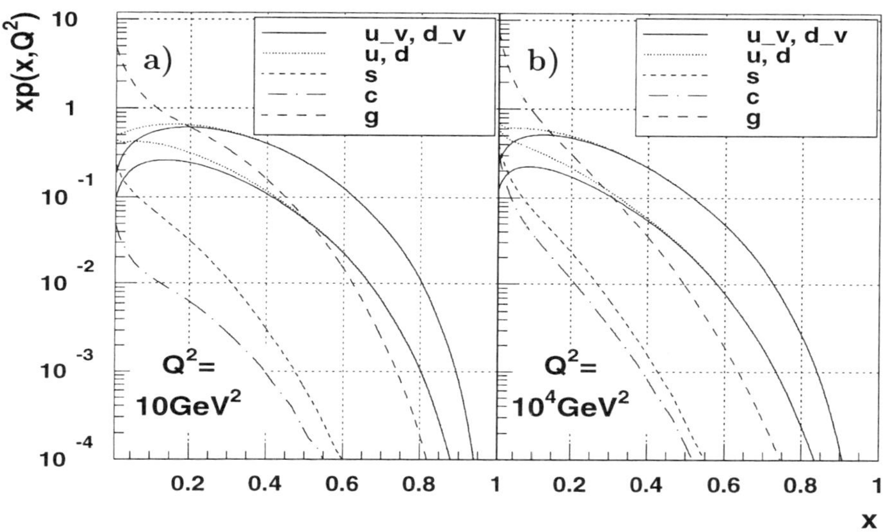

Fig. 2.8. Parton distribution functions according to the parameterization MRS-R1 (**a**) for $Q^2 = 10\,\text{GeV}^2$ and (**b**) for $Q^2 = 10^4\,\text{GeV}^2$. The larger of the two valence distributions represents u_v; the differences between the *dotted* and the *solid lines* indicate the u and d sea quark distributions

Figure 2.8 shows some selected PDFs as functions of x for two different values of Q^2. A few observations may be listed:

- Independently of Q^2, the valence distributions are clearly dominant for $x \gtrsim 0.4$. However, at lower x the sea-quark PDFs become comparable in size to the valence distributions and can hence not be neglected for investigations of cross sections at Q^2 values of some $100\,\text{GeV}^2$ and above (corresponding to $x \gtrsim 0.001$ at HERA).
- The ratio $r_\text{v} = d_\text{v}/u_\text{v}$ is significantly below the naïve expectation $r_\text{v} = 0.5$ at $x \gtrsim 0.3$ and actually approaches zero as $x \to 1$, a fact which is common

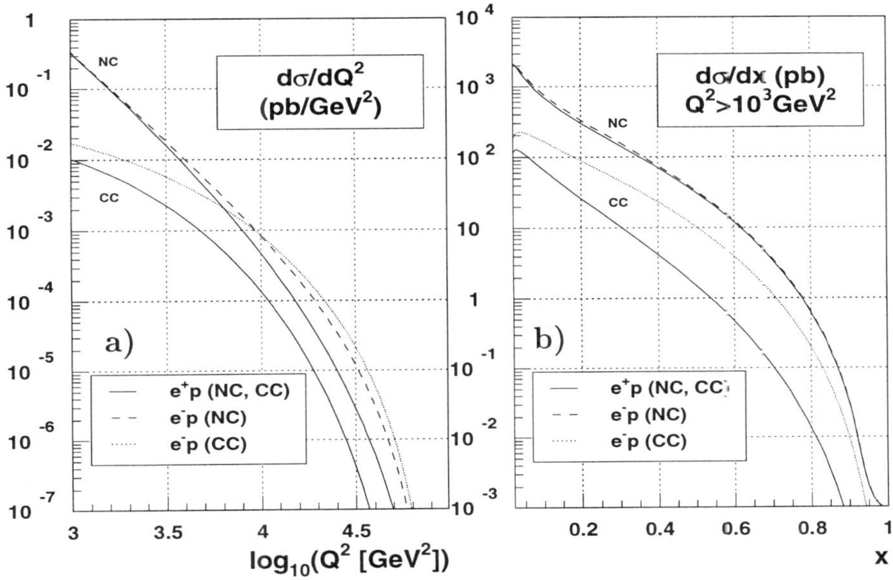

Fig. 2.9. Differential cross sections for $Q^2 > 10^3\,\mathrm{GeV}^2$, (**a**) $\mathrm{d}\sigma/\mathrm{d}Q^2$ and (**b**) $\mathrm{d}\sigma/\mathrm{d}x$. The cross sections for e^+p scattering (*solid lines*), e^-p NC scattering (*dashed lines*) and e^-p CC scattering (*dotted lines*) are indicated. All cross sections are calculated using the PDF set MRS-R1. Electroweak radiative corrections are neglected

to all modern PDFs which all have $b_{d_\mathrm{v}} > b_{u_\mathrm{v}}$ (see (2.42)). We will study the high-x limit of the ratio r_v in more detail in Sect. 2.1.6.

- As expected, the PDFs are shifted to lower x for increasing Q^2, and the fraction of the sea-quark contribution to the momentum integral (2.45) grows. However, for the HERA center-of-mass energy, the valence quark distributions are still responsible for the bulk of the cross section at $Q^2 \gtrsim 10^3\,\mathrm{GeV}^2$.

The cross sections $\mathrm{d}\sigma/\mathrm{d}Q^2$ and $\mathrm{d}\sigma/\mathrm{d}x$ for $Q^2 > 10^3\,\mathrm{GeV}^2$, obtained by integrating $\mathrm{d}^2\sigma/(\mathrm{d}x\,\mathrm{d}Q^2)$ over x and Q^2, respectively, are shown in Fig. 2.9 both for NC and for CC scattering. At $Q^2 \ll M_Z^2$, the NC cross section is dominated by the photon propagator, yielding a steep fall-off with $1/Q^4$, whereas the CC cross section exhibits a weaker Q^2 dependence. Once Q^2 rises to values of $\mathcal{O}(M_Z^2)$, both the NC and the CC cross sections become similar in magnitude, demonstrating the unification of electromagnetic and weak interactions. Note that at the same Q^2 scale the $\mathcal{F}_3^{\mathrm{NC}}$ contributions, which induce the difference between the NC e^+p and e^-p cross sections (see (2.4)), become important. In the CC case, both the sign of the $\mathcal{F}_3^{\mathrm{CC}}$ term and the flavor composition of the structure functions cause differences between the e^+p and e^-p cross sections. In the high-x regime, the latter is dominated by the u_v PDF which significantly exceeds the d_v PDF relevant for e^+p CC reactions.

Cross Section Uncertainties

This section is devoted to studies of *ep* cross section uncertainties. We first investigate the propagation of errors on the experimental input data to PDF fits into the HERA high-Q^2 cross sections. Thereafter we discuss neglected higher-order QCD effects and uncertainties of the d and u distributions at high x and of the heavy-quark contributions to the cross sections. Finally, we will quantify the effects of measurement errors of the electroweak parameters entering the cross section formulae. A recent survey of PDF and electroweak contributions to cross section uncertainties at HERA can be found in [186]

The largest sources of error in the PDF parameterizations are the experimental uncertainties of the input data summarized in Sect. 2.1.4. As yet, none of the PDF groups MRS(T), CTEQ and GRV have provided estimates of the PDF uncertainties arising from the propagation of these errors. Taking differences between different PDF sets as a measure of these errors is inappropriate, since all these sets use essentially the same input data and very similar fit procedures. In order to obtain a realistic error estimate, M. Botje from the ZEUS collaboration has performed a QCD fit – occasionally referred to as the *ZEUS QCD fit* in the following – using a restricted input data set (NC and CC structure function data from HERA and fixed-target experiments with $x > 0.001$ ($x > 0.1$ for CCFR data), $Q^2 > 3\,\mathrm{GeV}^2$ and $W^2 > 7\,\mathrm{GeV}^2$, and the E866 Drell–Yan data, see Sect. 2.1.4) [187]. Statistical and systematic uncertainties of the input data, as well as those of $\alpha_s(M_Z^2)$, of the strange-quark content of the proton, of nuclear corrections and of the charm quark mass are propagated through this fit. The corresponding variations of the fit results are added in quadrature to those obtained by modifying various fit parameters such as the Q^2 and W cuts or the starting point for the QCD evolution, Q_0^2. The resulting "total error", as shown in Fig. 2.10 for the reduced NC and CC cross sections,

$$\tilde{\sigma}^{e^+p\to e^+X} = \frac{xQ^4}{2\pi\alpha^2[1+(1-y)^2]}\,\frac{\mathrm{d}^2\sigma^{e^+p\to e^+X}}{\mathrm{d}x\,\mathrm{d}Q^2}\,, \tag{2.46}$$

$$\tilde{\sigma}^{e^+p\to \bar{\nu}_e X} = \frac{2\pi x}{G_\mathrm{F}^2}\left(\frac{Q^2+M_W^2}{M_W^2}\right)^2\frac{\mathrm{d}^2\sigma^{e^+p\to \bar{\nu}_e X}}{\mathrm{d}x\,\mathrm{d}Q^2}\,, \tag{2.47}$$

is dominated by the uncertainties of the experimental input data. The fit results are found to be stable against variation of the lower bound on W, indicating that higher-twist terms affect negligibly the PDF extrapolation to high Q^2. In the NC case, the PDFs induce a 3 – 6% error on the differential cross section, rising to about 8% at highest Q^2 and to 10 – 12% at $x \gtrsim 0.8$. For e^+p CC reactions the corresponding figure rapidly grows with x and exceeds values of 20% for $x \gtrsim 0.5$ and also at $Q^2 \gtrsim 20\,000\,\mathrm{GeV}^2$. For e^-p CC scattering the uncertainty is significantly smaller and close to the NC case. The large differences between the NC, CC e^-p and CC e^+p uncertainties are related to the precision with which the respective combinations of

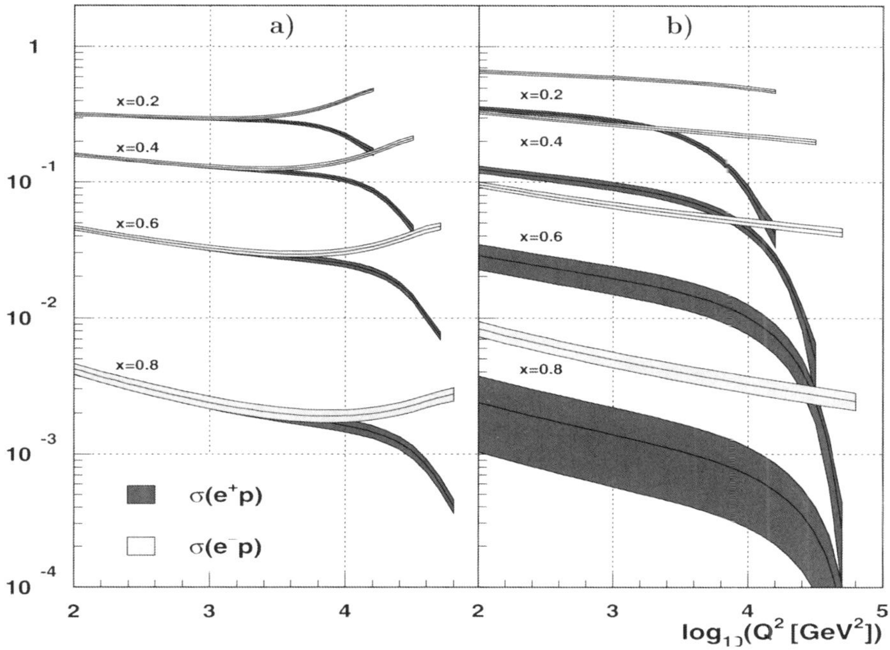

Fig. 2.10. Reduced cross sections for (**a**) NC and (**b**) CC ep DIS derived from the results of the QCD fit by M. Botje [187]. The *dark* (*light*) shaded bands indicate the total uncertainties for e^+p (e^-p) scattering (see text)

quark distributions are constrained by experimental data: $4u + d$ (relevant for NC) and, to some extent, also u (CC e^-p) are accurately determined by the charged-lepton fixed-target data, whereas d (CC e^+p) depends on the less accurate neutrino data.

A further source of error, unrelated to experimental data, is the unknown size of neglected higher-order QCD corrections. In order to estimate their impact, LLA and NLLA PDFs and cross sections (here the PDF sets CTEQ-4L and CTEQ-4H [185] were used) are compared in Fig. 2.11. Obviously the relative difference between the LLA and NLLA PDFs is of $\mathcal{O}(20\%)$ and significantly larger than the differences between the corresponding cross sections.[6] In the high-Q^2 domain, where α_s is small and the QCD perturbation series is expected to converge safely, the differences between two successive orders of accuracy are assumed to be larger than the sum of all neglected higher-order terms. Figure 2.11 can hence be interpreted as indicating a maximum uncertainty of a few percent due to pQCD contributions beyond the NLLA. Also, since changes of the factorization and renormalization scales induce modifications of the order of the next-higher pQCD order (see Sect. 2.1.2), scale

[6] Which illustrates that mixing different factorization schemes, or inserting LLA PDFs into NLLA formulae or vice versa, will yield wrong results (cf. Sect. 2.1.2).

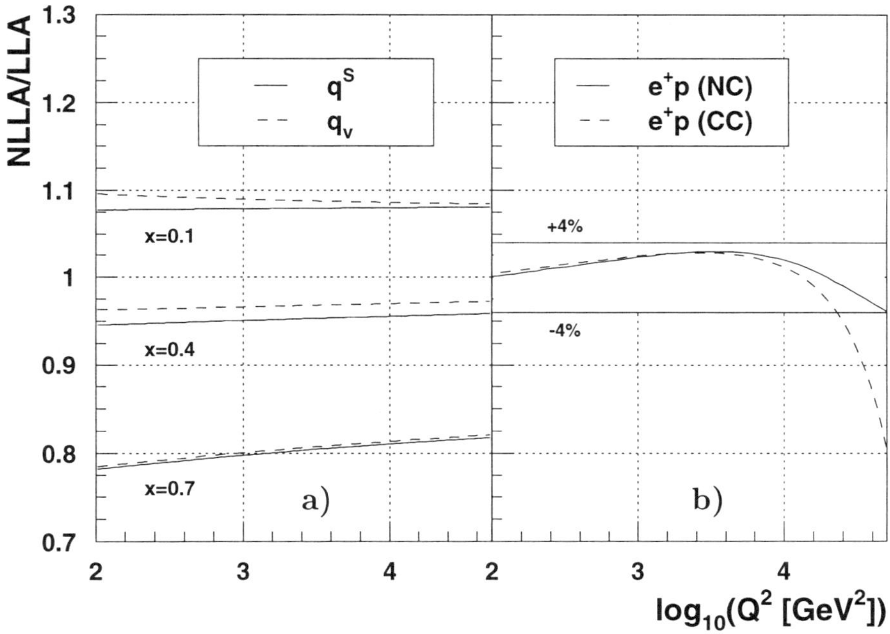

Fig. 2.11. Ratios of (**a**) the NLLA and LLA PDFs and (**b**) the NLLA and LLA differential cross sections for $Q^2 > 100\,\text{GeV}^2$. The PDF sets CTEQ-4L and CTEQ-4H have been used. For the PDFs, the singlet distribution q^{S} (see (2.26)) and the valence distribution $q_{\text{v}} = d_{\text{v}} + u_{\text{v}}$ are shown

uncertainties are estimated to be also small. An independent confirmation of this conjecture has been obtained in M. Botje's QCD fit [187] (see above), in which both scales have been varied in the range $Q^2/2 < \mu_{\text{r}}^2, \mu_{\text{f}}^2 < 2Q^2$. The resulting uncertainties of the reduced cross sections, as shown in Fig. 2.12 for NC and CC scattering, are substantially smaller than those resulting from the propagation of the experimental errors in Fig. 2.10.

In addition to neutrino data, NC DIS structure function measurements using proton and deuteron targets are used to constrain the ratio $r = d/u$, which in turn is the essential input for disentangling the d and u distributions in the valence-dominated high-x region. Extracting the neutron structure function $\mathcal{F}_2^n$ from the difference of the deuteron and proton data and using isospin symmetry between the proton and neutron, one obtains the LO high-x approximation

$$\frac{\mathcal{F}_2^n}{\mathcal{F}_2^p} \xrightarrow{x \to 1} \frac{4d_{\text{v}}/u_{\text{v}} + 1}{4 + d_{\text{v}}/u_{\text{v}}} . \tag{2.48}$$

Existing analyses are consistent with the assumption that the ratio $\mathcal{F}_2^n/\mathcal{F}_2^p$ approaches 0.25 for $x \to 1$ [189, 190], implying that r goes to zero in this limit. This paradigm is used for all standard PDF parameterizations. However,

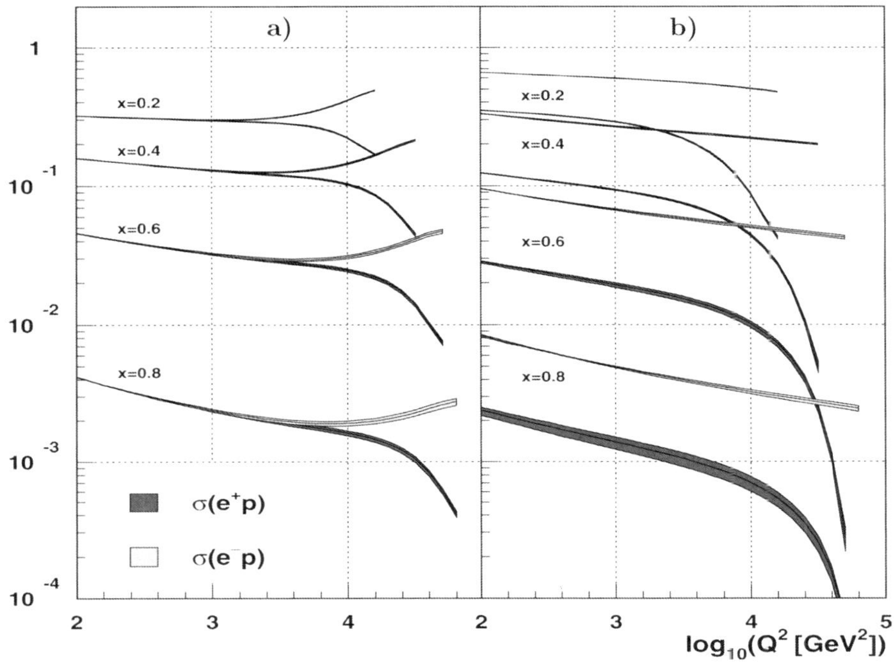

Fig. 2.12. Renormalization and factorization scale uncertainties derived from the results of the PDF fit by M. Botje [187] for (**a**) NC and (**b**) CC ep DIS. The *solid lines* indicate the reduced cross sections as functions of Q^2 for different fixed values of x; the *dark* (*light*) *shaded bands* represent the e^+p (e^-p) cross section modifications resulting from scale variations in the range $Q^2/2 < \mu_r^2, \mu_f^2 < 2Q^2$

recently it was pointed out [188] that a consistent relativistic treatment of Fermi motion and off-shell effects in the deuteron modifies the $\mathcal{F}_2^n$ results, and in particular induces an increase of $\mathcal{F}_2^n/\mathcal{F}_2^p$ at high x as shown in Fig. 2.13a. Consequently, and in agreement with QCD expectations [191], d/u assumes a value of about 0.2 for $x \to 1$, and the high-x tails of the parton distributions have to be modified. A recent discussion of these issues can be found in [192].

In a simple approach to modeling these changes we have adjusted the valence distributions to reproduce the $\mathcal{F}_2^n/\mathcal{F}_2^p$ ratio of Fig. 2.13a while at the same time requiring that $4u_v + d_v$ remains unaltered. The valence modifications were applied at $Q^2 = 20\,\mathrm{GeV}^2$ and fed into an LLA QCD evolution. It was checked that the modified PDFs and cross sections are still compatible with the neutrino DIS data [162, 193] and that the sum rules, (2.43)–(2.45), are only negligibly violated. Figure 2.13b shows the resulting PDF and structure function ratios, and demonstrates that the outcome is a significant increase in the CC cross section at high x. The cross section enhancement amounts to $\mathcal{O}(10\%)$ for $Q^2 \gtrsim 2.5 \times 10^4\,\mathrm{GeV}^2$. By construction, the NC cross sections (dominated by $4u + d$) are almost unaffected.

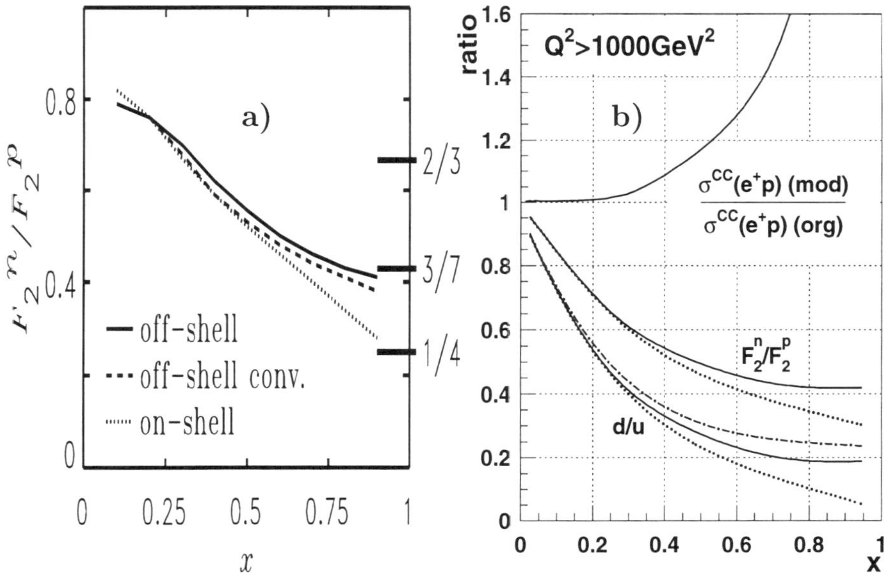

Fig. 2.13. Modification of the high-x PDFs and structure functions due to a revised treatment of Fermi motion and off-shell effects in the deuteron: (**a**) shows the modified ratio $\mathcal{F}_2^n/\mathcal{F}_2^p$ (*solid line*) compared with the result obtained in the free-nucleon approximation (*dotted line*) (plot reproduced from [188]). The result of a corresponding modification of the PDFs (see text) is shown in (**b**): the $\mathcal{F}_2^n/\mathcal{F}_2^p$ and d/u ratios both grow at high x (*dotted*: CTEQ-4L original; *solid*: modified as described in text; *dash-dotted*: d/u enhancement suggested by Bodek and Yang [126]). The CC e^+p cross section is significantly enhanced at high x (*top solid line*)

Similar considerations have led Bodek and Yang [126] to introduce an additive correction of the form $r \to r' = r+(0.1\pm0.01)(x+1)$ to the d/u ratio (see Fig. 2.13b). The so-modified PDF set MRS-R2 (NLLA, $\alpha_s(M_Z^2) = 0.120$) [183] is also compatible with $r \to 0.2$ at $x = 1$ and describes well a large spectrum of experimental data. However, the QCD fit by M. Botje [187] demonstrates that a similar increase of d/u at high x can also be obtained with standard parameterizations of the type (2.42) having $b_{u_v} > b_{d_v}$. As yet, such considerations have not been included in the standard PDF sets used for the HERA analyses, and, particularly for the e^+p CC cross sections at high x, modifications of the magnitude indicated in Fig. 2.13b may be envisaged (cf. Sect. 4.2.2).

As the last aspect of PDF-related uncertainties we will discuss the heavy-quark contributions to the high-Q^2 e^+p DIS cross sections. Figure 2.14 shows the relative contributions from charm and bottom quarks to $\mathrm{d}\sigma/\mathrm{d}x$ and $\mathrm{d}\sigma/\mathrm{d}Q^2$ at $Q^2 > 1000\,\mathrm{GeV}^2$, calculated in LO using the c and b PDFs taken from the set CTEQ-4L. Whereas the bottom contribution is small and falls below the PDF uncertainties mentioned above, charm adds significantly to

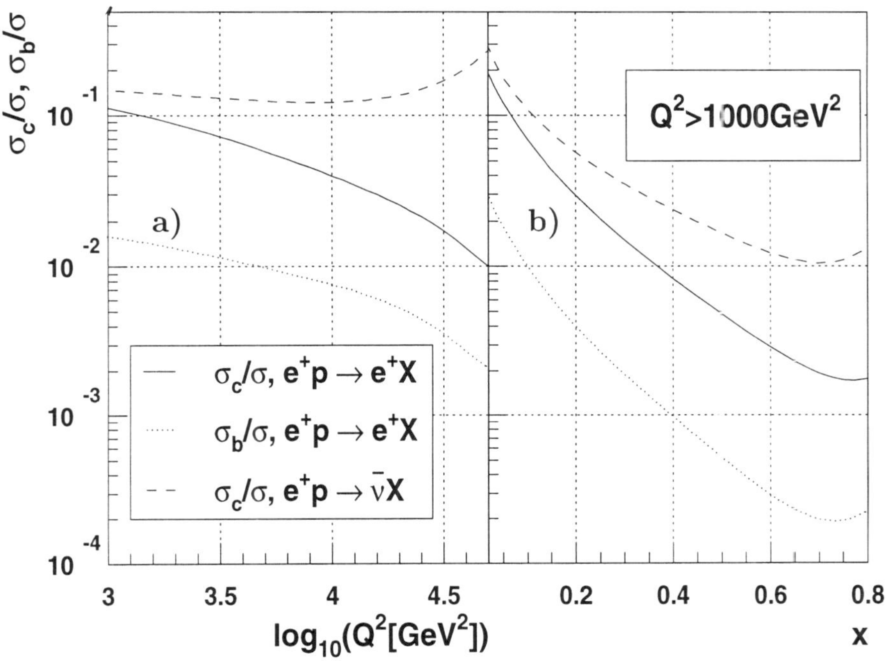

Fig. 2.14. Contributions from charm and bottom PDFs to the e^+p NC and CC DIS cross sections at $Q^2 > 1000\,\mathrm{GeV}^2$: (**a**) $\mathrm{d}\sigma/\mathrm{d}Q^2$ and (**b**) $\mathrm{d}\sigma/\mathrm{d}x$. The *curves* show the result of an LO calculation using the PDF set CTEQ-4L. According to the discussion in Sect. 2.1.1 there is no bottom contribution to CC scattering

the cross section, for both NC and CC processes. The rise of the charm contribution to the CC cross section with increasing Q^2 is caused by the suppression of $e^+d \to \bar{\nu}q$ due to the $(1-y)^2$ term in (2.12). At very high x, $x \gtrsim 0.6$, the size and x dependence of the heavy-quark contributions shown in Fig. 2.14 are entirely determined by the analytic form assumed for the PDF parameterizations (see (2.42)) and their QCD evolution, and are not supported by experimental data.

The method of calculation used for Fig. 2.14 assumes the VFNS (see Sect. 2.1.2), which is expected to yield accurate results if Q^2 is far above the threshold domain, $Q^2 \gg m_c^2, m_b^2$. The main source of uncertainty in the calculation of the heavy-quark cross sections is the treatment of the threshold region in the PDF fits. A comparison of the c and b structure functions obtained from the LO CTEQ set with the analysis by MRS [135], which includes a refined treatment of the heavy-quark QCD evolution, indeed indicates agreement to better than 10% for $Q^2 \gtrsim 10\,\mathrm{GeV}^2$ and $x \gtrsim 10^{-3}$. It may be concluded that the maximal relative uncertainty of the heavy-quark contributions is of $\mathcal{O}(10\%)$, amounting to an uncertainty in the overall cross sections at the percent level.

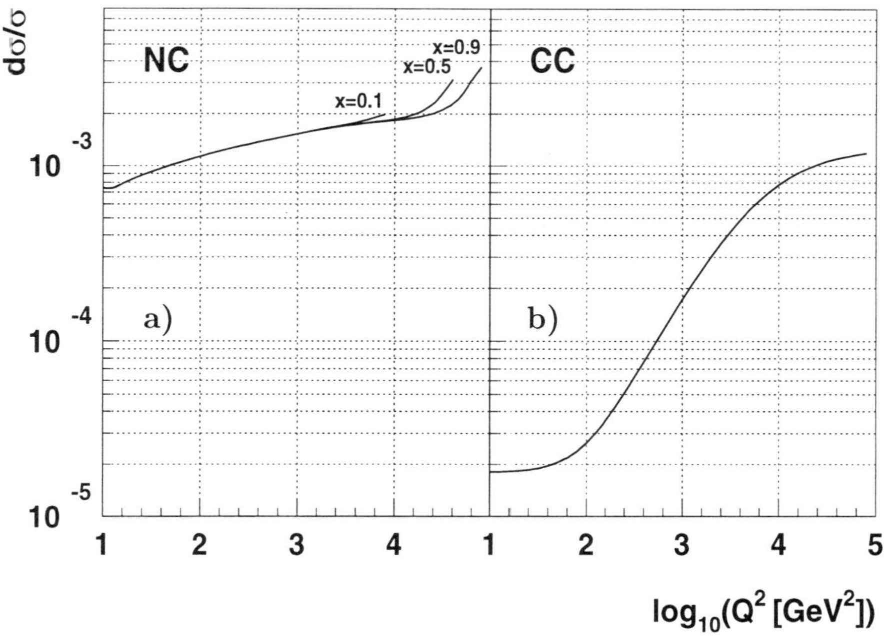

Fig. 2.15. Relative uncertainties of differential e^+p DIS cross sections due to errors in the relevant electroweak parameters, (**a**) for NC and (**b**) for CC scattering. In the NC case, the cross section uncertainty is dominated by the errors in α and $\sin^2\theta_W$, in the CC case by that in M_W

Finally, we want to leave the field of PDFs and QCD and come back to the electroweak parameters (see Sect. 2.1.4). Propagating the parameter uncertainties quoted in Table 2.1 in quadrature through the LO formulae (2.4) and (2.12) yields the relative errors shown in Fig. 2.15. The dominant contribution to the uncertainty of α comes from the relative uncertainty in $\alpha(M_Z^2)$, which has been assigned to the radiative correction, $\alpha(Q^2)-\alpha(0)$. The cross section uncertainties in Fig. 2.15 do not exceed 0.3% in the kinematic region of interest and can hence be safely neglected against the much larger errors caused by the PDF parameterizations.

To conclude, the set of ingredients required to calculate high-Q^2 ep differential cross sections is completely available and combines theoretical perception from different sectors of the SM into a consistent picture describing a wealth of experimental data from different reactions at wide ranges of energy scales. The uncertainties of the cross sections are dominated by the errors in the experimental input data to the PDF parameterizations and may, in particular in the CC case, acquire additional contributions from theoretical issues related to low-energy phenomena. In the region of interest, the PDFs induce a relative uncertainty of $\lesssim 6\%$ for NC reactions (up to 12% at $x \gtrsim 0.8$) and of $\lesssim 20\%$ for CC reactions with $Q^2 \lesssim 2 \times 10^4$ GeV^2 and $x \lesssim 0.5$.

2.1.7 Modeling the Hadronic Final State

The previous sections have been dedicated to questions concerning DIS cross sections and their uncertainties, but completely discarded the question "What happens to the partons emerging from the hard subprocess before they turn into observable particles?" Although the hadronic final state of DIS reactions is not subject of this book, a brief summary of the related concepts is useful to comprehend their effect on experimental aspects like event selection or kinematic reconstruction. More details can be found in two recent reviews covering the theory of jet production in DIS [194] and the QCD aspects of the hadronic final state in DIS [48].

Usually, one differentiates two stages of the final-state evolution: the *parton shower*, which can be viewed as QCD radiation of a cascade of partons each having a virtuality or transverse momentum which is larger than the typical hadronic energy scale of $\mathcal{O}(1\,\mathrm{GeV})$, and the *hadronization*, in which color-neutral hadrons are formed from the partons produced in this cascade. Whereas pQCD is not applicable at all to the hadronization, the parton shower is a process which is governed by pQCD but requires resummation techniques for a detailed description since full matrix-element calculations are not available for multiple-splitting processes. Both the parton shower and the hadronization stages are described by phenomenological models implemented in different Monte Carlo (MC) generators which are routinely used to simulate *ep* reactions at HERA.[7] Their basic ingredients are:

- *The parton shower.* In general the primary final-state partons emerge from the hard subprocess with some time-like virtuality V, which they reduce by successive radiation of further time-like or quasi-real partons. Equivalently, also the initial-state parton radiates and thus builds up space-like virtuality. Two approaches model these processes. The first is based on the DGLAP equations, which are applied to evolve V down to the hadronization scale. V is related to the scale of the hard subprocess either by an ad hoc assumption like $V = Q^2$ or by using the algorithm in conjunction with the $\mathcal{O}(\alpha_s)$ or $\mathcal{O}(\alpha_s^2)$ matrix elements [195]. The latter option is denoted MEPS and is used for event generation in the LEPTO program [196]. The second model groups the partons into pairs forming *color dipoles*, which radiate further partons, until some lower virtuality cutoff is reached [197]. This model as implemented in the ARIADNE program [198] is currently part of the standard procedure for DIS simulation.
- *Hadronization.* Once the virtuality of the partons in the QCD shower reaches the order of typical hadronic masses, the shower simulation stops and hands the parton state to the hadronization algorithm. Again, two different procedures are in common use. In the Lund string-fragmentation model [199], the partons are connected by color field strings, which break up by creating quark–antiquark pairs from the vacuum and finally produce

[7] See Sect. 3.2 for a discussion of MC simulations in the high-Q^2 analyses.

color-neutral hadrons (implemented in JETSET [200, 201]). In the cluster fragmentation model [202], the remaining gluons are split into quark–antiquark or diquark–antidiquark pairs, and each (di)quark is combined with its nearest partner in phase space with which it can form a color-neutral cluster, then decaying into hadrons (implemented in HERWIG [203]). In both hadronization models the energy–momentum and quantum number flow of the parton state is imaged by the corresponding hadronic distributions (*local parton–hadron duality* [204]), in accordance with experimental evidence.

By virtue of the factorization theorem (see Sect. 2.1.2), the concepts used for modeling the parton shower and hadronization are in principle applicable to different processes containing partons in their final states. However, parameter sets which have been tested and tuned, e.g. in e^+e^- reactions or in fixed-target DIS experiments, cannot simply be adopted for HERA physics for two reasons: (i) because HERA operates in a hitherto unexplored kinematic region, and (ii) because of conceptual difficulties which are induced by the fact that the proton remnant in *ep* scattering is an extended object which cannot be treated as pointlike and therefore requires special algorithms. A large number of studies have been performed to test the various approaches of simulating the hadronic final state at HERA against experimental results, both from DIS reactions and from photoproduction (the relevant ZEUS and H1 references are summarized in Table 2.2). In particular, event topologies like jet multiplicities and angular distributions, energy flow and global event shapes are particularly sensitive to parton shower properties, whereas single-particle distributions rather carry information on the hadronization process. The main conclusions are that the models described above satisfactorily reproduce many details of the hadronic final states of different reaction types at different energy scales. However, details like hadronic emission in the forward direction, i.e. close to the proton remnant, are not in all cases well simulated.

Table 2.2. H1 and ZEUS studies of the hadronic final state referring to hadronization and parton shower models. The tags in parentheses indicate the reaction type: PHP = photoproduction, DIS = deep inelastic scattering, DIS/dif. = diffractive DIS reactions. Some analyses appear in more than one slot

Type of analysis	H1	ZEUS
Jet topologies	[22] (PHP) [21, 24–26] (DIS)	[14, 15, 17, 19] (PHP) [13, 18, 51, 205] (DIS)
Energy flow, event shapes	[22] (PHP) [32, 206–209] (DIS) [32–34] (DIS/dif.)	 [28, 210] (DIS) [28] (DIS/dif.)
Single particles, multiplicities	[211–214] (PHP) [25, 39, 207, 213, 216–220] (DIS) [34, 35] (DIS/dif.)	[215] (PHP) [221–226] (DIS)

2.2 Modifications of the Standard-Model Scenario

As mentioned previously, in early 1997 the ZEUS and the H1 collaborations have both reported evidence for an excess of DIS events above predictions at $Q^2 \gtrsim 2 \times 10^4\,\mathrm{GeV}^2$, for NC as well as for CC reactions. Without intending to anticipate the detailed discussion of the HERA results in Chap. 4, it will be useful for the following considerations to keep in mind that the kinematic region of interest, where the surplus of events is located, is roughly defined by $Q^2 > 1.5 \times 10^4\,\mathrm{GeV}^2$ and $0.4 \lesssim x \lesssim 0.65$. The NC excess cross section in this region as inferred from the 1994–96 data [71, 72] amounts to about 0.4 pb and has been used in many of the related phenomenological studies. Including the 1997 data decreases this value to about 0.15 pb. Although the statistical significance of the excess is doubtful, and neither its firm experimental corroboration nor its "disappearance" can be expected in the forthcoming few years, it has stimulated an impressive collection of related theoretical work, considering both possible explanations within the SM and hypothesized physics beyond the SM. The recent developments are characterized by an intense symbiosis between experiment and theory, each benefiting from their respective progress and from the fact that many of the suggested explanations of the HERA effect would have observable consequences in LEP and Tevatron experiments and in low-energy phenomena.

In this section we want to summarize the theoretical scenarios which have been proposed and discuss how their parameters are constrained by existing experimental data, both from HERA and from other experiments. The intention is to provide a synopsis of the basic ideas and their relevant experimental implications rather than to strive for completeness or to go into details of particular models. The compatibility of the HERA high-Q^2 data with the different models will be discussed in Chap. 4. We will start by reviewing some approaches involving nonstandard PDFs, which, however, would not require extensions of the SM (Sect. 2.2.1), and then go on to models beyond the SM (Sect. 2.2.2). The latter include in particular the resonant production of heavy states which couple to both electrons and quarks, and interactions at much higher mass scales which are parameterized as effective four-fermion couplings (*contact interactions*, CIs). Comprehensive phenomenological studies of these scenarios can for example be found in [227–232]. A few ideas not included in these categories will, finally, be mentioned at the end of Chap. 4.

2.2.1 Nonstandard Parton Distribution Functions

When the Tevatron experiment CDF announced the observation of an excess of jets with high transverse momentum, $p_t^{\mathrm{jet}} \gtrsim 300\,\mathrm{GeV}$ [178], in early 1996, it only took a few months until the CTEQ group suggested a modification of the PDFs [185], specifically the gluon distribution, which could reasonably well explain the CDF data and still was in agreement with the usual input data to PDF fits. However, the gluon distribution only enters ep cross sections

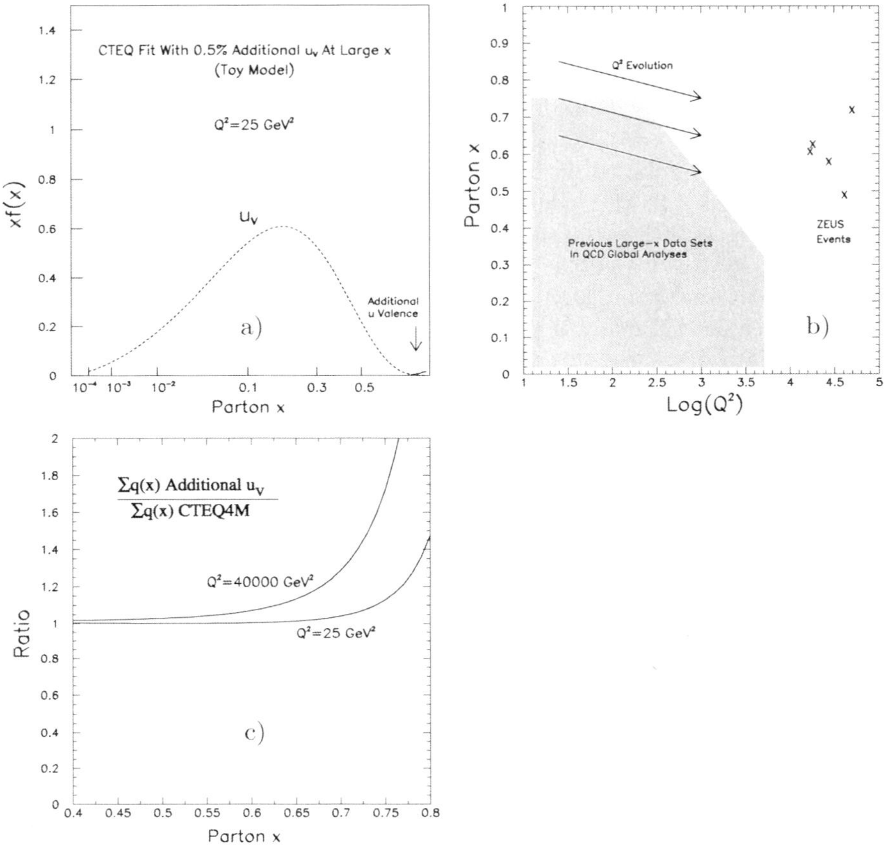

Fig. 2.16. Suggested modification of the u_v distribution to explain the observed excess of events in the HERA high-Q^2 regime: (**a**) the additional u_v contribution, (**b**) a sketch how this modification would affect the HERA region via QCD evolution, (**c**) the resulting relative change of the singlet distribution as a function of x. All plots are reproduced from [233]

via $\mathcal{O}(\alpha_s)$ corrections and cannot directly be made responsible for a possible HERA excess cross section. The question whether a PDF modification nevertheless could account for an anomalous HERA event rate at very high Q^2 was again taken up by members of the CTEQ collaboration: Kuhlmann et al. [233] introduced a small extra valence contribution, $\delta u_v = 0.02(1-x)^{0.1}$, which would enhance the cross section in the fixed-target Q^2 range only at $x \gtrsim 0.75$ (where experimental data are sparse and the theoretical predictions suffer from nonperturbative and higher-twist effects), but would migrate down to lower x in the relevant HERA Q^2 domain by virtue of the QCD evolution. Three figures from [233] demonstrating the proposed PDF modification are shown in Fig. 2.16.

We note that the cross section enhancement obtained by Kuhlmann et al. is only marginally compatible with the 1996 HERA excess events, which are concentrated at $Q^2 \lesssim 4 \times 10^4\,\mathrm{GeV}^2$ and $0.4 \lesssim x \lesssim 0.65$. In two subsequent publications a modification of quark PDFs was discussed in the context of experimental consequences. Babu et al. [234] argue that an enhancement of u or d quark densities would also induce significant changes to the CC cross sections at HERA, which in the case of an anomalous d component would already have been seen in the HERA e^-p CC data [50–54, 235, 236]. A u modification would cause a CC excess of about the same relative size as in the NC case. Rock and Bosted [237] analyze the SLAC data at very high x, $0.7 < x < 0.97$, and find that a scenario à la Kuhlmann et al. would necessitate the introduction of large negative higher-twist contributions and would not be compatible with the Bloom–Gilman duality [238], which relates the average cross section at low W to the perturbative DIS expectation.

The above discussion does not include any physical motivation for a PDF contribution which is much harder than the standard valence distributions (and has hence been classified as a toy model by the authors, cf. Fig. 2.16). One possible explanation could, however, be the presence of a small valence-type $c\bar{c}$ component of the proton, i.e. the flavor state $p = |uud\rangle$ would be changed to $p_{\mathrm{IC}} = (1-\beta)|uud\rangle + \beta|uudc\bar{c}\rangle$ with $\beta = \mathcal{O}(1\%)$ (*intrinsic charm*, IC). This hypothesis has been introduced by Brodsky et al. [240] and improves to some extent the agreement between theoretical predictions and e.g. the EMC $\mathcal{F}_2^c$ data [241] or measurements of Λ_c production in pp collisions [242]. In the context of the HERA results, IC has been discussed by Gunion and Vogt [243] and by Melnitchouk and Thomas [239]. The first authors assume that the IC contributions to the c and $\bar{c}$ PDFs are equal and do not cause a large effect for NC cross sections in the high-Q^2, high-x region. Melnitchouk and Thomas base their argument on a meson cloud model in which the nucleon can fluctuate to a virtual $\Lambda_c\overline{D}^0$ pair, thereby inducing asymmetry between c and $\bar{c}$ and allowing for a $\bar{c}$ contribution which is concentrated at large x. The relevant plots from [239] are reproduced in Fig. 2.17.

It becomes obvious that IC, within the bounds experimentally allowed, cannot easily account for an enhancement of the HERA NC cross sections at $0.4 \lesssim x \lesssim 0.65$ (corresponding to $190\,\mathrm{GeV} \lesssim M \lesssim 240\,\mathrm{GeV}$) by more than about 20% and generates modifications which are even smaller than those in the model of Kuhlmann et al. In contrast, the corresponding variation of the CC e^+p cross section is considerable and in the reach of experimental sensitivity at HERA. Note, however, that a possible IC signature in e^+p CC scattering at HERA will be hard to distinguish from the effect of a modified d/u ratio which may be of similar shape and magnitude (see Fig. 2.13).

We want to conclude this section with the observation that modifications of the PDFs which significantly improve the agreement with the HERA NC measurements lead to changes in the CC predictions at a level close to present experimental sensitivity.

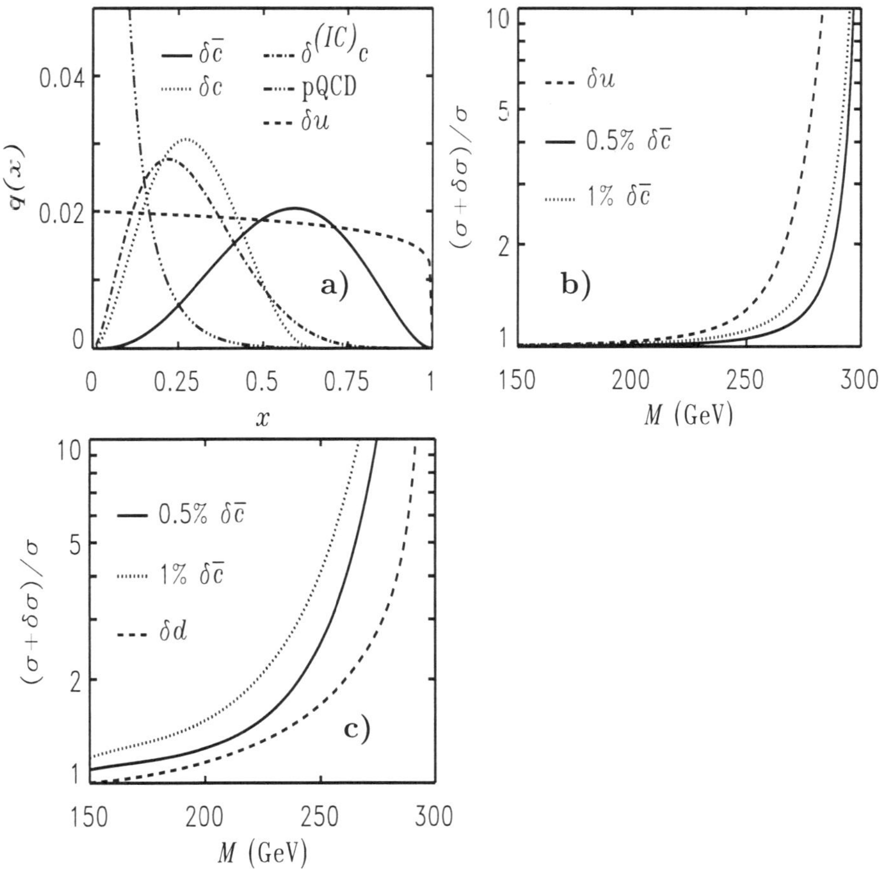

Fig. 2.17. Modification of the HERA high-Q^2 cross section due to intrinsic charm: (**a**) extra contributions to the parton distributions at high x according to the models of Melnitchouk and Thomas ($\delta c, \delta\bar{c}$), Gunion and Vogt ($\delta^{\mathrm{IC}}c$), and Kuhlmann et al. (δu). The resulting relative changes of (**b**) the e^+p NC and (**c**) the e^+p CC cross sections $\mathrm{d}\sigma/\mathrm{d}M$ at $Q^2 = 2 \times 10^4\,\mathrm{GeV}^2$ (with $M = \sqrt{xs}$) are shown according to the model of Melnitchouk and Thomas for two different normalizations β of the IC component (see text). All plots are taken from [239]

2.2.2 Theoretical Scenarios Involving New Physics

Hypothesized models involving particles or interactions not contained in the SM are generically denoted as *new physics*. Since HERA explores an as yet uncharted kinematic realm, it cannot be excluded that the first hints of such phenomena beyond the SM might show up and manifest themselves as modifications of the cross sections (and possibly other quantities) predicted by the SM. As a consequence, considerable excitement was stirred up by the HERA high-Q^2 results [71, 72] which indicate a possible discrepancy between prediction and observation. The relevant questions to be answered by phenomenological studies of any particular non-SM scenario of ep scattering are: (i) can the HERA observations be explained? (ii) are the underlying assumptions and implications for other processes in accordance with the corresponding experimental data? (iii) which predictions can be made which are subject to future experimental investigation, and how can the model be distinguished from other approaches which may result in similar experimental signatures?

Three classes of non-SM scenarios which have received particular attention in the context of the HERA data will be summarized and examined in the light of these questions in the following sections: the resonant production of leptoquarks and R_P-violating squarks, and contact interactions. Finally, a synopsis of other ideas will wind up the discussion of non-SM contributions to ep DIS cross sections.

Leptoquarks

Leptoquarks (LQs) are hypothesized particles coupling to both leptons and quarks and carrying both lepton and baryon number. They are expected to exist in different extensions of the SM, like composite models with quark and lepton substructure [244], technicolor [245], grand unified theories (GUTs) based on different gauge groups [246–248] or superstring-inspired models [249]. Some of the recent phenomenological studies of LQ production at HERA are collected in [227–231, 250, 251]; related theoretical work like model-building approaches including LQs can be found in [252–263]. LQ-type squarks occurring in supersymmetric models will be considered in the next subsection. We will for now restrict our attention to LQs with masses below the HERA center-of-mass energy and the unique possibility of their direct s-channel production in ep reactions (see Fig. 2.18). Virtual effects induced by heavier LQs will be covered by the discussion of contact interactions at the end of Sect. 2.2.2.

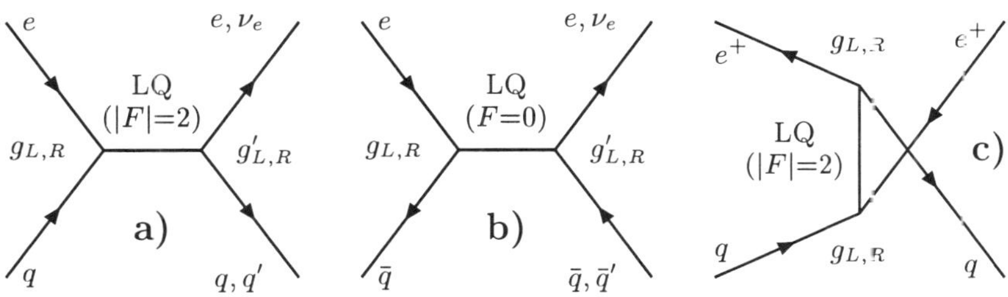

Fig. 2.18. Leading-order Feynman graphs for (**a**) s-channel LQ formation and decay in eq and (**b**) in $e\bar{q}$ scattering, and (**c**) for the u-channel LQ exchange in e^+q scattering (note that $e^+\bar{q}$ interactions require the exchange of an $F=0$ LQ). The LQ–quark–lepton Yukawa couplings g and g' carry helicity indices characterizing the spin state of the electron (see Table 2.3)

Table 2.3. Classification of LQs following [264]. The columns F, S, T and Q denote fermion number, spin, weak isospin and charge, respectively. The signs and factors of the chiral coupling constants are only relevant for relating the couplings of members of one multiplet to each other. The LQ name conventions from [229] are in widespread use in experimental LQ analyses and are also used in this book, but they differ from those in [264], which have been adopted as the PDG standard [108]. Note that the couplings of the LQs with $F=0$ are denoted $h_{\mathrm{L,R}}$ in [264]

LQ name [229]	LQ name [108, 264]	F	S	T	Q	Decay mode	Chiral coupling
S_0	S_1	2	0	0	$-1/3$	$e_{\mathrm{L}}u$	g_{L}
						$\nu_{\mathrm{L}}d$	$-g_{\mathrm{L}}$
						$e_{\mathrm{R}}u$	g_{R}
$\tilde{S}_0$	$\tilde{S}_1$	2	0	0	$-4/3$	$e_{\mathrm{R}}d$	g_{R}
S_1	S_3	2	0	1	$+2/3$	$\nu_{\mathrm{L}}u$	$\sqrt{2}g_{\mathrm{L}}$
					$-1/3$	$\nu_{\mathrm{L}}d$	$-g_{\mathrm{L}}$
						$e_{\mathrm{L}}u$	$-g_{\mathrm{L}}$
					$-4/3$	$e_{\mathrm{L}}d$	$-\sqrt{2}g_{\mathrm{L}}$
$V_{1/2}$	V_2	2	1	1/2	$-1/3$	$\nu_{\mathrm{L}}d$	g_{L}
						$e_{\mathrm{R}}u$	g_{R}
					$-4/3$	$e_{\mathrm{L}}d$	g_{L}
						$e_{\mathrm{R}}d$	g_{R}
$\tilde{V}_{1/2}$	$\tilde{V}_2$	2	1	1/2	$+2/3$	$\nu_{\mathrm{L}}u$	g_{L}
					$-1/3$	$e_{\mathrm{L}}u$	g_{L}
$S_{1/2}$	R_2	0	0	1/2	$-2/3$	$\nu_{\mathrm{L}}\bar{u}$	g_{L}
						$e_{\mathrm{R}}\bar{d}$	$-g_{\mathrm{R}}$
					$-5/3$	$e_{\mathrm{L}}\bar{u}$	g_{L}
						$e_{\mathrm{R}}\bar{u}$	g_{R}
$\tilde{S}_{1/2}$	$\tilde{R}_2$	0	0	1/2	$+1/3$	$\nu_{\mathrm{L}}\bar{d}$	g_{L}
					$-2/3$	$e_{\mathrm{L}}\bar{d}$	g_{L}
V_0	U_1	0	1	0	$-2/3$	$e_{\mathrm{L}}\bar{d}$	g_{L}
						$\nu_{\mathrm{L}}\bar{u}$	g_{L}
						$e_{\mathrm{R}}\bar{d}$	g_{R}
$\tilde{V}_0$	$\tilde{U}_1$	0	1	0	$-5/3$	$e_{\mathrm{R}}\bar{u}$	g_{R}
V_1	U_3	0	1	1	$+1/3$	$\nu_{\mathrm{L}}\bar{d}$	$\sqrt{2}g_{\mathrm{L}}$
					$-2/3$	$e_{\mathrm{L}}\bar{d}$	$-g_{\mathrm{L}}$
						$\nu_{\mathrm{L}}\bar{u}$	g_{L}
					$-5/3$	$e_{\mathrm{L}}\bar{u}$	$\sqrt{2}g_{\mathrm{L}}$

Many authors have recently reviewed the phenomenology of LQs. The most important issues and conclusions are summarized in the following.

LQ Classification. Using the natural assumptions that LQs conserve lepton and baryon numbers, respect the $SU(2)_L \times U(1)_Y$ symmetry, only couple to fermions and gauge bosons, and that their couplings to fermions are chiral, ten different LQ multiplets can be formed [264] from the left- and right-handed quarks and leptons of a given generation. The corresponding name conventions, quantum numbers and decay channels are summarized in Table 2.3 for the example of first-generation LQs, i.e. those which couple to u and d quarks, the electron and the electron neutrino.

We observe that LQs can carry either spin 0 (*scalar LQs*) or spin 1 (*vector LQs*), that they can have fermion number $F=0$ or ± 2, and that some couple to both eq and $\nu q'$ states, whereas others only have one coupling. In e^+p reactions, LQs with $F=0$ can be produced from quarks, whereas LQs with $F=-2$ require antiquarks in the initial state. Note that the full spectrum of LQs in Table 2.3 can in principle occur for any combination of lepton and quark generations, and that it is a priori not excluded that LQs might couple to different generations simultaneously.

In some models, and also in the context of the HERA high-Q^2 events, "unconventional" LQs have been considered which do not fit into the scheme of Table 2.3, i.e. these models explicitly break $SU(2)_L \times U(1)_Y$, involve right-handed neutrinos or exotic states beyond the SM, and/or have extended gauge structures [253, 255, 260, 261, 265].

LQ Production at HERA. For the moment, we want to postpone the discussion of existing constraints on LQ masses and couplings and instead investigate the experimental signatures of a hypothetical LQ production at HERA. In the case of s-channel production according to Figs. 2.18a,b, the LQ mass, M_{LQ}, is equal to the invariant mass of the initial electron–quark system and is therefore given by $M_{\mathrm{LQ}}^2 = xs$ if quark masses and QCD radiation effects inducing a nonzero initial-state virtuality are neglected. LQs of a given type would hence produce a resonance-like signal in x.

The cross section induced by any of the LQs in Table 2.3 receives contributions from the s-channel process (Figs. 2.18a,b), from the u-channel exchange (Fig. 2.18c) and from the interference of each of these processes with the SM DIS amplitude. For LQs with small couplings, $g^2/(4\pi) \ll 1$, the LQ decay width [264]

$$\Gamma_{\mathrm{LQ}} = \frac{1}{2+S}\,\frac{g^2}{8\pi}\,M_{\mathrm{LQ}}^2\,, \tag{2.49}$$

can be neglected in the cross section calculation (*narrow-width approximation*, NWA). Using the NWA and taking into account only the dominant s-channel but not the u-channel and interference contributions, the cross section σ^{LQ} for $ep \to \mathrm{LQ} + X$ with a subsequent decay $\mathrm{LQ} \to eq$ has been calculated in leading order [264]:

$$\frac{\mathrm{d}\sigma^{\mathrm{LQ}}}{\mathrm{d}y} = \frac{\pi\, g^2}{4s}\, \beta_e\, q(x, \mu_{\mathrm{LQ}}^2) \begin{cases} 1 & \text{if } S = 0 \\ 6(1-y)^2 & \text{if } S = 1 \end{cases}, \tag{2.50}$$

$$\sigma_{\mathrm{tot}}^{\mathrm{LQ}} = \int_0^1 \mathrm{d}y\, \frac{\mathrm{d}\sigma^{\mathrm{LQ}}}{\mathrm{d}y} = \frac{\pi\, g^2}{4s}\, (1+S)\, \beta_e\, q(x, \mu_{\mathrm{LQ}}^2)\,, \tag{2.51}$$

where q is the initial-state quark PDF (usually evaluated at a mass scale $\mu_{\mathrm{LQ}} = M_{\mathrm{LQ}}$), g is the appropriate coupling (taken to be the same for formation and decay), $\beta_{\ell,\nu} = \mathrm{BR}(\mathrm{LQ} \to \ell q, \text{`}\overline{\nu}\text{'} q)$ denotes the branching ratio for the LQ decay into the final state under consideration, and $M_{\mathrm{LQ}}$ and $S$ are the LQ mass and spin, respectively. The cross section for LQ production with a $\text{`}\overline{\nu}\text{'} q$ final state is given by (2.50) and (2.51) provided the appropriate branching ratio $\beta_{\text{`}\overline{\nu}\text{'}}$ is used. The leading QCD corrections to σ^{LQ} have been reported [251, 266] to be of $\mathcal{O}(20-50\%)$, rising with M_{LQ}.

The HERA LQ search analyses report exclusion limits on g as functions of M_{LQ} (see Sect. 4.4.1). In particular, if M_{LQ}^2 is close to s, or for LQ species produced from antiquarks or from quarks other than d or u, the coupling values corresponding to these limits do not necessarily fulfill the NWA, and the interference contributions (σ_i^{LQ}) as well as the u-channel term (σ_u^{LQ}) become important. Whereas σ_u^{LQ} is deselected by the searches since it does not produce a resonance-like x distribution, σ_i^{LQ} can contribute significantly to the signal close to the kinematic limit and even for $M_{\mathrm{LQ}} \gtrsim \sqrt{s}$ (more details can be found in Sect. 4.4.1).

Assigning the HERA high-Q^2 NC excess events from 1994–96 to LQ production yields $M_{\mathrm{LQ}} = \mathcal{O}(200\,\mathrm{GeV})$, and a coupling strength g which is in the range $0.01-0.08$ if the LQ is produced from a valence quark, and of $\mathcal{O}(0.25)$ in the case of an antiquark [227, 229, 267]. In either case, the resulting LQ width is limited to $\Gamma_{\mathrm{LQ}} < 1\,\mathrm{GeV}$ and hence smaller than the experimental resolution (see Sect. 3.3.3). It is interesting to note that $e^+\bar{q}$ fusion to LQs with $F = -2$ implies that the charge-conjugate process in e^-p reactions involves valence quarks (provided $q = d, u$), and hence has a much larger cross section. In fact the absence of an LQ signal in the $1.5\,\mathrm{pb}^{-1}$ of e^-p data collected at HERA in 1993 and 1994 [55, 56, 60, 63] and in $16.3\,\mathrm{pb}^{-1}$ of e^-p data from 1998 and 1999 analyzed preliminarily by ZEUS [268] almost completely rules out $|F| = 2$ LQs coupling to d or u quarks as possible explanations of the e^+p excess.

Figure 2.19 shows the modification of the NC cross sections $\mathrm{d}\sigma/\mathrm{d}Q^2$ and $\mathrm{d}\sigma/\mathrm{d}x$ by the presence of an LQ with $M_{\mathrm{LQ}} = 200\,\mathrm{GeV}$ and $\beta_e = 1$, which is assumed to couple to e^+u with a strength $g = 0.02$. Obviously, an LQ with these specifications would give a significant excess cross section at high Q^2, but the most striking and experimentally most distinctive property of LQ production is the spike in the x distribution. However, it is not only the experimental resolution in x which poses a problem in exploiting this signature, but also uncertainties related to QED and QCD radiative corrections

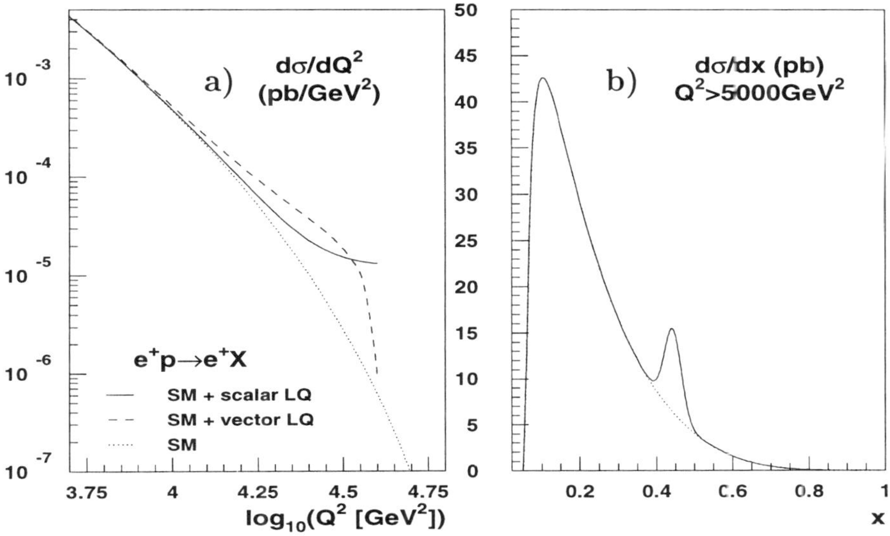

Fig. 2.19. Modification of the e^+p NC cross sections for $Q^2 > 5 \times 10^3\,\mathrm{GeV}^2$ due to the presence of an LQ with mass $M_{\mathrm{LQ}} = 200\,\mathrm{GeV}$ which couples to the positron and u quark with strength $g = 0.02$: (**a**) for $\mathrm{d}\sigma/\mathrm{d}Q^2$ and (**b**) for $\mathrm{d}\sigma/\mathrm{d}x$. In (**a**), the different Q^2 dependences of the LQ production are demonstrated for a scalar and a vector LQ, respectively. In (**b**), the LQ resonance in x is shown for the scalar LQ (the vector LQ would simply yield double the signal). The width of the LQ signal in x has been chosen to correspond to $\delta x/x = 0.05$ and roughly reflects the experimental resolution, which by far exceeds the intrinsic LQ decay width

[251, 266, 269, 270] and the QCD processes involved in the LQ fragmentation and decay[8] [271–274]. Both aspects have been analyzed in the quoted references, but as yet no Monte Carlo simulation containing NLO QCD matrix elements or specific LQ fragmentation models is available to the collaborations. The degree to which these issues affect the x or M_{LQ} resolutions obtained by the different kinematic reconstruction methods (see Sects. 3.3.3 and 4.4.1) is the subject of ongoing and future investigation.

Bounds on LQ Parameters. The existence of LQs with masses and couplings in the characteristic range discussed above would have consequences for a large variety of different processes, from low-energy phenomena like atomic parity violation (APV) and pion decays to $p\bar{p}$ reactions at the highest available energies. In many cases, LQ signatures are in the range of experimental sensitivity. A few examples will be collected in the following; detailed summaries can be found in [275].

[8] Since the typical lifetime of the LQs conjectured as an explanation for the HERA NC excess is long compared with hadronization timescales, these LQs may have the chance to fragment into color-neutral objects before they decay, thus changing significantly the pattern of the QCD cascade emitted by the decay quark and proton remnant.

- Currently, the only accelerator besides HERA where production of LQs with a mass of $\mathcal{O}(200\,\text{GeV})$ would be kinematically allowed is the Tevatron at Fermilab,[9] providing $p\bar{p}$ collisions with $\sqrt{s} = 1.8\,\text{TeV}$. The LQs would dominantly be pair-produced by $q\bar{q}$ or gg fusion with an intermediate s-channel gluon. For scalar LQs, the couplings in this process are completely determined by α_s and the $SU(3)_c$ gauge structure and in particular do not depend on the LQ Yukawa couplings $g_{L,R}$, rendering a pair-production cross section which depends on M_{LQ} as the only unknown parameter [278]. For vector LQs additional possible anomalous couplings come into play, but a lower bound on the LQ pair-production cross section which is independent of these unknowns has been calculated [279]. Using these predictions, the Tevatron experiments convert the observed numbers of candidates for the reaction $p\bar{p} \to \text{LQ}\,\overline{\text{LQ}}X$ with subsequent decays $\text{LQ} \to eq$ or $\text{LQ} \to \nu q'$ into 95% confidence level limits on M_{LQ} which are functions of β_e. For scalar LQs with $\beta_e = 1$ ($\beta_e = 0.5$), CDF [280] and DØ [281] quote lower limits of $M_{LQ} > 213\,\text{GeV}$ ($180\,\text{GeV}$) and $M_{LQ} > 225\,\text{GeV}$ ($204\,\text{GeV}$), respectively. A limit of $M_{LQ} > 242$ GeV for $\beta_e = 1$ has been obtained from the combined CDF and DØ data [282]. For lower β_e the limits are weaker since the searches include final states with neutrinos, which are experimentally less well constrained. The exclusion plot from DØ for scalar LQs is shown in Fig. 2.20. For vector LQs, DØ has reported limits of $M_{LQ} > 298\,\text{GeV}$ ($270\,\text{GeV}$) [283] which, however, still await publication. In any case, the vector LQ limits are significantly stronger than those for scalar LQs and rule out vector LQ interpretations of the HERA data for the relevant mass range, $M_{LQ} \lesssim 240\,\text{GeV}$.
- LQs with right- and left-handed couplings to ed and νu, respectively, would contribute to the helicity-suppressed decay $\pi^+ \to e^+\nu$. The precise experimental confirmation of e–μ universality in $\pi^\pm$ decays restricts the product of these couplings to $g_L g_R \lesssim 10^{-6}\,(M_{LQ}/200\,\text{GeV})^2$ [264], implying that LQs of the type $S_{1/2}$ or $V_{1/2}$ (see Table 2.3) cannot mediate $ed \to \nu u$ transitions at a level observable at HERA.
- Data from several rare decays and the absence of flavor-changing neutral currents imply that LQs must be dominantly flavor-symmetric, i.e. couple neither to quark mass-eigenstates nor to leptons from different generations simultaneously. Quantitative bounds and theoretical implications have been discussed in [275] and recently also in [228, 284]. Searches for LQs with lepton-flavor-violating couplings at HERA [59, 64, 84], and also for LQs coupling to second- or third-generation leptons at CDF [285, 286] and DØ [287, 288] did not yield evidence for the existence of such states.
- Several LQ constraints from other low-energy measurements can be found in [275]. In addition, the impact of precision measurements of atomic

[9] LQ production at the Tevatron and HERA is covered in a recent review [276]. The LQ discovery potential at future hadron colliders is discussed in [277].

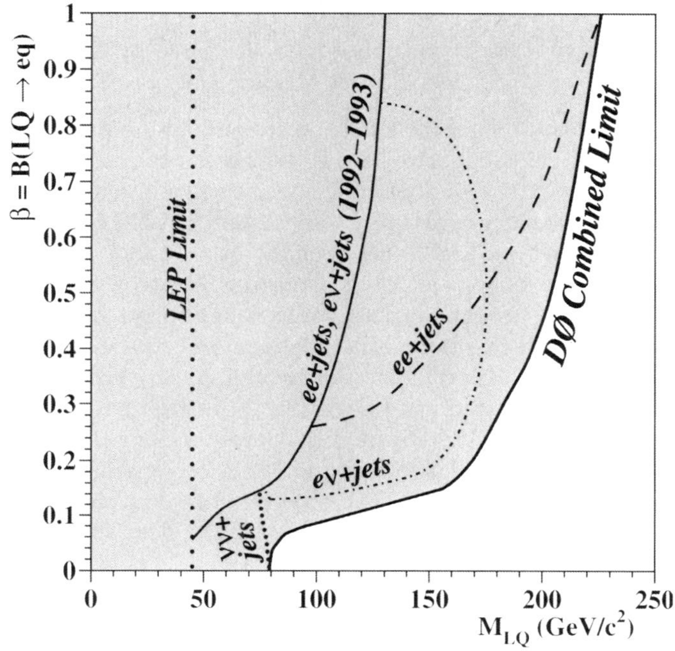

Fig. 2.20. Limits on scalar LQ parameters from DØ. The *shaded area* represents the region which is excluded at 95% C.L. The plot is reproduced from [281]

parity violation in cesium[10] which have recently become available [289] is discussed in [290, 291]. The LQ scenarios proposed as explanations for the HERA high-Q^2 excess events are compatible with these constraints.

- The question as to what extent virtual processes involving LQs would affect e^+e^- cross sections or the electroweak parameters has been addressed by [229, 292–294], the general conclusion being that the current experimental accuracy is not sufficient to exclude LQ scenarios which might explain the HERA excess events.

Combining the bounds obtained, we can now discard most of the LQ species in Table 2.3 as potential sources of a possible HERA excess: the LQs with $F = -2$ because of the HERA e^-p data and also because the required large couplings would be in conflict with low-energy bounds, and the vector LQs because of the Tevatron limits. From the remaining $S_{1/2}$ and $\tilde{S}_{1/2}$ LQs we know that – even if they coexist with the stringent Tevatron limits – they cannot couple to both e and ν (π decay). In particular, any contribution of LQ production to the CC cross section seems to be excluded. However, Babu et al. [292] argue that a mixing between different scalar LQs might produce CC-like

[10] See the discussion of contact interactions below for more details on APV.

events and would furthermore avoid the Tevatron limits by reducing β_e. Other scenarios with LQ decays into neutrinos have been discussed by Altarelli at al. [265], Hewett and Rizzo [255], Sekiguchi et al. [261] and Xiao [263].

To summarize, an explanation of the HERA excess of high-Q^2 events with LQs of the "standard" type summarized in Table 2.3 appears to be ruled out. On the one hand, the only two of the LQ species which are compatible with constraints from low-energy experiments and from the HERA e^-p data decay exclusively to an electron and quark. On the other hand, a solution with $\beta_e < 1$ would be necessary in order to avoid the Tevatron limits (see Fig. 2.20). One needs to go to extended or at least fine-tuned LQ models in order to find solutions which resolve this dilemma and are at the same time compatible with all other experimental constraints. As already indicated in Fig. 2.20, it has become customary to treat β_e as a free parameter in experimental LQ search analyses, thereby implicitly presupposing that the set of assumptions leading to the LQ classification in Table 2.3 may be violated. One theoretical scenario containing LQ-type objects with alternative decay modes (and therefore with $\beta_e < 1$) will be discussed in the next subsection.

R_P-Violating Squarks

Supersymmetry (SUSY) is considered to be one of the most promising scenarios to extend the SM and to overcome some of its inherent theoretical difficulties (see [295] for a detailed and comprehensive overview). In SUSY, a fermionic (bosonic) partner[11] is assigned to each SM boson (fermion), except in the Higgs sector, which is more complex. SM and SUSY particles differ in their R-parity, which is defined as

$$R_P = (-1)^{3B+L+2S} = \begin{cases} +1 & \text{for SM particles} \\ -1 & \text{for SUSY particles} \end{cases}, \tag{2.52}$$

where B, L and S denote the baryon and lepton number and the spin of the particle.

In the minimal supersymmetric model (MSSM) R_P is required to be a conserved quantity, with the immediate consequences that SUSY particles can only be produced in pairs, and that the lightest SUSY particle (LSP) must be stable. This scenario does not offer a straightforward explanation of anomalous DIS cross sections (see, however, [257, 296] for an R_P-conserving extension of the MSSM which could produce anomalous high-Q^2 cross sections at HERA). However, R_P conservation is not an indispensable ingredient of SUSY and can be abandoned, allowing one to add Yukawa-type terms to the superpotential which include squark–quark–lepton couplings and thus

[11] SUSY partners of fermions are denoted by a prefix *s*, like *sfermion*, whereas partners of bosons are characterized by the ending *-ino*, like *photino*.

allow for LQ-like production and decay mechanisms of squarks. The relevant terms in the Lagrangian are given by [297]

$$\begin{aligned}\mathcal{L}_{\not R_P} = -\lambda'_{ijk} \Big[& \tilde{u}^j_{\rm L} \bar{d}^k_{\rm R} e^i_{\rm L} + \bar{\tilde{d}}^k_{\rm R} (\bar{e}^i_{\rm L})^c u^j_{\rm L} + \tilde{e}^i_{\rm L} \bar{d}^k_{\rm L} u^j_{\rm L} \\ & - \tilde{\nu}^i_{\rm L} \bar{d}^k_{\rm R} d^j_{\rm L} - \tilde{d}^j_{\rm L} \bar{d}^k_{\rm R} \nu^i_{\rm L} - \bar{\tilde{d}}^k_{\rm R} (\bar{\nu}^i_{\rm L})^c d^j_{\rm L} \Big] + \text{h.c.} , \qquad (2.53)\end{aligned}$$

where the tilde indicates sfermions (which are scalar), the superscript c denotes charge-conjugate spinors and the i, j, k are generation indices. Actually, the superpotential acquires two further contributions with couplings λ_{ijk} and λ''_{ijk}, which, however, would induce lepton number violation (λ_{ijk}) or baryon number violation (λ''_{ijk}) [298] if they were present simultaneously with the terms in (2.53), and are hence assumed to be negligible. If the λ_{ijk} terms are nonzero, they also induce contributions to e^+e^- scattering and allow for single-slepton production in e^+e^- or $p\bar{p}$ reactions which both might be in reach of experimental sensitivity at LEP or the Tevatron [299, 300].[12] The phenomenology of direct production of R_P-violating squarks at HERA has recently been covered in a series of publications [227, 229–232, 265, 302–311]. A compilation of current upper limits on the λ_{ijk}, λ'_{ijk} and λ''_{ijk} derived from various experimental constraints can be found in [312]. Different theoretical aspects of SUSY scenarios with R_P violation and their embedding in fundamental gauge symmetries are discussed in [313–318] in the light of corresponding interpretations of the HERA data.

If the relevant couplings λ'_{ijk} are nonzero, (2.53) allows for production and subsequent decay of squarks in e^+p reactions via

$$\begin{aligned}&(e^+)^i d^k \to \tilde{u}^j \to e^+ d \, , \\ &(e^+)^i \bar{u}^j \to \bar{\tilde{d}}^k \to e^+ \bar{u}, \, \bar{\nu} \bar{d} \, , \qquad (2.54)\end{aligned}$$

according to the LO Feynman graphs in Fig. 2.21. The generation indices i, j, k have only been indicated for the initial-state particles and the squarks in (2.54). In general, the flavor indices of the final state can differ from those of the initial state, as indicated in Fig. 2.21. At HERA, obviously $i = 1$ is required, and reactions involving valence quarks are only possible for $k = 1$ and d quarks. Note that in this case no CC-type final state can be produced.

The squarks in (2.54) act exactly like scalar LQs (where $\tilde{u}^k$ and $\bar{\tilde{d}}^j$ correspond to the left-handed states of $\tilde{S}_{1/2}$ and S_0, respectively, in Table 2.3 [59], with the one important exception that they will in general have competing R_P-conserving decay modes and can hence avoid the Tevatron limits and the dilemma pointed out for LQs at the end of the previous subsection. In all other respects, the statements made about scalar-LQ production at HERA

[12] The presence of λ_{ijk} or λ'_{ijk} terms might also affect the interactions of ultrahigh-energy cosmic neutrinos in a way which could be observed in large-volume neutrino telescopes [301].

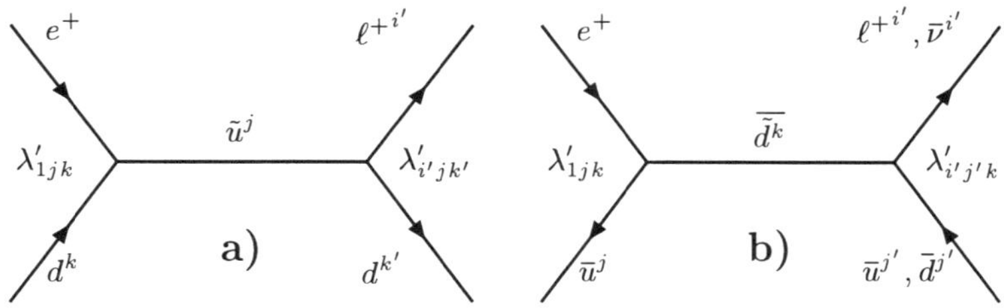

Fig. 2.21. Leading-order Feynman graphs for s-channel squark formation and decay in (**a**) ed and (**b**) $e\bar{u}$ scattering. The squark–lepton–quark vertices are characterized by Yukawa couplings λ'_{1jk} (initial state) and $\lambda'_{i'jk'}$ or $\lambda'_{i'j'k}$ (final state). The quark flavors d and u are used here to denote d- and u-type quarks of any generation

(and also at the Tevatron) are directly applicable to the case of R_P-violating squarks. In particular, the cross section formulae (2.50) and (2.51) remain valid after replacing LQ $\rightarrow \overset{\leftrightarrow}{\tilde{q}}$ and $g, g' \rightarrow \lambda'_{ijk}, \lambda'_{i'j'k'}$.

The major results of several careful studies of the phenomenological aspects of the squark scenario which were published before [299, 319–321] and after [227, 229–232, 265, 302–311] the advent of the HERA high-Q^2 excess can be summarized as follows:

- If more than one of the couplings λ'_{ijk} is nonzero for any particular squark, lepton flavor violation or flavor-changing neutral currents are induced at tree level. One therefore usually assumes that all but one of the λ'_{ijk} are zero (an approach with several nonzero λ'_{ijk} is discussed by Belyaev and Gladyshev [303]). However, some residual flavor-changing potential is left since the flavor states in (2.53) are the weak eigenstates which are connected to the mass eigenstates via the nondiagonal CKM matrix [227].
- The relevant couplings λ'_{1j1} and (for initial-state sea quarks) λ'_{1j2}, λ'_{11k} or λ'_{12k} are severely constrained by low-energy phenomena. Note that – in contrast to the LQ case – any nonzero λ'_{ijk} implies that all six terms in (2.53) are present and contribute to a large spectrum of processes. The most restrictive bounds are derived from limits on neutrinoless double-beta decay (which effectively rule out contributions from λ'_{111}, cf. Sect. 4.4.1), from charged-current universality (eliminating λ'_{11k}) and from limits on the electron-neutrino mass (deleting λ'_{122} and λ'_{133} from the list). As in the LQ case, $e^+\bar{u} \rightarrow \bar{\tilde{d}}^k$ is also almost certainly excluded by the absence of a signal from the charge-conjugate process in the HERA e^-p data.
- The remaining possibilities which might explain the HERA high-Q^2 excess events are

$$e^+d \rightarrow \tilde{c} \rightarrow e^+q \qquad (ijk) = (121)\,, \tag{2.55}$$

$$e^+d \rightarrow \tilde{t} \rightarrow e^+q \qquad (ijk) = (131)\,, \tag{2.56}$$

$$e^+s \rightarrow \tilde{t} \rightarrow e^+q \qquad (ijk) = (132)\,. \tag{2.57}$$

- The reaction channels with stop ($\tilde{t}$) production have attracted particular attention since they involve the additional subtlety of $\tilde{t}_{\rm R}$–$\tilde{t}_{\rm L}$ mixing, which is expected to be more pronounced for the stop than for other squarks owing to the high top mass. They were investigated by Ellis et al. [307], Kim and Ko [309], Asakawa, Kamoshita and Sugamoto [302], and Kon and Kobayashi [310], who argue that the existence of two mixing stops with similar masses in the HERA range leads to a broad x spectrum of the signal and could thus explain the absence of a sharp peak in the data. However, it has also been claimed recently [322] that the SUSY parameter space in which reaction (2.57) is allowed is severely constrained by the experimental limit on the electron-neutrino mass in conjunction with corresponding theoretical contributions due to the sneutrino vacuum expectation value.
- As for the LQs, none of the preferred solutions involves νq final states, and there is hence no natural explanation for a CC cross section anomaly in the R_P-violating SUSY scenario. Several authors [265, 304, 311] have investigated the question whether R_P-violating SUSY would provide reaction mechanisms which mimic the experimental CC signatures and would be interpreted as a CC excess. They confirm that νq final states are excluded, but argue that R_P-conserving cascade decays of the stop may entail multiquark final states with large missing transverse momentum due to neutrinos or LSPs. However, although such events would enter the CC selections of ZEUS and H1, they in general contain multiple hadronic jets and should easily be distinguishable from typical high-Q^2 DIS events like the one shown in Fig. 1.3.

As already indicated above, the experimental situation in an R_P-violating SUSY scenario is characterized by the decay of the squarks into a variety of R_P-conserving channels in addition to the R_P-violating eq mode. The combined R_P-conserving branching ratio typically exceeds 50%, thus providing an additional handle for squark searches. A detailed account of possible decay channels and their experimental signatures at HERA has been worked out in [323]. Recently, it has been suggested that scenarios involving squarks might explain the H1 observation of an excess of events characterized by missing transverse momentum, an isolated lepton (e or μ) with high transverse momentum and a hadronic jet [88]. This event topology (denoted by $\not{P}_{\rm t}\ell j$ in subsequent chapters) is compatible with a decay chain $\tilde{t} \to \tilde{b} + W \to d\nu + \ell\nu$ (equivalently for $\tilde{c}$). However, as yet none of the direct searches for production of R_P-violating squarks in e^+p scattering (by ZEUS [78] and H1 [65, 85]), in e^+e^- reactions (by ALEPH [324] and OPAL [325]) and in $p\bar{p}$ scattering (by CDF [326] and DØ [327]) has yielded positive evidence in any of the decay channels investigated.

To summarize, there are scenarios of R_P-violating SUSY which could accommodate an anomalous high-Q^2 NC cross section compatible with the recent HERA measurements. However, no excess of CC-type events with the typical DIS (1+1)-jet topology is expected in these models.

Contact Interactions

In the last two sections we have dealt with the direct production of conjectural LQs or squarks at HERA. We will now turn to scenarios in which the high-Q^2 *ep* cross sections are modified by virtual effects of new interactions at a much higher mass scale than the HERA center-of-mass energy. One can describe a broad spectrum of such processes by parameterizing their low-energy manifestations as an effective four-fermion interaction, the so-called contact interaction (CI). In the following, we will discuss general aspects of CIs and their impact on *ep* and other cross sections, and collect the constraints due to existing experimental data. Finally, we will classify those scenarios which avoid these constraints and might be in the reach of experimental sensitivity in the investigations of ZEUS and H1 (see Sect. 4.4.2).

CI Cross Sections. The effective diagram for electron–quark CIs and Feynman graphs of some processes which it might represent are shown in Fig. 2.22. In particular, CIs can parameterize virtual effects of LQs or R_P-violating squarks, of additional gauge bosons, or of composite leptons and quarks. However, CIs are good approximations for these scenarios only if their characteristic mass scales are large compared with the HERA center-of-mass energy. Also, for interactions of composite quarks with pointlike electrons or vice versa, the *ep* cross section is modified by a form-factor term and CIs are inappropriate [328]; we will shortly address this case at the end of this section.

The most general effective Lagrangian for *eeqq* CIs is given by

$$\begin{aligned}\mathcal{L}_{\mathrm{CI}}^{eeqq} = \epsilon\frac{g^2}{\Lambda^2}\sum_q &\left[\eta_{\mathrm{LL}}^q(\overline{e_{\mathrm{L}}}\gamma_\mu e_{\mathrm{L}})(\overline{q_{\mathrm{L}}}\gamma^\mu q_{\mathrm{L}}) + \eta_{\mathrm{RR}}^q(\overline{e_{\mathrm{R}}}\gamma_\mu e_{\mathrm{R}})(\overline{q_{\mathrm{R}}}\gamma^\mu q_{\mathrm{R}})\right.\\ &\left.+\,\eta_{\mathrm{LR}}^q(\overline{e_{\mathrm{L}}}\gamma_\mu e_{\mathrm{L}})(\overline{q_{\mathrm{R}}}\gamma^\mu q_{\mathrm{R}}) + \eta_{\mathrm{RL}}^q(\overline{e_{\mathrm{R}}}\gamma_\mu e_{\mathrm{R}})(\overline{q_{\mathrm{L}}}\gamma^\mu q_{\mathrm{L}})\right]\\ &+\text{scalar terms} + \text{tensor terms}\,, \end{aligned} \tag{2.58}$$

where the subscripts L and R denote the left- and right-handed helicity projections of the respective fermion fields, g is the overall coupling, Λ is the mass scale and ϵ is an overall sign which determines whether the interference with the SM amplitude is constructive or destructive. A prominent historical example of a CI-type effective theory is the description of weak interactions in terms of the Fermi constant, G_{F}, which is the exact equivalent of g^2/Λ^2. A specific combination of ϵ and the coefficients η_{mn}^q ($m,n=$L, R) is referred to as a "CI scenario" in the following; any given theoretical model inducing CIs is related to one particular such scenario. In most CI searches, each of the η_{mn}^q is assumed to be either ±1 or zero. Once the η_{mn}^q are fixed, the CI strength is given by the ratio of g and Λ and is usually quantified in terms of an effective mass scale defined by

$$\frac{g^2}{\Lambda^2} \overset{\text{convention}}{=} \frac{4\pi}{(\Lambda_{\mathrm{eff}})^2}\,. \tag{2.59}$$

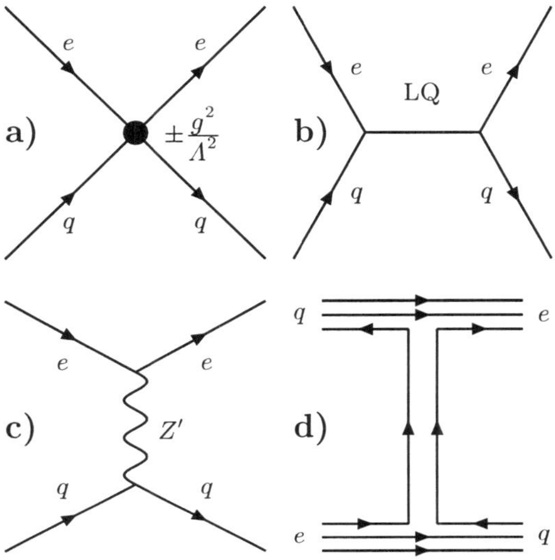

Fig. 2.22. The effective contact interaction diagram (**a**) and three examples of leading-order Feynman graphs it might represent: (**b**) s-channel formation of an LQ; (**c**) t-channel exchange of a massive neutral gauge boson and (**d**) exchange interaction of composite fermions

In the following, the superscript "eff" will be omitted and Λ will always denote the effective scale. The scalar and tensor terms in (2.58) are strongly constrained by data on helicity-suppressed decays [275] and will not be considered further here. In various publications [227, 265, 329–335] the phenomenology of CIs in ep reactions at HERA has recently been investigated. Some authors have also related the interpretation of the HERA high-Q^2 excess events in terms of CIs to higher-dimension gauge symmetries [336–338] and also to the possible existence of extra neutral gauge bosons [339].

The LO[13] ep cross section including CIs is given by the SM formula, (2.4), if A_q and B_q from (2.7) and (2.8) are replaced by [340]

$$
\begin{aligned}
A_q(Q^2) &\longrightarrow \frac{1}{2}\left[(V_q^{\mathrm{L}})^2 + (V_q^{\mathrm{R}})^2 + (A_q^{\mathrm{L}})^2 + (A_q^{\mathrm{R}})^2\right] , \\
B_q(Q^2) &\longrightarrow V_q^{\mathrm{L}}\, A_q^{\mathrm{L}} - V_q^{\mathrm{R}}\, A_q^{\mathrm{R}} , \\
V_q^m &= Q_q - (v_e \pm a_e)\, v_q\, P_Z + \frac{\epsilon Q^2}{2\alpha\Lambda^2}(\eta_{m\mathrm{L}}^q + \eta_{m\mathrm{R}}^q) , \\
A_q^m &= \quad - (v_e \pm a_e)\, a_q\, P_Z + \frac{\epsilon Q^2}{2\alpha\Lambda^2}(\eta_{m\mathrm{L}}^q - \eta_{m\mathrm{R}}^q) .
\end{aligned}
\tag{2.60}
$$

In (2.60), the superscript m is L or R; the plus (minus) sign in the definitions of V_q^m and A_q^m is for $m=\mathrm{L}$ ($m=\mathrm{R}$). Three observations can be made. (i) Q^2/Λ^2 enters linearly into the SM×CI interference terms but quadratically

[13] Note that CIs are an effective, nonrenormalizable theory. In particular, virtual loop corrections are divergent. Consequently, CIs are always used in leading order; see Sect. 4.4.2 for a description of how the dominant radiative corrections due to photon emission are taken into account.

into the CI×CI terms. In fact both contributions are important for typical Λ scales under discussion for HERA. (ii) On first sight, the CIs seem to modify only $d\sigma/dQ^2$ but not $d\sigma/dx$. However, the different y dependence of the $\mathcal{F}_2$ and $\mathcal{F}_3$ contributions effectively induces an x dependence of the ratio $\sigma_{SM+CI}/\sigma_{SM}$ of the doubly differential cross sections. An additional x dependence of the integrated cross section $d\sigma/dx$ is of course introduced by the fact that the average x is a function of Q^2. (iii) CI effects for e^+p and e^-p NC scattering differ by the relative sign of the $\mathcal{F}_2$ and $\mathcal{F}_3$ terms. The pattern of cancellations between the four principal contributions to the CI-induced cross section (SM×CI and CI×CI terms in $\mathcal{F}_2$ and $\mathcal{F}_3$) therefore depends both on ϵ and on the charge of the initial-state lepton. Many combinations of the η^q_{mn} yield significantly different experimental sensitivities for $\epsilon = +1$ and for $\epsilon = -1$; typically, this "asymmetry" is reversed between e^+p and e^-p scattering. CI searches in e^+p and e^-p data are hence complementary, and the HERA results reported in this book will have to be combined with CI searches in the e^-p data to reach optimal sensitivity to all CI scenarios.

In Fig. 2.23 the effect of a CI on the e^+p DIS cross section is shown for the case that all η_{mn} are equal (denoted "VV" below, see Table 2.4) and that $\Lambda = 3\,\text{TeV}$.

As a consequence of crossing symmetry, the presence of an $eeqq$ CI will also affect hadron production in e^+e^- scattering and Drell–Yan processes in

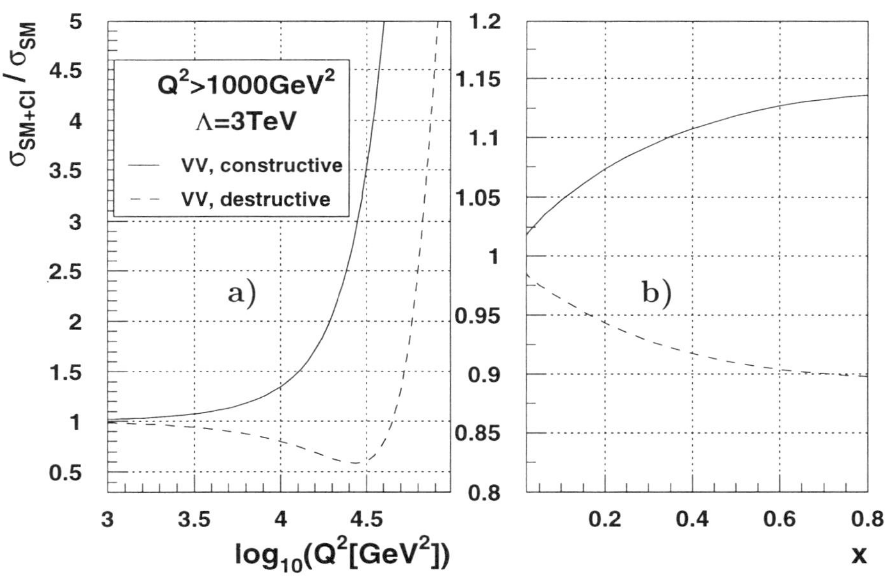

Fig. 2.23. Modification of the NC e^+p DIS cross sections by a CI of type VV (see text) with $\Lambda = 3\,\text{TeV}$. Shown are the ratios (**a**) of the cross sections $d\sigma/dQ^2$ and (**b**) of $d\sigma/dx$ with and without CI, for constructive (*solid lines*) and for destructive (*dashed lines*) SM×CI interference

$p\bar{p}$ collisions. It is, however, important to note that the relative sign of the SM and CI amplitudes changes under crossing, with the consequence that constructive interference in ep scattering corresponds to destructive interference in the other two reactions and vice versa. Also, for e^+e^- and $p\bar{p}$ scattering the LL and RR contributions are strongest, whereas in e^+p scattering these are suppressed by $(1-y)^2$ with respect to the LR and RL terms.

So far, we have only considered $eeqq$ CIs, which are relevant for NC DIS at HERA. The CC equivalents are $e\nu qq'$ CIs, which are governed by a Lagrangian that is identical to (2.58) except that the $\eta_{\rm RL}$ and $\eta_{\rm RR}$ terms are absent since they would involve right-handed neutrinos.[14] Again, scalar and tensor terms, but also the $\eta_{\rm LR}$ contribution, are ruled out by pion decay data [333]. The remaining $\eta_{\rm LL}$ modifies the CC cross section by a weakly Q^2-dependent factor [342] (see (2.19), (2.20)):

$$\frac{G_{\rm F}^2}{2\pi}\left(\frac{M_W^2}{Q^2+M_W^2}\right)^2 \longrightarrow \frac{1}{\pi}\left(\frac{G_{\rm F}}{\sqrt{2}}\frac{M_W^2}{Q^2+M_W^2} \mp \frac{\pi}{2\Lambda^2}\right)^2 . \tag{2.61}$$

One obvious way to relate $eeqq$ and $e\nu qq'$ CIs is to require ${\rm SU(2)_L \times U(1)_Y}$ invariance, whence [330]

$$\begin{aligned} \eta_{\rm RL}^{eedd} &= \eta_{\rm RL}^{eeuu} , \\ \eta_{\rm LL}^{eedd} - \eta_{\rm LL}^{eeuu} &= \eta_{\rm LL}^{e\nu ud} , \end{aligned} \tag{2.62}$$

where now the superscript of the η's denotes the fermions participating in the contact interaction. In addition to $e\nu qq'$ CIs, ${\rm SU(2)_L \times U(1)_Y}$ symmetry also implies the existence of $\nu\nu qq$ terms which have experimental consequences for neutrino–nucleon reactions.

Experimental Constraints on CIs. In addition to those from HERA (see Sect. 4.4.2), experimental constraints on CIs come from high-energy accelerator data (CDF [343], DØ [344] and CCFR [345] at the Tevatron, ALEPH [346, 347], DELPHI [348], L3 [349] and OPAL [350, 351] at LEP), from the investigation of virtual effects in Z production at LEP [352] and from low-energy measurements. A recent global CI analysis taking into account all these data and their statistical and systematic uncertainties can be found in [353].

Of the low-energy experimental data (see [290, 305, 330] for details), the atomic-parity-violation (APV) results turn out to be particularly restrictive for CIs. They are sensitive to a parity-violating mixing of S and P states in atomic cesium, from which the "weak charge" of the nucleus, $Q_{\rm w}$, can be inferred.[15] $Q_{\rm w}$ receives a well-defined contribution from the SM and would be modified by CI according to

[14] Limits on CIs including heavy right-handed neutrinos have been derived from the supernova SN1978A data [341].

[15] Cesium is particularly suited for these measurements since its atomic wave function is accurately known and $Q_{\rm w}$ can therefore be derived with high accuracy ($\mathcal{O}(1\%)$) from the experimental results [354].

$$\begin{aligned}\Delta^{\mathrm{CI}}Q_{\mathrm{w}} = \; & \frac{2Z+N}{\sqrt{2}G_{\mathrm{F}}}\frac{4\pi\epsilon}{\Lambda^2}\left(\eta^u_{\mathrm{LL}}+\eta^u_{\mathrm{LR}}-\eta^u_{\mathrm{RL}}-\eta^u_{\mathrm{RR}}\right)\\ & + \frac{Z+2N}{\sqrt{2}G_{\mathrm{F}}}\frac{4\pi\epsilon}{\Lambda^2}\left(\eta^d_{\mathrm{LL}}+\eta^d_{\mathrm{LR}}-\eta^d_{\mathrm{RL}}-\eta^d_{\mathrm{RR}}\right)\end{aligned} \tag{2.63}$$

(Z and N are the numbers of protons and neutrons in the nucleus). On the basis of a recent measurement [289], Q_{w} has lately been redetermined with improved precision [355], yielding [291]

$$\Delta^{\mathrm{exp}}Q_{\mathrm{w}} = 1.28 \pm 0.46\,. \tag{2.64}$$

This result translates into lower limits on Λ of $8-20\,\mathrm{TeV}$, provided only one of the η^q_{mn} is nonzero. However, any CI which is parity-conserving will have $\Delta^{\mathrm{CI}}Q_{\mathrm{w}} = 0$ and hence avoid the APV constraint. Furthermore, mutual cancellations of several CI-like processes may occur [356]. If taken at face value, (2.64) implies the existence of a non-SM contribution to Q_{w} at the 2.5σ significance level, which could e.g. be a CI with $\Lambda = \mathcal{O}(10\,\mathrm{TeV})$. Other scenarios beyond the SM resulting in a nonzero ΔQ_{w} are discussed in [291, 357].

Several authors have analyzed the compatibility of the aforementioned data with a CI interpretation of the HERA high-Q^2 results for NC [227, 329, 330, 334, 358] and CC [265, 332, 335] reactions. We will not reproduce a detailed synopsis of all relevant data and theoretical considerations here, but instead conclude that the existing HERA data are not sensitive to CC CIs of a size corresponding to the exclusion limits of other experiments. In the NC case, typical $eeqq$ CIs which have been proposed as explanations of a significant increase of the HERA NC cross sections at high Q^2 are parity-conserving combinations with scales in the range $\Lambda = 3-4\,\mathrm{TeV}$. However, most such scenarios are severely constrained by existing data from e^+e^-, $p\bar{p}$ and low-energy experiments. As an illustrative example, Fig. 2.24 (reproduced from [330]) demonstrates how the CDF Drell–Yan data restrict the CIs which would fit the 1996 combined H1 and ZEUS Q^2 distributions best. Nevertheless, the overall conclusion certainly depends on the exact properties of the HERA high-Q^2 data as studied in the ZEUS and H1 CI analyses (see Sect. 4.4.2).

Classification of CI Scenarios. Although a priori there are no constraints relating the different η^q_{mn} to each other, most experimental CI analyses search for signals corresponding to specific linear combinations of them. This procedure allows us easily to impose physics constraints on the CI scenarios under study (e.g. requiring parity conservation) and simplifies the statistical treatment. The typical CI scenarios investigated in experimental analyses are listed in Table 2.4 (each row in this table corresponds to two CI scenarios, differing in their overall sign). As indicated, all but the "purely chiral" scenarios (LL, LR, RL, RR) avoid the APV constraints, and only some of the U1–U6 combinations violate the $\mathrm{SU(2)_L}$ symmetry or would induce $e\nu qq'$ CIs

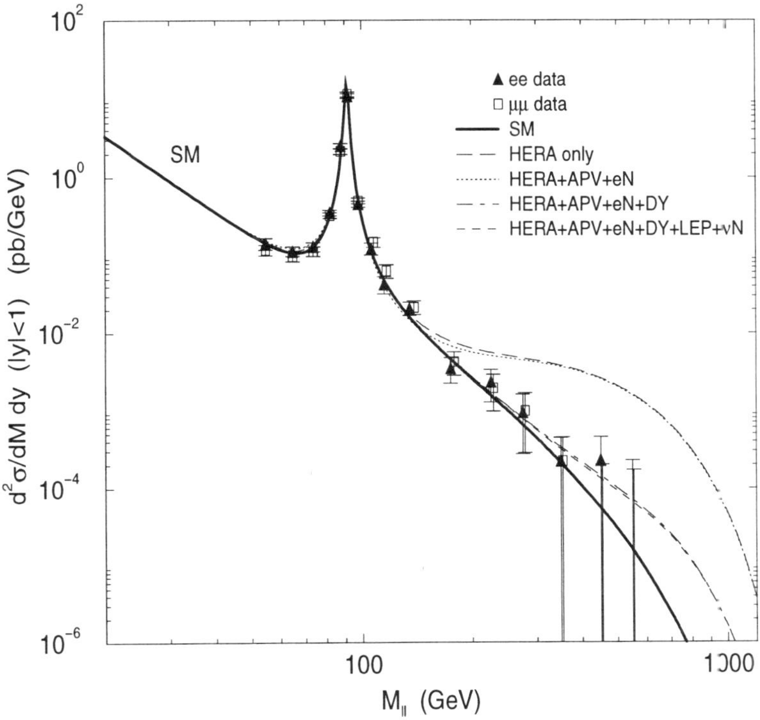

Fig. 2.24. Drell–Yan lepton-pair production cross section measured by CDF as a function of the invariant mass of the lepton pair (*points with error bars*) compared with the SM prediction and various CI scenarios. The *lines* represent results of CI fits including different selections of data sets. The figure is reproduced from [330]

if this symmetry holds. These Un terms (which could be constructed for the d quarks as well) are studied at HERA since the corresponding SM cross sections are dominated by u quarks and the resulting CI–SM interference terms are therefore the largest to be expected for a given Λ. A CI signal from this source could therefore manifest itself at HERA while avoiding strong $SU(2)_L$ breaking effects e.g. at LEP.

Note that the CI classification in Table 2.4 involves first-generation fermions only. Whereas this is obviously correct for the lepton (which is forced to be an electron by the choice of the initial state or by the signal definition), the situation is more complicated for the quark sector. In ep and $p\bar{p}$ interactions, at least one of the quarks has to be present in the initial state, implying that effects of second- or third-generation quarks are strongly suppressed. In contrast, quark–antiquark pairs of all generations can be produced in e^+e^- reactions, and CI results derived e.g. from the total hadronic cross section at LEP therefore depend on assumptions relating the η^q_{mn} for different quark generations.

Table 2.4. CI combinations considered in experimental analyses. The η^q_{mn} represent the coefficients of the respective helicity terms in the vector part of the Lagrangian (2.58). Plus or minus signs are used to denote +1 and −1, respectively. Each row corresponds to two CI scenarios differing in the overall sign, ϵ. The columns marked P, $\mathrm{SU(2)_L}$ and CC indicate whether the CI scenarios conserve parity (i.e. avoid the APV limits, cf. (2.63)), whether they conserve $\mathrm{SU(2)_L}$ symmetry and whether they would also induce $e\nu qq'$ CIs if $\mathrm{SU(2)_L}$ symmetry holds (see (2.62). The last column indicates recent phenomenological publications in which the particular CI was discussed. The X3 (X4) combination is sometimes denoted by LL + RR or V0 (LR + RL or A0) in the literature

CI	η^u_{LL}	η^u_{LR}	η^u_{RL}	η^u_{RR}	η^d_{LL}	η^d_{LR}	η^d_{RL}	η^d_{RR}	P	$\mathrm{SU(2)_L}$	CC	Discussed in
VV	+	+	+	+	+	+	+	+	$\surd$	$\surd$	—	[330, 334, 358]
AA	+	−	−	+	+	−	−	+	$\surd$	$\surd$	—	[330, 334, 336, 358]
VA	+	−	+	−	+	−	+	−	$\surd$	$\surd$	—	[334, 338]
X1	+	−	0	0	+	−	0	0	$\surd$	$\surd$	—	[358]
X2	+	0	+	0	+	0	+	0	$\surd$	$\surd$	—	
X3	+	0	0	+	+	0	0	+	$\surd$	$\surd$	—	
X4	0	+	+	0	0	+	+	0	$\surd$	$\surd$	—	[333, 337, 358]
X5	0	+	0	+	0	+	0	+	$\surd$	$\surd$	—	
X6	0	0	+	−	0	0	+	−	$\surd$	$\surd$	—	[333, 358]
U1	+	−	0	0	0	0	0	0	$\surd$	$\surd$	$\surd$	[358]
U2	+	0	+	0	0	0	0	0	$\surd$	—		
U3	+	0	0	+	0	0	0	0	$\surd$	$\surd$	$\surd$	
U4	0	+	+	0	0	0	0	0	$\surd$	—		[329, 339]
U5	0	+	0	+	0	0	0	0	$\surd$	$\surd$	—	
U6	0	0	+	−	0	0	0	0	$\surd$	—		[358]
LL	+	0	0	0	+	0	0	0	—	$\surd$	—	[342]
LR	0	+	0	0	0	+	0	0	—	$\surd$	—	
RL	0	0	+	0	0	0	+	0	—	$\surd$	—	
RR	0	0	0	+	0	0	0	+	—	$\surd$	—	

Other Approaches

The phenomenological scenarios developed in the preceding sections represent particular approaches for extensions of the SM. Although the large majority of publications related to the high-Q^2 excess of e^+p events at HERA elaborate on these approaches, there are also a few other ideas which potentially yield modifications of the high-Q^2 cross sections. In the following we briefly summarize four classes of such ideas. It is beyond the scope and intention of this book to judge whether all of these suggestions are self-consistent and compatible with existing experimental data and theoretical constraints.

Compositeness. Compositeness denotes a class of models in which the fundamental particles of the SM are compound objects, built of pointlike "preons". Complementary to the contact term approach, which covers the phenomenology of compositeness at large mass scales, some authors [328, 359, 360] speculate that the HERA data could be due to the direct production of excited states or heavy partners of the SM fermions. Whereas Adler [359] conjectures an object decaying into e^+g, which would be consistent with the observed event topologies, the other authors assume final states containing a fermion and an electroweak gauge boson, which do not reflect the experimental observation but could potentially pass the ZEUS and H1 DIS event selection.

It should be noted that both ZEUS [66–68, 80] and H1 [60, 70, 87] have searched for the production of excited electrons and neutrinos and their subsequent decay to $e\gamma$, eZ, νW or $\nu\gamma$, eW, νZ, respectively. No indication of a signal has been found in any of the decay channels investigated.

Particular experimental signatures are expected if the mass scale is much larger than the HERA center-of-mass energy and exchange graphs of the type Fig. 2.22d are prohibited, either because only leptons or only qarks are composite, or because they don't carry common constituents. In this case, the modification of the DIS differential cross sections can be approximated by a form-factor term [361],

$$\sigma = \sigma_{\mathrm{SM}} \left(1 - \frac{1}{6} R^2 Q^2\right)^2 , \tag{2.65}$$

where R is the effective radius of the composite fermion.[16] Equation (2.65) is a valid approximation only if $R^2Q^2 \ll 1$. On the basis of data taken up to 1994 (1995), H1 [61] (ZEUS [362]) have derived upper limits of $R < 2.6 \times 10^{-16}$ cm (1.4×10^{-16} cm) at 95% C.L., corresponding roughly to the expected spatial resolution of $1/\sqrt{Q^2_{\mathrm{max}}}$ (see Sect. 2.1.2), where $Q^2_{\mathrm{max}} \approx 10^4\,\mathrm{GeV}^2$ is the maximum Q^2 covered by those data. Recently, H1 has reported a limit of $R < 1.7 \times 10^{-16}$ cm [86] based on the 1994–97 data. However, one has to keep in mind that (2.65) predicts a decrease of the cross section at high Q^2, where actually an excess is observed; limits on R derived from these data must hence be interpreted with care, irrespective of whether the excess is caused by a fluctuation or has a physics origin.

Extra Spatial Dimensions. Recently, a theoretical framework containing n extra space dimensions of size R_{g} has been suggested [363] in the context of models featuring quantum gravity. Striking properties of this scenario would be the existence of a characteristic mass scale for gravitation, M_{D}, which is related to the Planck mass, M_{P}, by $M_{\mathrm{P}}^2 = 1/G_{\mathrm{N}} \sim R_{\mathrm{g}}^n\, M_{\mathrm{D}}^{2+n}$ (G_{N} being the gravitational constant), and a violation of Newton's law at distances smaller than R_{g}. However, since gravitation has not been tested at submillimeter distances, such scenarios can be consistent with existing experimental constraints if $n > 2$ and $R_{\mathrm{g}} \lesssim 100\,\mu\mathrm{m}$ [364].

In the related phenomenology [365], the gravitons are massless spin-2 objects propagating in $4+n$ dimensions, whereas the SM particles are constrained to the usual $3+1$ space–time dimensions. The projections of the

[16] If both leptons and quarks are composite, R^2 is the quadratic sum of both radii.

gravitons onto four dimensions produce a spectrum of Kaluza–Klein excitations coupling to the energy–momentum tensors of the SM particles and thus inducing non-SM contributions to the *eq* and also to the *eg* scattering amplitudes. Summing over the spectrum of Kaluza–Klein excitations up to masses of M_{D}, one obtains modifications of the *ep* cross sections that depend on an effective coupling $\lambda_{\mathrm{g}}/M_{\mathrm{D}}^4$, where λ_{g} is expected to be of the order of unity.

The experimental signature for extra dimensions at HERA would be similar to that of CIs, although the exact kinematic dependence of the cross section modification would be different. In a search for corresponding deviations from the SM prediction in the 1994–97 e^+p data [86], H1 found no significant evidence for extra spatial dimensions and placed 95% C.L. limits of $M_{\mathrm{D}} > 0.48\,\mathrm{TeV}$ ($M_{\mathrm{D}} > 0.72\,\mathrm{TeV}$) for $\lambda_{\mathrm{g}}=+1$ ($\lambda_{\mathrm{g}}=-1$). Exclusion limits of similar magnitude have been reported by the LEP experiments [351, 366, 367].

Alternative Higgs Sector. White [368] constructs a theory which is SM-like, except that the Higgs sector is replaced by a flavor doublet of color-sextet quarks and the electroweak symmetry breaking is a strong-interaction effect. In this scenario, the *ep* cross sections at Q^2 above the weak scale, i.e. exceeding $M_{W,Z}^2$, may receive extra contributions from instanton interactions involving the color-sextet quark doublet.

Tensor Admixture to the $V-A$ Currents. Chizhov [369] has argued that a small tensor admixture to the SM $V-A$ currents can explain both increased HERA NC and CC cross sections and anomalous tensor form factors measured in rare π and K decays. In this model, the tensor current is mediated by new heavy bosons which are described by an antisymmetric second-rank tensor field. Possible constraints from low-energy processes other than π and K decays are not discussed.

3. Experimental Issues

The survey of theoretical and phenomenological aspects of ep DIS at high Q^2 in the previous chapter related measured cross sections and distributions to interpretations in terms of underlying models and the determination of their parameters. We now want to discuss the experimental issues relevant to deriving theoretically interpretable measured quantities from the ep collisions provided by HERA. We will emphasize the aspects which are important for the high-Q^2 analyses.

Two multipurpose hybrid detector systems select and register ep reactions at HERA. They have been designed and built and are maintained and operated by the international collaborations ZEUS and H1. Although quite different in details of technical solutions, both detectors accomplish similar design principles, which are enforced by the particular accelerator environment at HERA and by the common physics goals.

Section 3.1 is devoted to a comparative description of both detectors with special emphasis on the central components. An account of the methods, quality and limitations of the Monte Carlo (MC) event and detector simulation and a survey of the MC generators used for the high-Q^2 analyses are given in Sect. 3.2. The offline reconstruction of event topology and kinematic variables from the raw detector data is discussed in Sect. 3.3, and the selection of the final NC and CC high-Q^2 DIS event samples, including studies of background sources and selection efficiencies, is detailed in Sect. 3.4. Finally, some figures characterizing the data samples used for the ZEUS and H1 high-Q^2 analyses are presented in Sect. 3.5.

3.1 The ZEUS and H1 Detectors

The general outline of both the ZEUS and the H1 detectors is dictated by the goal to register as completely and accurately as possible the final-state particles in many different ep reaction categories. As a consequence, both detectors are hermetic with the exception of holes to accommodate the beam pipe in the forward and backward directions, providing a coverage of more than 99.6% of the full solid angle. Also, they exhibit a marked forward–backward asymmetry, adapted to optimal measurement of the final states of

ep reactions with a center of mass which moves into the forward (i.e. proton) direction with an overall velocity of $0.935c$ for the 1994–97 beam energies.

The accelerator environment at HERA is characterized by the high bunch-crossing frequency of 10.4 MHz, which necessitates fast detection and readout techniques, and by a large background-to-signal ratio caused by high rates of interactions of beam protons with residual gas molecules in the beam pipe and with beam-line elements. Both aspects require fast multilevel trigger systems based on pipelined storage buffers to reduce the overall interaction rate of $\mathcal{O}(10^5\,\mathrm{Hz})$ by roughly four orders of magnitude before writing reaction data to mass storage.

We will, in the following, summarize the design and structure of both detectors (Sect. 3.1.1) and then focus on those components which are of particular importance for the high-Q^2 analyses.

3.1.1 Overview

A schematic overview of the ZEUS detector is shown in Fig. 3.1; the equivalent for the H1 detector in the pre-1995 configuration is presented in Fig. 3.2. Detailed technical reports describing the design, test and initial operational experience of the detectors are available both for ZEUS [370] and for H1 [371, 372]. We will summarize the detector setups below, and also mention their most relevant changes since 1993. Thereafter the components which are of particular importance for the high-Q^2 analyses will be presented in more detail: the calorimeters (Sect. 3.1.2), the central tracking detectors (Sect. 3.1.3), the luminosity monitors (Sect. 3.1.4) and the triggers (Sect. 3.1.5).

The ZEUS Detector. Enclosing the nominal interaction point (IP), a system of wire drift chambers provides precise track measurements. The principal component is the cylindrical central drift chamber (central tracking device, CTD) [373], covering an angular range of approximately $15° < \theta < 164°$ (θ being the polar angle with respect to the proton beam direction). An additional cylindrical wire chamber serving as a vertex detector (VXD) [374] was not used for analysis purposes after 1993 and was removed in the 1995/1996 winter shutdown. It will be replaced by a silicon microstrip detector scheduled to be operational from 2001 onwards. The central tracking detectors operate in a longitudinal magnetic field of 1.43 T, provided by a superconducting thin solenoid surrounding the CTD. A planar drift chamber end cap in the rear direction (RTD), and a forward system of planar drift chambers [375] interleaved with transition radiation detectors [376] (FDET) complement the central tracking components.

A sampling calorimeter (CAL) [377] with alternating layers of depleted uranium as an absorber and a scintillator as an active material surrounds the tracking volume and the solenoid. It consists of planar forward (FCAL) and rear (RCAL) parts and of a cylindrical central part (barrel calorimeter, BCAL). Each of these parts is further segmented into independent readout cells, thus providing position in addition to energy measurement. The

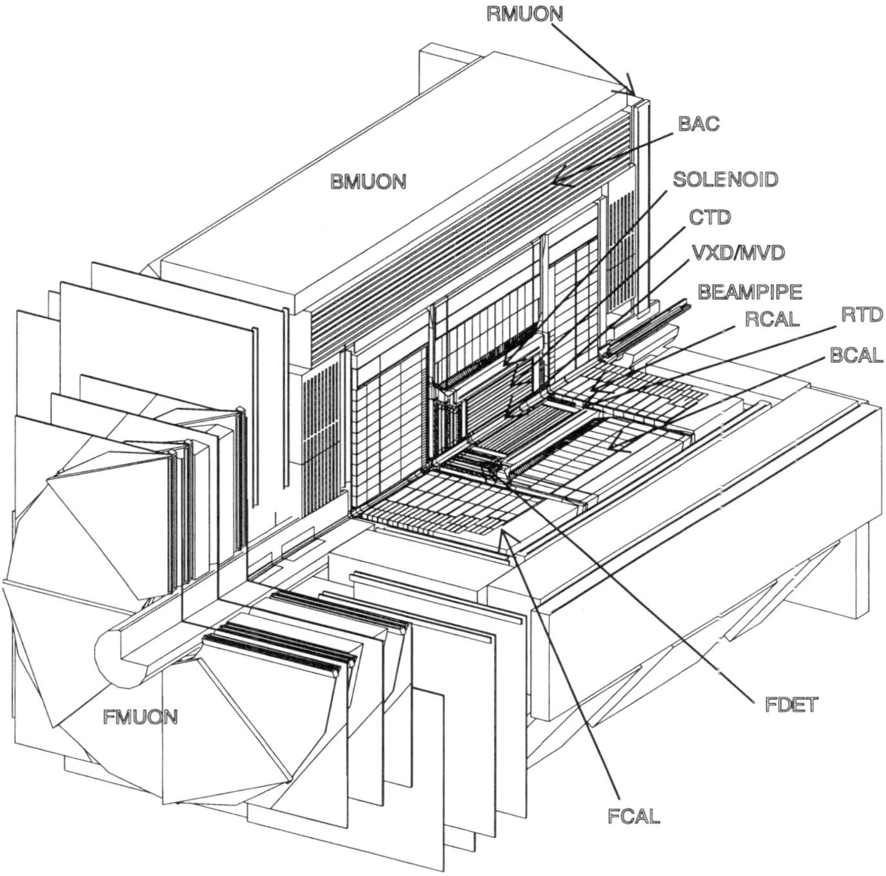

Fig. 3.1. The ZEUS detector. See text for a discussion of the components

CAL parameters are designed such as to ensure equal response to hadronic and electromagnetic showers (*compensation*), compromising the electromagnetic resolution to some extent. Since 1998, CAL has been complemented by a lead–scintillator sandwich calorimeter covering the very forward region, $0.6° \lesssim \theta \lesssim 6°$ (forward plug calorimeter, FPC [378]). In the rear direction, a hodoscope of crossed scintillator strips covers the inner surface of the RCAL close to the beam pipe (small-angle rear tracking detector, SRTD) [379] and supports position as well as energy measurement of scattered electrons in NC DIS reactions with $Q^2 \lesssim 70\,\mathrm{GeV}^2$. Since the 1995 data-taking period, RCAL and FCAL have been equipped with one-layer scintillator preshower detectors (presampler, RSAM and FSAM) [380], which are used to correct for energy loss of electrons and photons in the inactive material in front of the calorimeter. The barrel equivalent has been operational since 1999. Also, both FCAL and RCAL incorporate a layer of silicon detectors, which are

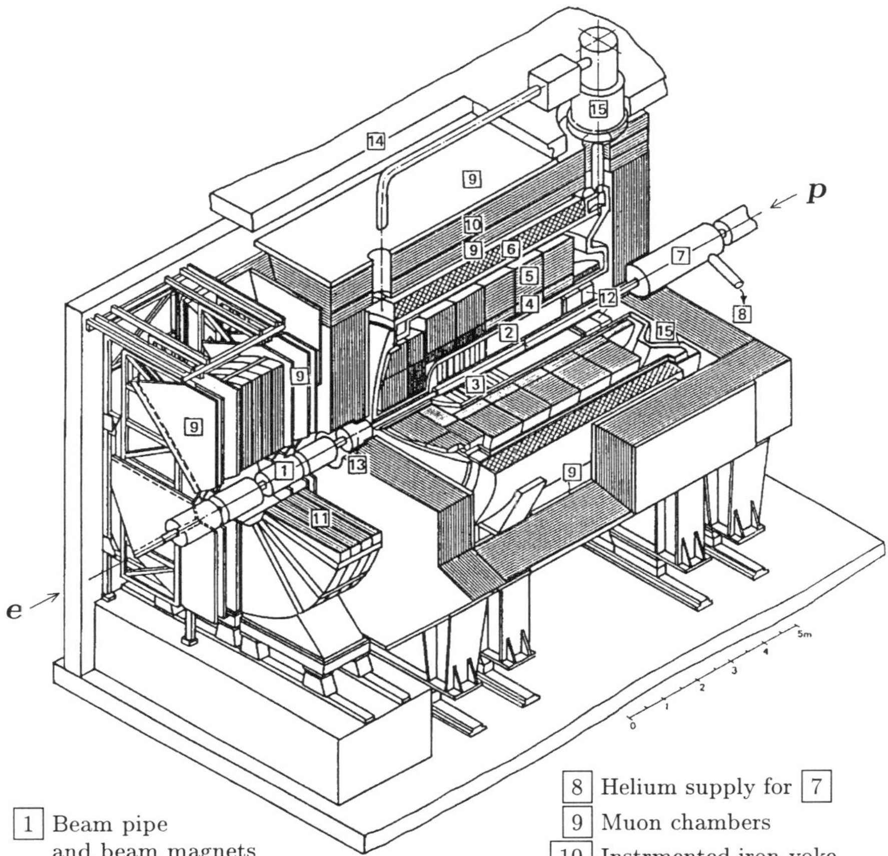

1 Beam pipe and beam magnets	8 Helium supply for 7
2 Central tracking device	9 Muon chambers
3 Forward tracking device	10 Instrmented iron yoke
4 Electromagnetic LAr calorimeter	11 Forward muon toroid
5 Hadronic LAr calorimeter	12 Backward electromagnetic calorimeter (BEMC)
6 Superconducting coil (1.15 T)	13 PLUG calorimeter
7 Compensating magnet	14 Concrete shielding
	15 Liquid-argon cryostat

Fig. 3.2. The H1 detector in the 1994 configuration. See text for a discussion of the components

positioned at a depth corresponding to the average longitudinal maximum of electromagnetic showers and serve to distinguish hadrons and electrons (hadron–electron separator, HES) [37, 381].

Two further components close the central detector. The backing calorimeter (BAC) [382], consisting of instrumented iron layers forming the return yoke of the solenoid, measures shower leakage out of the CAL and contributes to muon identification. A system of muon detectors [383], equipped with limited-streamer tubes, is installed in the rear (RMUON), central (BMUON) and forward (FMUON) directions. It serves to identify muons and, using the return yoke magnetization and a toroidal magnetized iron setup in the forward direction, to measure their momenta.

A series of additional detector components is situated at larger distances from the IP, mostly for detection of the scattered electron in very low-Q^2 reactions (like the beam pipe calorimeter, BPC, which was mentioned in Sect. 2.1.4), or the leading baryon in diffractive-type interactions. The only outer detector component which will be discussed here is the luminosity monitor, comprising two electromagnetic calorimeters [384] which are placed downstream in the e-beam direction and are designed to measure electrons (LUMI-e) and photons (LUMI-γ) from Bethe–Heitler processes $ep \to ep\gamma$.

The H1 Detector. As in the case of ZEUS, the volume around the IP is occupied by tracking detectors: two central, concentrically cylindric jet chambers (CJC) [385–388] are complemented by a silicon-strip microvertex detector operational since 1995 [389], by a planar arrangement of drift chambers and passive transition radiators in the forward direction (FTD) [390–392], and by a backward proportional chamber (BPC), which was replaced by a planar drift chamber (BDC) in 1995. Additional multiwire proportional chambers are used for triggering purposes [386, 393]. The rear, the central, and parts of the forward tracking detectors are situated in a longitudinal magnetic field of about 1.15 T, produced by a superconducting solenoid, which – in contrast to the situation in ZEUS – also encloses the calorimeter.

The H1 main calorimeter [394–396] covers the central and forward region ($4^\circ < \theta < 153^\circ$) and uses liquid-argon technology, with lead absorbers in the electromagnetic section and stainless-steel absorbers in the hadronic section. The forward cone ($0.7^\circ < \theta < 3.3^\circ$) is covered by a copper–silicon sandwich calorimeter (PLUG) [397]. In the rear direction, a lead–scintillator calorimeter (BEMC) [398] covering the angular range $151^\circ < \theta < 176^\circ$ was originally installed for the measurement of electrons in NC DIS reactions with Q^2 up to about 100 GeV^2. In 1995, BEMC was replaced by a lead–scintillating-fiber calorimeter (SPACAL) [399] with extended coverage and an improved resolution of $\delta E/E \approx 7\%/\sqrt{E\,(\mathrm{GeV})}$ for electromagnetic showers.

Similarly to ZEUS, H1 is equipped with a backing calorimeter (tail catcher, TC), a muon system [400, 401] and several external components in the electron or proton beam direction. The luminosity measurement is based on the same principle as in ZEUS.

3.1.2 Calorimetric Energy and Position Measurements

We will see below that neither the reconstruction of kinematic event variables nor the efficient selection of sufficiently pure event samples is possible without making extensive use of calorimetric measurements. The calorimeters are therefore the key detector components for the high-Q^2 (as well as for other) analyses, and a detailed understanding of their performance is crucial for reliability and accuracy of physical results. In the following we first discuss the design of the ZEUS uranium–scintillator calorimeter and the H1 liquid-argon calorimeter,[1] and then summarize their operation characteristics obtained from test beam measurements and *ep* data taken at HERA.

Design of the ZEUS Uranium Calorimeter (CAL). Energy measurement in the ZEUS CAL is performed using the light yield of scintillators, sandwiched between uranium absorber layers, as a measure of the integrated charged-particle flow traversing them, which in turn is proportional to the energy deposited in the electromagnetic or hadronic shower initiated by the primary particle. The compensating response of the ZEUS calorimeter has been optimized by varying the thicknesses d_{U}, d_{Sc} of the depleted-uranium (DU) and scintillator layers. The final design uses $d_{\mathrm{Sc}} = 2.6\,\mathrm{mm}$ and $d_{\mathrm{U}} = 3.3\,\mathrm{mm}$ (corresponding to one radiation length), with some additional material due to stainless-steel cladding of the DU. The DU–scintillator stacks are assembled in modules and read out via wavelength-shifting light guides running along the modules' lateral sides to photomultipliers (PMs) positioned at the outer CAL surface. In RCAL and FCAL the modules are installed vertically, each module spanning the full height and depth of the CAL and being 20 cm wide. The 32 modules of the BCAL are wedge-shaped, span its full length and depth and are installed at an azimuthal tilt of 2.5° with respect to the radial direction.[2] The BCAL and FCAL (RCAL) modules are subdivided in depth into one electromagnetic and two (one) hadronic readout layers, denoted EMC and HAC, respectively. The readout for each such layer is further partitioned into single cells with transverse sizes of about $20 \times 20\,\mathrm{cm}^2$ (HAC), $10 \times 20\,\mathrm{cm}^2$ (EMC in RCAL) and $5 \times 20\,\mathrm{cm}^2$ (EMC in BCAL and FCAL). The BCAL EMC cells point projectively in polar angle to the nominal IP.

The CAL consists of 5918 cells altogether, each being read out by two PMs, one connected to each of the cell's sides. This arrangement provides measurement redundancy, backup for single nonoperational PMs and position information via signal sharing. The PM signals are fanned out to the first-level trigger, the data acquisition chain [370] and a branch which records the time-integrated average used for calibration purposes. Owing to its segmentation, the CAL also allows precise measurements of the spatial position of energy deposits. This is particularly important for the determination of the scattering angle of the final-state electrons in DIS which are outside the tracking-chamber acceptance. The position measurement exploits the distri-

[1] The H1 rear calorimeters, BEMC and SPACAL, are of lesser importance for the high-Q^2 studies.

[2] This tilt has been designed to keep particles with high transverse momentum (i.e. the scattered electrons in *ep* DIS events) from traveling along the module cracks. Owing to the magnetic field, this objective is accomplished more effectively for e^-p than for e^+p scattering.

bution of the signal between the two PMs of each CAL cell, and between adjacent cells in extended showers.

Since all processes between energy deposition in the ZEUS CAL and arrival of the signals at the PMs are "fast", i.e. are characterized by velocities of $\mathcal{O}(c)$, a time measurement with a resolution of $\mathcal{O}(1\,\mathrm{ns})$ is possible [402].

The energy calibration of the ZEUS CAL [403–405] is based on test beam measurements, on laser light injection and on the signal caused by the uranium radioactivity. The latter is monitored as a function of time for each PM and allows one to calibrate the absolute energy scale with an accuracy which is stable to within less than 0.2% with time, to intercalibrate single cells to within 1% and 2% for the EMC and the HAC, respectively, and to detect malfunctioning PMs. Typically, 2 – 3% of all CAL channels are excluded from the readout owing to hardware problems. However, since each CAL cell has two readout channels, the affected cells still provide energy measurement with somewhat reduced resolution. Generally only a few cells (on average less than four in the 1996 and 1997 data taking periods) are completely inactive.

Design of the H1 Liquid-Argon Calorimeter (LAC). The measurement principle realized in the H1 LAC uses the ionization produced by charged particles crossing liquid argon placed in the gaps between absorber plates. A voltage of 625 V/mm is applied across the gaps, and the negative charge is collected by pads on the ground side of the gaps, from where it is fed into the readout chain. The inner part of the calorimeter is optimized for measuring electromagnetic showers and consists of 2.4 mm thick lead plates with 2.35 mm wide argon-filled gaps between them, and is read out in independent cells of transverse size $10 - 100\,\mathrm{cm}^2$, with a three- to four-fold subdivision in depth. The hadronic part is organized in layers of a total of 19 mm stainless steel with double gaps of 2.4 mm argon and is read out in units of transverse size $50 - 2000\,\mathrm{cm}^2$ with four to six subdivisions in depth. The whole LAC sits in a cryogenic pressure vessel, where it has been maintained at liquid-argon temperature (about 90 K) since 1991 [406]. It is subdivided into eight wheels, with the absorber plates arranged in concentric layers around the beam direction in the rearmost three wheels, and perpendicular to the beam in the others. The inhomogeneities arising from the mechanical structure are designed such as to minimize acceptance loss.

The 44 352 LAC readout channels feed the trigger and the data acquisition chain and allow for measurements with much finer granularity than in the ZEUS case. Since the LAC response to electromagnetic and hadronic showers differs ($e/h \approx 1.3$), i.e. LAC is not compensating, the small cell sizes are required for offline algorithms correcting for this effect on the basis of shower topologies (see [372] and references therein). The position reconstruction also profits from the fine granularity, but on the other hand suffers from the shower distortion in the magnetic field. Also, since the drift of ionization charges in the argon requires $\mathcal{O}(200\,\mathrm{ns})$, timing measurements using the LAC signals do not provide accuracies much better than the bunch-crossing period, $T = 96\,\text{ns}$.

Monitoring and electronics calibration of the LAC are performed using charge injection techniques [396]. In addition, the argon purity is regularly monitored by dedicated probes.

Performance and operation parameters of the ZEUS and H1 calorimeters, and the calibration of the absolute energy scales, have been determined and optimized in various test beam measurements and in test runs using cosmic rays (see [403, 404, 407, 408] for ZEUS and [395, 396, 409] for H1). The relevant figures are summarized in Table 3.1. However, it has to be kept in mind that not all test beam results are directly transferable to the HERA environment, since (i) the tested setups differ in details from the final configurations; (ii) the response to multiparticle final states (jets) has not been assessed in detail in test beam measurements; (iii) the test measurements were performed without a magnetic field; (iv) the effects of detector inhomogeneities on calorimetric measurements depend on many parameters, not all of which can be scanned in a test setup; (v) the inactive material between the IP and calorimeter cannot be taken into account in detail; (vi) the precision of the detector alignment with respect to the point of particle incidence is different.

Table 3.1. Operation parameters of the ZEUS and H1 calorimeters as obtained in test beam measurements and from analyses of cosmic-ray and ep data. The $\oplus$ operator indicates addition in quadrature. The momenta of test beam particles are denoted by p, the measured energies of calorimeter deposits by E. Parameters characterizing electromagnetic and hadronic showers are indicated by "elm." and "had.", respectively

Quantity	ZEUS	H1
$\delta E/E$ (elm.)	$18\%/\sqrt{E\,(\text{GeV})} \oplus 1\%$	$12\%/\sqrt{E\,(\text{GeV})} \oplus 1\%$
$\delta E/E$ (had.)	$35\%/\sqrt{E\,(\text{GeV})} \oplus 1\%$	$50\%/\sqrt{E\,(\text{GeV})} \oplus 2\%$
Linearity	Better than 1% in B/FCAL for $p < 120\,\text{GeV}$ and in the RCAL for $p < 50\,\text{GeV}$	Better than 1% for e and high-momentum π's, better than 2% for π's with $p \lesssim 20\,\text{GeV}$
E scale uncertainty	elm.: 2% / 1% / 3% for RCAL / BCAL / FCAL had.: 3%	elm.: 0.7% (central) to 3.0% (forward) had.: 2%
e/h	1.00 ± 0.01 for $p \gtrsim 5\,\text{GeV}$; $\gtrsim 0.9$ for $p \gtrsim 2\,\text{GeV}$	Approximately 1.3
Time resolution	$\approx 2\,\text{ns}/\sqrt{E\,(\text{GeV})} \oplus 0.5\,\text{ns}$	$\mathcal{O}(100\,\text{ns})$
Polar-angle resolution of elm. showers	3 mrad	< 3 mrad

As a consequence, both collaborations are performing extensive studies with the goal to check and verify the parameters from Table 3.1 using *ep* data from the first years of operation. For example, ZEUS verified the overall EMC CAL scale in the BCAL by comparison of track momenta and CAL energies to within 5%. Electron energy scale studies using the scattered electrons in high-Q^2 NC reactions will be presented in Sect. 3.3.4. Many such investigations are not yet completed since they have just become feasible with the full statistics of the available data samples or even need more data.

3.1.3 Track Measurements

Tracking information is used in the DIS analyses for several purposes like identification of the scattered electron in NC reactions, measurement of its scattering angle, determination of the event vertex (to which calorimetric event variables are very sensitive, see Sect. 3.3.1), background rejection and triggering. The principal tracking detector component of ZEUS is the CTD. In the case of H1, both the CJC including its auxiliary components and the FTD contribute to the high-Q^2 analyses. In the following, we describe the design and functionality of these detectors and thereafter summarize their operation parameters. The remaining tracking components currently do not have a significant impact on the high-Q^2 analyses and will not be discussed in detail here.

Design of the ZEUS Central Tracking Chamber. The ZEUS central tracking device (CTD) is a cylindrical wire drift chamber with an active volume extending from $z = -100\,\mathrm{cm}$ to $z = 105\,\mathrm{cm}$ and from 18.2 cm to 79.4 cm in radius (the coordinate origin is defined to be the nominal IP). The sense wires are organized in nine superlayers (SLs) with eight wire layers each, and are parallel to the beam axis in SLs $1, 3, \ldots, 9$ ("axial layers") and tilted by stereo angles between $\pm 4.6°$ and $\pm 5.8°$ in the remaining SLs ("stereo layers"). The sense wires and the field-shaping ground and cathode wires of an SL are arranged in planes rotated by 45° in the radial direction in order to compensate for the Lorentz drift angle, i.e. to guarantee azimuthal drift of electrons for operation at nominal voltages and in the ZEUS magnetic field of 1.43 T.

The axial and stereo layers allow accurate measurements of charged-particle trajectories in the projection perpendicular to the beam direction. The polar-angle information is obtained from the stereo layers, to which transverse track segments from at least one adjacent axial layer can be extrapolated. In addition, the wires of the first SL and every second wire of the third and fifth SLs provide an individual z measurement based on the timing difference of the signals read out at each end. This *z-by-timing* information is used for trigger purposes and for offline track reconstruction.

The analog signal readout furthermore allows the measurement of the energy loss, $\mathrm{d}E/\mathrm{d}x$, of charged particles with a precision of about 8%.

Design of the H1 Central and Forward Tracking Chambers. The H1 central tracker is a hybrid device consisting of coaxial cylindrical chambers surrounding each other: two jet drift chambers (CJC1 and CJC2) [385] with altogether 56 layers of sense wires[3] parallel to the beam direction measure the $r\phi$ coordinates of tracks, two cylindrical layers of drift chambers with azimuthal wires (CIZ [388] and COZ [387]) provide z sensitivity and two cylindrical proportional chambers (CIP and COP) [386] are used for trigger purposes. With increasing radius, these components are arranged in the order CIP–CIZ–CJC1–COZ–COP–CJC2. They enclose a sensitive volume extending from 15.4 cm to 84.4 cm in radius and from −112.5 cm to 107.5 cm in z.

Both for the jet and for the z chambers the position information along the direction of a wire is determined from the charge division between the analog signals read out at each end of the wire, with a precision of 1% of the wire length in the CJC and 2% in the z chambers. Also, $\mathrm{d}E/\mathrm{d}x$ can be measured in the CJC with a relative uncertainty of about 7%.

The central tracking chambers provide a measurement of the time offset at which an event occurred with respect to the HERA clock (resolution $\sigma_t \approx$ 1.6 ns), which is used for background rejection in the offline analyses.

The forward tracking detector (FTD) [390–392] is a combination of drift chambers, multiwire proportional chambers (FWPCs) and inactive transition radiators. It consists of three identical supermodules, each organized into three layers of planar drift chambers with parallel wires (*planars*), an FWPC plane, a radiator plane for electron identification [391] and a drift chamber with radial wires (*radials*). The three planars of each supermodule are rotated with respect to each other by 60°. The radials measure up to twelve space points per supermodule crossed by a track. The overall active volume extends from $r = 15$ cm to $r = 79$ cm in radius and from $z = 134$ cm to $z = 254$ cm, providing angular coverage down to about 5°. The track reconstruction in the FTD [392] yields a polar-angle resolution of better than 1 mrad, and a momentum resolution of roughly $\delta p/p \approx 15\%/\,p\,(\mathrm{GeV})$.

The FTD contributes to the trigger and is essential for the vertex determination, in particular for events without central tracks.

Some performance parameters of the ZEUS and H1 tracking detectors are summarized in Table 3.2. Note that event-related quantities like the resolution of the vertex position, which depend strongly on the event topology and on other detector components, will be covered in Sect. 3.3.

3.1.4 Luminosity Measurements

Both collaborations use very similar detector systems to measure the integrated luminosity provided by the HERA collider [370, 372]. The Bethe–Heitler process, $ep \rightarrow ep\gamma$ [410], is used as a reference reaction, with a

[3] The sense wire planes are rotated with respect to the radial direction by 30° to account for the Lorentz angle.

Table 3.2. Operation parameters of the ZEUS and H1 tracking detectors. The angular acceptances are quoted with respect to the nominal IP. The $\oplus$ operator indicates addition in quadrature

	CTD (ZEUS)	
Polar-angle acceptance	$15° < \theta < 164°$ (≥ 2 superlayers)	
Point resolutions	230 μm ($r\phi$) 1 mm (z, stereo layers) 4.5 cm (z-by-timing)	
σ_{p_t}/p_t	$0.0058 \times [p_t\,(\mathrm{GeV})] \oplus 0.0065 \oplus 0.0014/[p_t\,(\mathrm{GeV})]$ (9 SLs)	
$\delta(\mathrm{d}E/\mathrm{d}x)$	ca. 8%	
	CJC (H1)	**FTD (H1)**
Polar-angle acceptance	$25° < \theta < 155°$ (at least CIZ–CJC1–COZ)	$5° < \theta < 25°$ (at least 1 supermodule)
Point resolutions	170 μm ($r\phi$) 200 – 500 μm (z, CIZ/COZ) <3.5 cm (z by charge division)	210 μm (planars, ⊥ wires) 180 μm (radials, azimuthal) 3 – 4 cm (radials, r)
σ_{p_t}/p_t	$0.009\,p_t\,(\mathrm{GeV}) \oplus 0.015$	20% ($0.5\,\mathrm{GeV}<p_t<1\,\mathrm{GeV}$) 8% ($1\,\mathrm{GeV}<p_t<3\,\mathrm{GeV}$) 5% ($p_t>3\,\mathrm{GeV}$)
$\delta(\mathrm{d}E/\mathrm{d}x)$	ca. 7%	—

cross section which has a theoretical uncertainty of less than 0.5% [411]. The final-state electron and photon emerge at small scattering angles of $\mathcal{O}(m_e/E_e) \approx 20$ μrad (i.e. at $\theta \approx 180°$ in the detector coordinate systems) and are detected by external calorimeters (lead–scintillator sandwich calorimeters with additional position detectors in the case of ZEUS, total-absorption Čerenkov crystal hodoscopes with PM readout for H1) positioned about 35 m (electrons) and 105 m (photons) from the IP in the e beam direction. The photons leave the beam pipe through thin exit windows after the electron beam is bent off by beam-line dipole magnets, whereas the electrons originating from Bethe–Heitler processes are separated from the e beam and deflected to their exit windows by accelerator quadrupole magnets. As a consequence, the detector acceptances cover almost the full energy range for the photons (with a lower cutoff due to inactive material and shielding against synchrotron radiation) but only a limited range of the electron energy.

The luminosity is determined from the rate of $e\gamma$ coincidences (H1) or from the γ rate alone (ZEUS, H1) after subtraction of the background, which is caused mainly by bremsstrahlung reactions of the electron beam with residual gas molecules in the beam pipe, $eN \to eN\gamma$, and can be quantified using unpaired e bunches provided by HERA with well-known currents. The recorded Bethe–Heitler rates are sufficiently large to guarantee that statistical errors are negligible. The experimental systematic uncertainties of the luminosity measurement are related to the background subtraction, to pile-up effects and to the energy calibration, linearity and acceptance of the calorimeters. Further uncertainties are caused by the fact that off-bunch collisions occur (*satellite bunches*) which contribute to the luminosity measurement but can partly not be used for the analyses owing to displaced event vertices. H1 assigns a luminosity uncertainty of $1-2\%$ to the satellite bunches, whereas ZEUS applies looser constraints on the vertex position and is therefore less sensitive to this effect. For the data taken in the period 1994–97, ZEUS and H1 quote overall systematic luminosity errors of 1.6% and 1.5%, respectively.

In addition to providing the integrated luminosities for the offline analyses, the luminosity detectors are also used for online luminosity monitoring (utilized e.g. for optimization of HERA operation parameters), and for detecting the electron in photoproduction reactions and the photon in reactions with initial-state radiation (see Sect. 2.1.3).

3.1.5 Triggering

The task of the triggers is to distinguish ep reactions with predefinable signatures from the background caused mainly by interactions of beam particles with residual gas molecules in the beam pipe (*beam–gas reactions*) or with elements of the beam line (*beam–wall reactions*), but also by cosmic rays or by beam-halo particles. The rate of interactions leaving signals in calorimeters and tracking chambers depends strongly on the beam quality and currents and on the pressure in the beam pipe and is typically of $\mathcal{O}(10-100\,\mathrm{kHz})$. This is to be contrasted with the rate of ep reactions, which (at design luminosity, $\mathcal{L}_{\mathrm{des}} = 1.5 \times 10^{31}\,\mathrm{cm}^{-2}\,\mathrm{s}^{-1}$) spans a range descending from about 200 Hz (inclusive photoproduction) through 0.03 Hz (NC DIS with $Q^2 > 1000\,\mathrm{GeV}^2$) to 1 per hour ($e^+p$ CC DIS with $Q^2 > 1000\,\mathrm{GeV}^2$).

Both collaborations use multilevel trigger systems, with the first level being the fastest but coarsest and with the higher levels being successively slower but more selective. Detailed descriptions can be found in [370, 372] and the references cited therein. Over the past years of operation, the trigger configurations were repeatedly adapted to the increasing instantaneous luminosity and to the developments of physics priorities. The following short overview of the design and functionality of both trigger systems refers to the 1997 data taking and does not reflect changes prior to this date.

The ZEUS Trigger. The first-level trigger (FLT) uses programmable logic to combine the individual component trigger signals, which arrive about $1.0-2.5\,\mu\mathrm{s}$ after the bunch crossing. The FLT decision is issued after $4.4\,\mu\mathrm{s}$, i.e. with a delay of 46 bunch crossings. The component data are stored in pipelined buffers during this time and are transferred to the second-level trigger (SLT) upon an FLT-accept. The FLT decision is mainly based on various combinations of the following information:

- calorimeter data [412]: global sums like EMC energy in different CAL parts, transverse energy or missing transverse momentum, but also local topology-dependent information, which is used in particular to tag isolated electromagnetic energy deposits;
- CTD data [373]: number of track candidates, subdivided into those that are compatible with coming from the vicinity of the nominal IP and others;
- signals recorded in the muon detectors.

The FLT is designed to reduce the event rate to below 1 kHz and was running with output rates of typically 200 – 600 Hz in recent years.

The second-level trigger (SLT) uses an asynchronous pipeline of transputer processors to operate on the fully digitized data of the events accepted by the FLT. The individual components participating in the SLT (including CAL and CTD) recalculate the quantities used for the FLT on a higher level of precision and provide additional variables. On the basis of this input, the FLT cuts are refined, and additional constraints like calorimeter timing cuts are applied. The SLT reduces the FLT rate by a factor of about ten.

From the SLT, the data are transferred to an array of transputers (the *event builder* [413]), where they are combined and formatted and finally passed on to the third-level trigger (TLT) upon an SLT-accept. The TLT is a workstation processor farm running a reduced version of the offline reconstruction and analysis code, providing information like electron and jet identification, track and event vertex reconstruction, and event classification. The TLT applies selection cuts which are similar to, but looser than, the standard offline analysis cuts and thus reduces the rate of non-*ep* background reactions significantly below the *ep* rate. Also, the TLT allows one to prescale the rate of single physics branches. The TLT output rate is adjusted to be in the range of 5 – 10 Hz.

A further pseudo-trigger level is introduced at the event reconstruction stage (see Sect. 3.3), where preselection cuts for the various physics analysis branches are applied and the selected events are correspondingly marked for offline analysis by so-called *DST bits*. However, all events (also those without DST-bit selection) are recorded and can be used for systematic checks and trigger studies.

The H1 Trigger. The H1 first-level trigger is a system of fast pipelines which independently receive and process information from several detector components and finally pass the results to the central level-1 trigger (CL1), which calculates subtriggers and issues a decision within 2.5 μs. The trigger decision is mainly based on

- vertex information obtained from the CIP, the COP and the FWPC data by applying histogramming techniques [393];

- track candidates in the CJC [414];
- time-of-flight measurements from two dedicated rear scintillator planes;
- the LAC signals, which are subdivided into 256 towers, from which overall sums such as transverse energy or missing transverse momentum are calculated, and signatures like isolated electromagnetic clusters or spatial coincidences with proportional-chamber hits are recognized;
- data from the rear calorimeter [398] and the muon system [401].

Upon a CL1-keep signal, the pipelines are stopped and read out. Two further trigger levels operate in the resulting dead time. The second-level trigger assesses the event topology using custom-designed hardware systems (including neural-network approaches). If the event is kept, the full readout procedure is activated and the level-3 trigger further refines the selection criteria, with a resulting output rate of less than 50 Hz (prior to 1995, the level-2 and level-3 systems were not used for trigger decisions and the quoted output rate was achieved by CL1 alone).

The level-4 trigger is, similarly to the TLT in the ZEUS case, a computer farm operating on the complete event data, which are merged and formatted after the level-3 accept. The level-4 output rate is about 8 Hz.

H1 applies an additional "level-5" offline event selection (*event classification*) equivalent to the ZEUS DST-bit flagging. Deselected events are stored on slow-access mass storage and remain accessible for offline analysis.

Table 3.3. Selection criteria and performance of the ZEUS and H1 trigger systems as used for the high-Q^2 analyses. The global event variables E_t (transverse energy), P_t (transverse momentum) and $E - P_z$ are defined in Sect. 3.3.1; note that they are evaluated with the characteristic precision of the individual trigger levels

NC	ZEUS	H1
Main trigger requirements	FLT: RCAL/BCAL signal consistent with e and jet, E_t SLT: Cuts on E_t, $E - P_z$ and CAL timing TLT: e identification, cuts on CAL timing	CL1: electron trigger (one big tower) Further trigger levels: refined cosmic and beam–gas rejection, cuts on vertex position and $E - P_z$
Trigger acceptance	$> 99.9\%$ for $Q^2 > 1200\,\mathrm{GeV}^2$, $> 99\%$ for $Q^2 > 400\,\mathrm{GeV}^2$	$> 99.5\%$ for $E'_e > 11\,\mathrm{GeV}$ (E'_e = energy of scattered e)
CC		
Main trigger requirements	FLT: P_t, E_t, CTD tracks SLT: cuts on timing, P_t, E_t, $P_t/\sqrt{E_t}$, FCAL energy TLT: cuts on P_t, tracks, vertex, timing	CL1: P_t Further trigger levels: refined cosmic and beam–gas rejection, cut on vertex position
Trigger acceptance	$> 99.5\%$ for $Q^2 > 400\,\mathrm{GeV}^2$	$\sim 98\%$ for $P_t > 25\,\mathrm{GeV}$, falling to $\sim 50\%$ for $P_t = 12\,\mathrm{GeV}$

The main selection criteria and the performance of the ZEUS and H1 trigger branches used for the high-Q^2 DIS analyses are summarized in Table 3.3. The details of the complicated trigger logic will not be described here. The offline selection procedures and their acceptances are discussed in Sect. 3.4.

3.2 Event and Detector Simulation

An essential prerequisite for the analysis of reactions recorded by complex detector systems is a detailed simulation, both of the interactions under study and of the detector response to the reaction products. Computer-based Monte Carlo (MC) techniques are used for sampling instances of reactions of a given type. Usually the simulation includes the following three independent stages.

Event Generation. Reactions including complete specifications of their final state are generated with distributions of kinematic variables which reflect differential cross sections inferred from measurements or theoretical models. The underlying algorithms are independent of the experimental situation, except that they know about the initial-state conditions.

The MC programs used by ZEUS and H1 for generating DIS reactions, potential sources of background and events involving R_P-violating squark or LQ production are summarized in Table 3.4. In the case of DIS, the simulation is based on the theoretical concepts discussed in Sects. 2.1.1 to 2.1.3.

Table 3.4. MC generators used by ZEUS and H1 for simulating DIS reactions, background processes and LQ or squark production. Generators marked with an asterisk (*) have been adapted by ZEUS or H1 for application in their environments

<table>
<tr><th>Reaction</th><th>ZEUS</th><th>H1</th></tr>
<tr><td>DIS:
$ep \to eX$ (NC)
$ep \to \nu X$ (CC)</td><td colspan="2">HERACLES [148, 415]
DJANGO [416]
LEPTO/ARIADNE [193]/[198]</td></tr>
<tr><td>Photoproduction:
$\gamma p \to$ jets</td><td>HERWIG [203]
PYTHIA [201]</td><td>PYTHIA [201]</td></tr>
<tr><td>High-E_{t} prompt γ's:
$\gamma p \to \gamma X$</td><td>HERWIG [203]</td><td>PYTHIA [201]</td></tr>
<tr><td>W and Z production:
$ep \to Z\,X,\ W^{\pm}\,X$</td><td colspan="2">EPVEC* [417]</td></tr>
<tr><td>$\ell^+\ell^-$ production:
$ep \to \ell^+\ell^-\,eX$</td><td colspan="2">LPAIR* [418]</td></tr>
<tr><td>LQ production:
$ep \to \mathrm{LQ}\,X$</td><td>PYTHIA [201]</td><td>LEGO [60, 63, 64, 419]</td></tr>
<tr><td>R_P-violating squark production:
$ep \to \tilde{q}\,X$</td><td>SPYTHIA* [420]</td><td>LEGO, SUSSEX [419]</td></tr>
</table>

The partonic final state of the hard subprocess, including photon radiation to $\mathcal{O}(\alpha)$ and the full set of one-loop electroweak corrections, is generated by HERACLES [148, 415] and subsequently handed over to the interface package DJANGO [416], which feeds it into programs supplementing the parton shower. Either the MEPS option of LEPTO [196] or the color-dipole model as implemented in ARIADNE [198] is used for this purpose (see Sect. 2.1.7). The generators used for LQ and squark production implement initial- and final-state QCD parton showers and an approximate treatment of QED ISR, but no higher-order QCD matrix elements; LQ processes simulated by PYTHIA only include s-channel LQ diagrams. Contact interaction signals are simulated by reweighting DIS MC sets appropriately (see Sect. 4.4.2).

The hadronization of partonic final states is simulated according to the cluster fragmentation model for applications of HERWIG, and using string fragmentation as implemented in JETSET [200, 201] for all other generators.

Detector Simulation. Both ZEUS and H1 employ detailed detector simulations based on the GEANT package [421] to investigate the detector response to reactions provided by the event generators. The primary final-state particles as well as all secondaries produced by interactions and decays are traced through the detector, and their energy deposits in active detector volumes are recorded and finally transformed into simulated "measured signals", which are filled into the same offline data structures as used for experimental data.

Although GEANT includes basically all relevant reaction mechanisms, the detector simulation is subject to several approximations and omissions:

- For reasons of CPU time consumption and coding complexity, the exact details of the detector geometry are usually not reproduced in GEANT. An example is the sandwich structure of a calorimeter, which is replaced by a fictitious homogeneous material carrying the appropriate average physical properties.
- The detailed simulation of electromagnetic and hadronic shower development is abridged by treating the shower tails in a parameterized way. This approach requires careful tuning and testing of parameters which are unique for a given detector type and geometry. As a cross-check, H1 also uses a version of the detector simulation including the detailed calorimeter geometry and full shower simulation, thereby increasing the average processing time per event by a factor of about 10 [372].
- Processes related to signal generation, like charge drift in media or photon production and propagation in scintillators, are not simulated by GEANT and have to be included in the MC programs in a parametric way. Typical examples of relevant effects for the ZEUS calorimeter are statistical fluctuations of the number of photons reaching the PMs for a given energy deposit, or the experimental resolution of timing measurements. Also, the MC simulation does not "know" about hardware defects or faults in the readout chain.

- Signals not related to the primary event, such as detector noise and overlays of cosmic rays or background interactions, are not covered by GEANT.
- Accelerator-dependent quantities, e.g. the distribution of ep vertex positions, vary from run to run and have to be derived summarily from real data and fed into the MC simulation. Correlations between accelerator effects and detector performance are usually neglected.

The above points limit the applicability of MC simulations to investigations of experimental resolutions and migration effects. However, the agreement of real and simulated distributions is in general sufficiently good to keep the related uncertainties of analysis results below the level of statistical errors or systematic effects of other origin. Some MC-data comparisons for the high-Q^2 samples are presented in Sect. 3.5.

Trigger Simulation, Reconstruction and Analysis. After the trigger decision is simulated by dedicated software modules, the MC data are processed by the same analysis chain as the experimental data. In particular, the same reconstruction algorithms and event selection procedures (see Sects. 3.3 and 3.4) are applied.

The comparison of the "true" quantities of the primary reactions as produced by the event generators and their "measured" equivalents resulting from the reconstruction step yields insight into the mechanisms of experimental smearing and the response functions of the detector. Furthermore, this information is used to study detector acceptance, data selection efficiencies and migration of events in the kinematic space due to measurement errors. Methods to correct for losses and migrations are usually based on MC studies. The corresponding details for the high-Q^2 analyses can be found in Sects. 3.4 and 3.5.

3.3 Event Reconstruction

Once an event has been recorded by the detector, it still has to undergo a long series of offline operations before it finally may contribute to physics analyses. In an initial step, the raw detector data are calibrated and converted to kinematic quantities like track momenta or energies and positions of calorimeter deposits. This step is performed centrally and the resulting data are written to mass-storage memory,[4] where they are available for individual physics analyses.

The relevant quantities characterizing single events are determined from the DST data. We will discuss the related analysis and correction methods and experimental resolutions in the following three subsections, which cover the determination of global event variables, the identification of the scattered

[4] Traditionally called *data summary tapes* (DSTs) although tape recording is not in all cases used.

electron in NC DIS reactions and the reconstruction of kinematic variables. Finally, new results of both collaborations on the offline electron energy calibration will be presented in Sect. 3.3.4 to demonstrate the level of precision which has meanwhile been reached in the high-Q^2 analyses. Unless stated otherwise, the descriptions, figures and results in the following sections refer to the latest analyses [73–75].

3.3.1 Global Event Variables

The term *global variables* is used here to distinguish a set of quantities which can be derived from the primary detector information from those variables which require the identification of the event as belonging to a particular reaction class. For DIS analyses, the most relevant global variables are

$$E_{\text{tot}} = \sum_i E_i \ , \tag{3.1}$$

$$E_{\text{t}} = \sum_i E_i \sin\theta_i \ , \tag{3.2}$$

$$P_{\text{t}} = \sqrt{\left(\sum_i E_i \sin\theta_i \cos\phi_i\right)^2 + \left(\sum_i E_i \sin\theta_i \sin\phi_i\right)^2} \ , \tag{3.3}$$

$$E - P_z = \sum_i E_i (1 - \cos\theta_i) \ , \tag{3.4}$$

$$\boldsymbol{v} = (v_x, v_y, v_z) = \text{event vertex position} \ , \tag{3.5}$$

where the sums run over all reconstructed energy deposits, E_i, above noise level in the calorimeter, and the angles are calculated from the positions of these calorimeter objects and of the event vertex. Before the above sums are evaluated, the calorimeter data are usually processed by algorithms (partly in the primary reconstruction, partly at analysis level) which combine energy deposits in adjacent cells (*clustering*) and apply corrections for detector effects like calorimeter inhomogeneities, inactive material or albedo from reactions of final-state particles with other detector volumes (see Sect. 3.3.3 for more details). The quality of these corrections is limited by inaccuracies of the MC simulation of the detector response and of event characteristics like the forward energy flow (see Sect. 2.1.7). Some MC-data comparisons of distributions of global variables will be shown in Sect. 3.5.

The vertex position is determined individually for each event from the reconstructed tracks, with typical resolutions in ZEUS (H1) of 1.5 mm (3 mm) longitudinally and <1 mm (1 mm) transversely. For CC candidate events without tracks, ZEUS applies an algorithm based on the CAL timing measurements to determine the longitudinal vertex position with a resolution of better than 10 cm. Events without a reconstructed vertex are usually excluded from the analyses. Note, however, that there are events for which a vertex reconstruction is faked by tracks coming from interactions of secondary particles in the beam pipe and inner detector volumes, by spurious tracks or by overlays of background reactions or cosmics. The fractions of events with fake vertices in the final data samples depend strongly on the reaction type and event topologies and are estimated to be less than 1% in the high-Q^2 analyses.

3.3.2 Electron Identification

The most distinctive signature of NC DIS reactions is the scattered electron, which, in the case of high Q^2, in general has high transverse momentum (above $\mathcal{O}(10\,\text{GeV})$) and does not overlap with the hadronic system. The corresponding detector signals, an isolated electromagnetic shower in the calorimeter and a high-momentum track pointing to it, are exploited to identify events with electron candidates. This task is particularly important and difficult for high-Q^2 NC investigations because the electrons are scattered into the central and forward parts of the detectors (see Fig. 1.2) where potential background from hadronic activity is concentrated. Furthermore the signal cross section falls rapidly with Q^2, yielding a tiny signal rate for $Q^2 \gtrsim 1000\,\text{GeV}^2$ as compared with potential background sources, which have much higher cross sections.

ZEUS uses two independent approaches for electron finding. The first one, which is used for the analysis of the high-Q^2 data [71, 73], operates on loosely preselected electron candidates which are required to be isolated, and to have a matching track if they are in the geometrical acceptance of the CTD; the actual cuts are summarized in Table 3.5. For each such candidate, characteristic calorimeter and tracking variables are determined. The distributions of these variables in samples of electrons selected by other algorithms are used to infer for each of the variables the likelihood that its value is compatible with originating from an electron. The product of these likelihoods is finally required to exceed a threshold value determined from MC and data studies.[5] Acceptance of the electron identification and the remaining backgrounds are discussed in Sect. 3.4. The second electron-finding algorithm, which is used for lower-Q^2 analyses and serves for systematic checks at high Q^2, is based on an MC-trained feed-forward neural network [423] operating on the calorimeter topology of candidate clusters preselected by loose requirements on their energy, "electromagneticity" (i.e. the energy fraction in the electromagnetic CAL sections) and transverse size.

H1 identifies electromagnetic clusters in the LAC using lateral and longitudinal shower profiles which have been investigated in detail in test beam measurements [395]. In order to be selected for the high-Q^2 NC analysis, candidate clusters have to fulfill additional requirements on isolation and energy, and have to have a matching track if they are inside the acceptance of the CJC, CIZ and COZ. The exact requirements are shown in Table 3.5.

[5] Recently, a neural network has been designed as alternative to this likelihood-based decision step [422].

Table 3.5. Electron identification requirements applied in the ZEUS and H1 high-Q^2 analyses. The following variables are used in the table: θ_e = polar angle of electron candidate calculated from calorimeter position and reconstructed vertex; p_{tr} = reconstructed track momentum; $\mathrm{DCA}_{\mathrm{tr}}$ = distance of closest approach between track extrapolation and calorimeter cluster; E_{clus} = total energy of candidate cluster; $E_{\not e}(R)$ = energy not belonging to candidate cluster in η–ϕ cone with radius R around its center; $E_{\mathrm{t}}(e)$ = transverse energy of e candidate

	ZEUS	H1
Shower topology	Longitudinal and transverse shower size in CAL	Longitudinal and transverse shower properties in LAC (using test beam results)
Matching track	For candidates with $\theta_e > 17.2°$: $p_{\mathrm{tr}} > 5$ GeV; $\mathrm{DCA}_{\mathrm{tr}} < 10$ cm; $p_{\mathrm{tr}}/E_{\mathrm{clus}}$ and direction match	For candidates with $\theta_e > 35°$: $\mathrm{DCA}_{\mathrm{tr}} < 12$ cm
Isolation	$E_{\not e}(R{=}0.8) < 5$ GeV	$E_{\not e}(R{=}0.25)/E_{\mathrm{clus}} < 0.02$
Energy, $E_{\mathrm{t}}(e)$	$E_{\mathrm{clus}} > 10$ GeV $E_{\mathrm{t}}(e) > 30$ GeV if $\theta_e < 17.2°$	$E_{\mathrm{clus}} > 11$ GeV
Fiducial cuts	Remove candidates in R/BCAL and B/FCAL transition regions	Remove candidates in LAC cracks and outside LAC (with $\theta_e \gtrsim 153°$)

The main difficulties of electron identification are (i) fake candidates coming mainly from hadronic energy deposits (which may look "electron-like" if they are dominated by photons, e.g. from π^0 decays) or from isolated photons with high transverse momentum (see [424]), and (ii) events containing more than one acceptable candidate. In the latter case decision algorithms are applied using the quality assessment of the electron identification procedure and the transverse momentum of the candidate.

3.3.3 Reconstruction of Kinematic Variables

Once the global variables are determined and electrons identified, events can be classified as NC or CC candidates, or be discarded for the high-Q^2 analyses (see the following section for details on selection requirements). The next step is now to determine the kinematic event variables.

For the moment, we want to postpone the discussion of ISR[6] a little and pretend that it can be neglected. Under this assumption, only two of the three basic kinematic variables (x, y, Q^2) are independent (see Sect. 1.3.1), and any combination of two defining quantities suffices to determine them. For NC events, the energies and scattering angles of the electron and the hadronic system provide four such quantities which are experimentally measurable,

[6] Photon radiation of one of the fermion lines in the hard subprocess has been discussed in Sect. 2.1.3.

and correspondingly there are several independent reconstruction methods. In the CC case, the missing knowledge of the scattered neutrino only leaves one possibility. An investigation of the different options can be found in [425–427]. In the following, we will focus on those aspects which are important for the high-Q^2 analyses. The resolutions of the kinematic variables will be summarized at the end of this section in Table 3.7.

Whereas the electron variables are directly determined from the calorimetric information and, where available, from the electron track, the hadronic quantities are defined in terms of calorimeter sums which exclude the electron cluster in NC reactions (denoted by primed sum symbols):

$$(P_{\rm t})_{\rm had} = \sqrt{\left(\sum_i{}' E_i \sin\theta_i \cos\phi_i\right)^2 + \left(\sum_i{}' E_i \sin\theta_i \sin\phi_i\right)^2} \,, \quad (3.6)$$

$$\Sigma = (E - P_z)_{\rm had} = \sum_i{}' E_i(1 - \cos\theta_i) \,, \quad (3.7)$$

$$\cos\gamma_{\rm had} = \frac{(P_{\rm t})^2_{\rm had} - (E - P_z)^2_{\rm had}}{(P_{\rm t})^2_{\rm had} + (E - P_z)^2_{\rm had}} \,, \quad (3.8)$$

$$E_{\rm had} = \frac{(P_{\rm t})_{\rm had}}{\sin\gamma_{\rm had}} \,. \quad (3.9)$$

The summation is over the same type of calorimeter deposits as in (3.1)–(3.4) and implies the same corrections described there. Note that the definition of the hadronic scattering angle, $\gamma_{\rm had}$, in (3.8) does not involve topology studies of the hadronic final state (like jet searches) and is independent of global scaling factors applied to the energies, i.e. of the absolute energy scale calibration. It can be shown that in the case of (2→2)-type hard subprocesses, corresponding to (1+1)-jet event topologies, $\gamma_{\rm had}$ actually is an unbiased estimate of the quark scattering angle. Furthermore, $\gamma_{\rm had}$ allows the determination of kinematic variables irrespective of the final-state topology (see below).

As mentioned in Sect. 3.3.1, the hadronic sums in (3.6)–(3.9) have to be corrected for effects of energy loss in inactive material, redirected energy flow due to secondary interactions in material between the vertex and calorimeter or by backsplash from the calorimeter (albedo), calorimeter noise, particle loss through the beam pipe and problems caused by position-migrated energy deposits of the proton remnant jet. In particular at decreasing $\gamma_{\rm had}$, i.e. for small y (see Fig. 1.2), the reconstruction of $\gamma_{\rm had}$ becomes increasingly difficult. As an example of the corrections applied, Fig. 3.3 illustrates the removal of albedo energy in ZEUS: $\gamma_{\rm had}$ is reconstructed from the energy clusters in CAL, and all clusters with energy below 3 GeV and a polar angle larger than $\gamma_{\rm max}$ are removed, where $\gamma_{\rm max}$ is an MC-determined function of $\gamma_{\rm had}$. The procedure is iterated until it converges. After this and other corrections (mainly an MC-based compensation for energy loss in inactive material), the

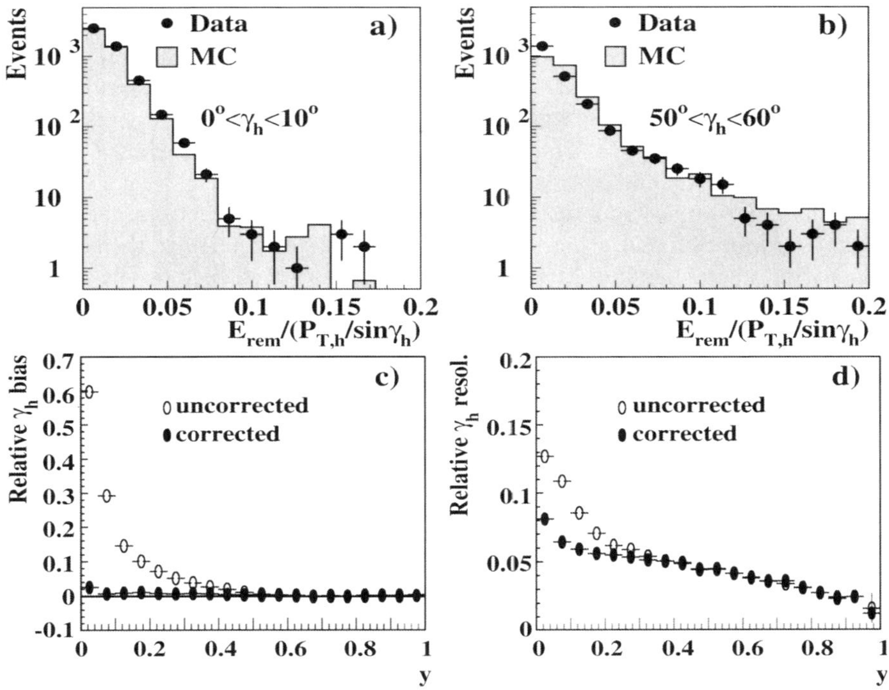

Fig. 3.3. Correction for albedo effects and reconstruction of $\gamma_{\rm had}$ in a ZEUS study of NC reactions: distributions of the ratio of removed albedo energy to the remaining hadronic energy (projected into the reconstructed $\gamma_{\rm had}$ direction), for (**a**) small and (**b**) intermediate values of $\gamma_{\rm had}$. The *points with error bars* denote the experimental data; the *shaded histogram* is the MC simulation. The lower plots show (**c**) the average relative shift and (**d**) the relative R.M.S. resolution with respect to the true $\gamma_{\rm had}$. The *open symbols* denote the results obtained by summing over uncorrected energies in all calorimeter cells; the *full circles* demonstrate the improvement obtained by clustering and applying corrections for energy loss in inactive material and for albedo effects. The figures are reproduced from [73]

reconstructed values of $\gamma_{\rm had}$ and $(P_{\rm t})_{\rm had}$ are sufficiently unbiased to allow for precise kinematic event reconstruction in the high-Q^2 analyses.

NC Kinematic Reconstruction. For determining the kinematic variables of NC reactions, different methods are used by ZEUS and H1, and each collaboration applies the other method(s) for systematic checks and also for offline calibration purposes. We will summarize the different reconstruction methods in the following and present the corresponding measurement accuracies in Table 3.7.

- The *double-angle (DA) method* uses the two angular variables θ_e and γ_h as primary input. The kinematic variables are calculated as[7]

[7] H1 sometimes uses the subscript 2α to denote the double-angle method.

$$x_{\mathrm{DA}} = \frac{E_e}{E_p} \frac{\sin\gamma_{\mathrm{had}}}{1-\cos\gamma_{\mathrm{had}}} \frac{\sin\theta_e}{1-\cos\theta_e}, \tag{3.10}$$

$$y_{\mathrm{DA}} = \frac{\sin\theta_e(1-\cos\gamma_{\mathrm{had}})}{\sin\gamma_{\mathrm{had}}+\sin\theta_e-\sin(\gamma_{\mathrm{had}}+\theta_e)}, \tag{3.11}$$

$$Q^2_{\mathrm{DA}} = s\,x_{\mathrm{DA}}\,y_{\mathrm{DA}} = 4E_e^2 \frac{\sin\gamma_{\mathrm{had}}(1+\cos\theta_e)}{\sin\gamma_{\mathrm{had}}+\sin\theta_e-\sin(\gamma_{\mathrm{had}}+\theta_e)}. \tag{3.12}$$

This method is used by ZEUS as the standard algorithm and by H1 for systematic checks and for the offline calibration of the electron energy measurement (see Sect. 3.3.4). Note that the only energies entering these formulae directly are those of the electron beam (E_e) and of the proton beam (E_p), whence the results are unaffected by detector energy scale calibration uncertainties. The DA method is applicable and stable over the full kinematic region of interest, as demonstrated in Fig. 3.4, showing the DA Q^2 resolutions obtained in an MC study by ZEUS.

- The *electron method (e)* determines the variables from the energy and scattering angle of the electron:

$$x_e = \frac{E_e}{E_p} \frac{E'_e(1+\cos\theta_e)}{2E_e - E'_e(1-\cos\theta_e)}, \tag{3.13}$$

$$y_e = 1 - \frac{E'_e}{2E_e}(1-\cos\theta_e), \tag{3.14}$$

$$Q^2_e = s\,x_e\,y_e = 2E_eE'_e(1+\cos\theta_e). \tag{3.15}$$

The e method, which was used by H1 for the 1996 analysis [72] and serves ZEUS for cross-checks, has the intrinsic disadvantage that its results are susceptible to uncertainties of the absolute calorimeter energy scales, and that it becomes unstable at low y. As can be seen from (3.14), low y implies that the denominator in (3.13) is given by a small difference of comparatively large numbers, of which one carries an experimental error. On the other hand, the e method has the advantage of being independent of the reconstruction of a partonic scattering angle from a multiparticle final state, and of not requiring a hermetic acceptance and understanding of hadronic energy measurements.

- The *Σ method (Σ)* [426, 427, 429], which is only used by H1, combines the electron variables with $\Sigma = (E-P_z)_{\mathrm{had}}$ as defined in (3.7):

$$y_\Sigma = \frac{\Sigma}{\Sigma + E'_e(1-\cos\theta_e)}, \tag{3.16}$$

$$Q^2_\Sigma = \frac{(E'_e\sin\theta_e)^2}{1-y_\Sigma}, \tag{3.17}$$

$$x_\Sigma = \frac{Q^2_\Sigma}{s\,y_\Sigma}. \tag{3.18}$$

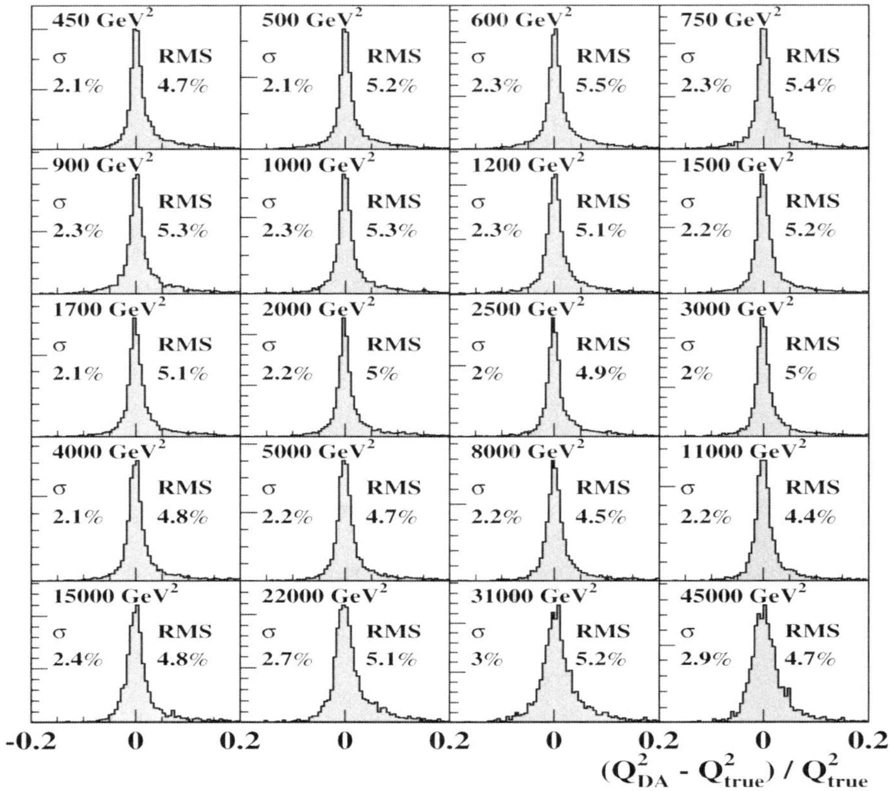

Fig. 3.4. Resolution of Q^2_{DA} determined in a ZEUS MC study. The individual plots show the distributions of the relative deviation of the measured from the true Q^2 values in the range $Q^2 > 400\,\mathrm{GeV}^2$ (arbitrary vertical scales). The widths of the peaks correspond to resolutions of 2 – 3%. The tails extending to the right ($Q^2_{\mathrm{DA}} > Q^2_{\mathrm{true}}$) are caused by ISR. The figure is reproduced from [428]

In contrast to the e method, the Σ reconstruction avoids the low-y instability and is furthermore less sensitive to ISR than other methods because y_Σ and Q^2_Σ do not depend on the electron beam energy.

- The *$e\Sigma$ method ($e\Sigma$)* [426, 427] combines advantages of the e and Σ methods by defining

$$y_{e\Sigma} = y_\Sigma \frac{Q^2_e}{Q^2_\Sigma} = y_\Sigma \frac{2E_e}{\Sigma + E'_e(1-\cos\theta_e)} \,, \tag{3.19}$$

$$Q^2_{e\Sigma} = Q^2_e \,, \tag{3.20}$$

$$x_{e\Sigma} = x_\Sigma \,. \tag{3.21}$$

This method yields good resolution of x and Q^2 in the complete kinematic range of the high-Q^2 analyses and is used by H1 in [75].

- The ω *method* (ω) [429] is an alternative approach which has been devised for high-Q^2 NC analyses in H1. It is based on the Σ method and exploits the redundancy of the information given by the electron variables and the hadronic system to calculate on an event-by-event basis individual corrections which reduce the influence of ISR.

CC Kinematic Reconstruction. In the CC case, the determination of kinematic variables entirely relies on the measured properties of the hadronic system. The standard formulae as proposed by Jacquet and Blondel [430] are (*JB method*)

$$y_{\mathrm{JB}} = \frac{(E-P_z)_{\mathrm{had}}}{2E_e} , \tag{3.22}$$

$$Q^2_{\mathrm{JB}} = \frac{(P_{\mathrm{t}})^2_{\mathrm{had}}}{1-y_{\mathrm{JB}}} , \tag{3.23}$$

$$x_{\mathrm{JB}} = \frac{Q^2_{\mathrm{JB}}}{y_{\mathrm{JB}}\, s} . \tag{3.24}$$

Although all corrections of the hadronic variables described for the NC case are applied in the CC analyses as well, the JB method is much less accurate than the NC kinematic reconstruction algorithms discussed above. It is unstable both at low y_{JB} (since $(E-P_z)_{\mathrm{had}}$ is small and hence particularly sensitive to detector effects) and at high y_{JB} (where small variations of y_{JB} induce large changes of Q^2_{JB}, see (3.23)). Furthermore, $(P_{\mathrm{t}})_{\mathrm{had}}$, which is one of the primary measured quantities, depends on the hadronic calorimeter resolution (which is worse than the electromagnetic one, see Table 3.1) and is affected by particles escaping full calorimetric measurement (muons, neutrinos, low-energy charged particles spiraling in the magnetic field). Note that Q^2_{JB} depends quadratically on $(P_{\mathrm{t}})_{\mathrm{had}}$ and is therefore particularly sensitive to the absolute hadronic energy scale, thus inducing the dominant experimental systematic uncertainty in the CC analyses (see Sect. 4.2.2).

In addition to purely experimental resolution effects, the emission of photons in the approximate direction of the beam electron (ISR) contributes to variable misreconstruction and event migration in the kinematic space. If the photon is nearly collinear with the e beam, the ep kinematics can be approximated by a decreased beam energy $\tilde{E}_e = E_e - E_\gamma$, corresponding to a reduction of s by a factor of

$$z_\gamma = 1 - E_\gamma / E_e . \tag{3.25}$$

The distribution of z_γ in unbiased event samples follows approximately the typical bremsstrahlung spectrum $\propto 1/z_\gamma$. It is important to note that z_γ is restricted by the event selection procedures for NC events to $z_\gamma \gtrsim 0.7$ (see next section); no such limitation exists in the CC case. If the photon escapes detection, the variables are falsely reconstructed since the nominal s is assumed, with different consequences for the various reconstruction methods

Table 3.6. Mismeasurement of kinematic variables due to undetected initial-state photon radiation. The factors quoted in the table are the ratios of measured values, obtained by assuming the nominal s, and the approximate true values resulting from a reduced value $z_\gamma s$. The y values appearing in the ratios are the measured ones. Note that no overall factors can be defined for the ω method since it involves event-by-event corrections

Reconstruction method	Ratio measured/true		
	x	y	Q^2
Double-angle (DA)	$1/z_\gamma$	1	$1/z_\gamma^2$
Electron (e)	$\dfrac{y_e-(1-z_\gamma)}{y_e z_\gamma}$	$\dfrac{y_e z_\gamma}{y_e-(1-z_\gamma)}$	$1/z_\gamma$
Σ	z_γ	1	1
$e\Sigma$	z_γ	$1/z_\gamma$	$1/z_\gamma$
Jacquet–Blondel (JB)	$\dfrac{z_\gamma-y_{\rm JB}}{z_\gamma(1-y_{\rm JB})}$	z_γ	$\dfrac{z_\gamma-y_{\rm JB}}{z_\gamma(1-y_{\rm JB})}$

Table 3.7. Experimental resolutions of x, y and Q^2 obtained in the high-Q^2 DIS analyses of ZEUS and H1. The H1 resolutions are taken from [72] and refer to the very high-Q^2 regions ($Q^2 > 10^4\,\mathrm{GeV}^2$ for NC and $P_{\rm t} > 50\,\mathrm{GeV}$ for CC). All CC resolutions are given for $y < 0.9$. The relative uncertainties of the electromagnetic and hadronic energy scales are denoted by $\delta_{\rm sca}E_{\rm elm}$ and $\delta_{\rm sca}E_{\rm had}$, respectively, and are given in Table 3.1. The entry denoted by (e, σ) shows the width of a Gaussian fit to the distribution of (measured−true)/true values; otherwise the R.M.S. widths of these distributions are quoted. The $\oplus$ operator is used here to indicate the addition of systematic contributions which are correlated between individual events

NC	ZEUS	H1
$\delta x/x$	$\approx 9\%$ (DA, for $0.03 < x < 0.2$), rising to 12% at high x [73]	$3\% \oplus \delta_{\rm sca}E_{\rm elm}/y_e$ (e, σ) [72]
$\delta y/y$	5% (DA), independent of $y_{\rm DA}$ [73]	$(1-4\%) \oplus \delta_{\rm sca}E_{\rm elm}\dfrac{y_e-1}{y_e}$ [72]
$\delta Q^2/Q^2$	$\approx 5\%$ (DA, $Q^2_{\rm DA} > 400\,\mathrm{GeV}^2$), independent of $Q^2_{\rm DA}$ [73]	$(2-7\%) \oplus \delta_{\rm sca}E_{\rm elm}$ (e) [72]
CC		
$\delta x/x$	$8-20\% \oplus \dfrac{\delta_{\rm sca}E_{\rm had}}{1-y}$, largest at $x < 0.02$ [74]	$20\% \oplus \dfrac{\delta_{\rm sca}E_{\rm had}}{1-y}$ [72]
$\delta y/y$	$8-11\% \oplus \delta_{\rm sca}E_{\rm had}$, largest at $y < 0.1$ [74]	$10\% \oplus \delta_{\rm sca}E_{\rm had}$ [72]
$\delta Q^2/Q^2$	$20\% \oplus \delta_{\rm sca}E_{\rm had}\dfrac{2-y}{1-y}$ [74]	$28\% \oplus \delta_{\rm sca}E_{\rm had}\dfrac{2-y}{1-y}$ [72]

[144]. The effect of ISR on the variable reconstruction can be quantified by the factors given in Table 3.6. On average, these systematic shifts amount to a few percent; for $0.4 < x < 0.6$ and $y > 0.25$ ZEUS finds $\delta_{\rm ISR} x_e = -2.5\%$, $\delta_{\rm ISR} Q_e^2 = +1.8\%$, $\delta_{\rm ISR} x_{\rm DA} = +1.7\%$ and $\delta_{\rm ISR} Q_{\rm DA}^2 = +2.5\%$ [431].

The experimental resolutions of kinematic variables obtained in the ZEUS and H1 high-Q^2 analyses are summarized in Table 3.7. Effects of ISR are included. Note that the kinematic regions for which the resolutions are quoted differ significantly between ZEUS and H1.

Both process-dependent and detector-specific systematic effects tend to induce biases in the kinematic reconstruction, which can be accounted for either by applying suitable corrections to the kinematic variables or by using algorithms to reverse distortions of the resulting distributions caused by event migrations in the kinematic space (*unfolding*). A discussion of these methods is deferred to Sect. 3.4.3.

3.3.4 Electron Energy Calibration

The redundancy of the NC kinematic reconstruction opens a very powerful possibility for calibrating the electromagnetic energy scale of the central and forward calorimeter parts using the high-Q^2 NC data. The basic idea is to compare the measured electron energy, E_e' (after corrections for energy loss in inactive material), with the corresponding value inferred from the kinematic variables reconstructed by the double-angle method ($E_{\rm DA}$) or the ω method (H1 only), which are independent of energy-scale variations. Using the extrapolation of the associated track into the calorimeter allows one furthermore to study the calorimeter response as a function of the position of particle impact and thus to investigate it across inhomogeneities.

Both collaborations have used this approach to devise position-dependent calibration methods, to test the quality of the MC simulation and to quantify the electromagnetic-energy-scale uncertainties quoted in Table 3.1. In Fig. 3.5 (from ZEUS), the mean ratio $\langle E_e/E_{\rm DA}\rangle$ is shown as a function of the z coordinate of the track extrapolation to the calorimeter, before and after applying an inhomogeneity correction. The modulation caused by the 5 cm cell structure is clearly visible in the raw data. After calibration, agreement to within better than 2% is found between the data and MC, and between $\langle E_e\rangle$ and $\langle E_{\rm DA}\rangle$. The equivalent result for H1 is shown in Fig. 3.6, indicating an even better agreement. In the transition regions between different LAC modules small discrepancies between the data and MC are observed, whence electron candidates in these regions are excluded from the analysis (see Sect. 3.3.2).

3.4 Event Selection and Correction Procedures

In this section we discuss the procedures applied to select those data samples from which cross sections and other physical quantities are eventually deter-

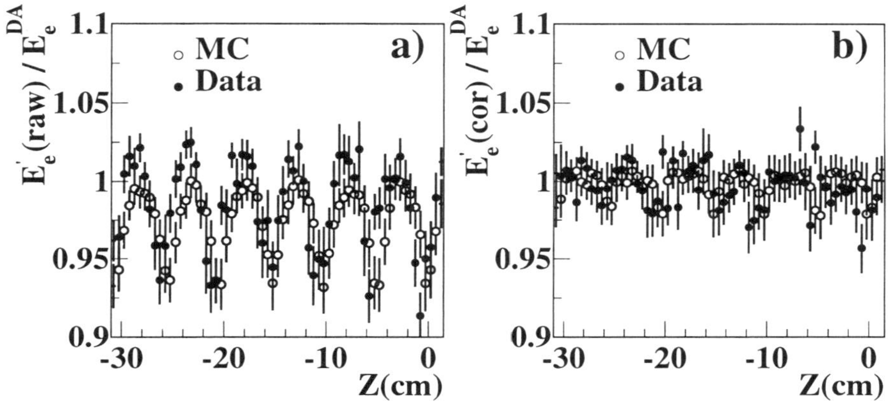

Fig. 3.5. Mean ratio $E'_e/E_{\rm DA}$ of BCAL electrons in ZEUS NC events as a function of the z position of the electron cluster: (**a**) uncorrected data, (**b**) after calibration of E_e. Data (*full circles*) and MC simulation (*open circles*) are both shown. The periodic structure in the raw data is caused by the 5 cm calorimeter cell structure. The plots are reproduced from [73]

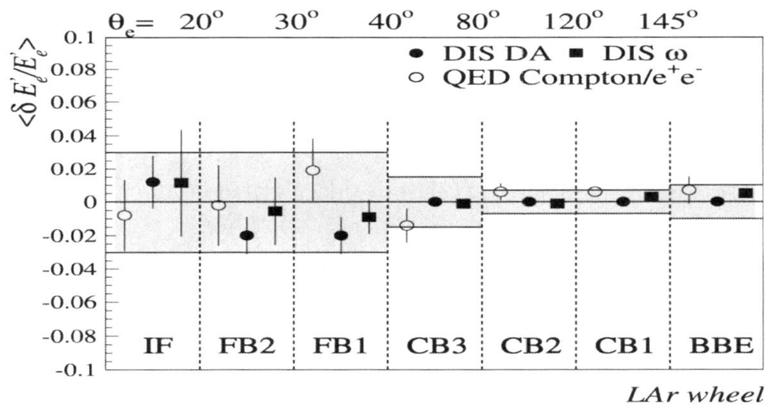

Fig. 3.6. Mean shift $\langle \delta E'_e/E'_e \rangle$ for LAC electrons of NC events in H1 as a function of the electron cluster polar angle, θ_e. The difference $\delta E'_e$ between the measured and predicted values of E'_e is determined using different calibration methods. BBE, CB1–CB3, FB1, FB2 and IF denote different LAC wheels. The *shaded band* demonstrates the effect of an energy scale variation in the MC according to the determined scale uncertainties of $\pm 0.7\%$ in CB1 and CB2, $\pm 1\%$ in BBE, $\pm 1.5\%$ in CB3 and $\pm 3\%$ in FB1–IF. The plot is reproduced from [75]

mined, and to correct them for remaining background contaminations and for acceptance losses and event migrations. In the following we first summarize the requirements for NC and CC events (Sect. 3.4.1) and the potential sources of background which have been investigated (Sect. 3.4.2), and finally discuss corrections applied to the resulting distributions (Sect. 3.4.3).

Both the estimate of remaining background contributions and the acceptance corrections are based on MC studies. It is therefore of central importance to validate the reliability of the MC simulation in each selection step, and for each of the quantities involved. In general, good agreement is found between the data and MC distributions, and even details of the detector response are reproduced satisfactorily by the simulation. We will not go into details of MC–data comparisons in the following, except for selected illustrative examples in Sect. 3.5.

3.4.1 Selection Criteria and Acceptances

The basic signatures of NC and CC reactions and the respective sources of potential backgrounds are distinctly different and necessitate independent selection procedures, which will be summarized in the following. For both the NC and the CC studies, the primary samples of candidate events are those passing the appropriate trigger branches (see Sect. 3.1.5). The offline selection criteria have been designed with the goal to find an optimal compromise between high acceptance for signal events and efficient rejection of remaining background in the primary samples.

For *NC events*, the following selection criteria are used (the actual cut values are summarized in Table 3.8):

- The event vertex has to be reconstructed from the tracking-chamber data and has to be consistent with coming from an *ep* reaction.
- The scattered electron has to be identified by the algorithms described in Sect. 3.3.2 and has to fulfill the requirements listed in Table 3.5. In particular, the associated calorimeter cluster has to lie in fiducial volumes defined such as to exclude regions in which the calorimeter response is not fully understood or is unreliable, like the ends of the BCAL for ZEUS or module boundaries in the LAC for H1 (see Sect. 3.3.4).

Table 3.8. Selection criteria for the ZEUS and H1 high-Q^2 NC analyses of the 1994–97 data [73, 75]. Note that the selection requirements are slightly different from those in the previous analyses [71, 72] and in preliminary studies [428, 431–435]

	ZEUS	H1
Vertex position	$\|v_z\| < 50\,\mathrm{cm}$	$\|v_z\| < 40\,\mathrm{cm}$
Electron identification	See Sect. 3.3.2, Table 3.5	
$E-P_z$	$\theta_e > 17.2°$: $38\,\mathrm{GeV} < E-P_z < 65\,\mathrm{GeV}$ $\theta_e \leq 17.2°$: $44\,\mathrm{GeV} < E-P_z < 65\,\mathrm{GeV}$	$E-P_z > 35\,\mathrm{GeV}$
$P_t/\sqrt{E_t}$	$P_t/\sqrt{E_t} < 4\,\mathrm{GeV}^{1/2}$	—
Kinematic cuts	$Q^2_{\mathrm{DA}} > 400\,\mathrm{GeV}$ $y_e < 0.95$	$Q^2_e > 150\,\mathrm{GeV}$ $y_e < 0.9$

The e-finding efficiency has been determined by both collaborations using MC studies and the results of independent algorithms, based e.g. on a selection of events having isolated tracks with high transverse momenta. Inside the fiducial volumes, ZEUS finds inefficiencies below 4% [73], with an error of $\pm 0.3\%$, rising to $\pm 1.5\%$ at the lowest electron energies. H1 quotes an electron detection efficiency of $> 90\%$ in [72], and an uncertainty of $\pm 2\%$ (rising to $\pm 4\%$ at $Q^2 < 500\,\mathrm{GeV}^2$ and $y > 0.5$) in [75].

- Several cuts are applied to reject background events, the most important being the requirement of longitudinal event containment. By energy–momentum conservation, the final state of any ep reaction has $E - P_z = 2E_e$ ($= 55\,\mathrm{GeV}$ for the HERA e beam energy). The measured quantity (see (3.4)) is affected by experimental resolution and by particles which escape detection. In particular, the requirement of a minimum measured value, $(E - P_z)_{\mathrm{min}}$, allows one to reject events with significant energy flow into the rear beam hole and thus to remove most of the photoproduction background (where the electrons are deflected into the rear beam pipe) and events with undetected high-energy ISR. Up to measurement errors, the factor z_γ defined in (3.25) is constrained to $z_\gamma \gtrsim (E - P_z)_{\mathrm{min}}/(2E_e)$. Further antibackground requirements are that the timing be consistent with an ep interaction and upper limits on $P_{\mathrm{t}}/\sqrt{E_{\mathrm{t}}}$ to remove cosmic-ray events. In addition, elastic QED-Compton events ($ep \to e\gamma p$) are discarded.
- Finally, the kinematic region under investigation is constrained by requirements on y_e (thereby rejecting remaining photoproduction events with fake, low-energy forward electron candidates) and the reconstructed value of Q^2.

The overall NC selection efficiency is about 80% [73, 433] for events with a true value of Q^2 above the selection threshold.

For *CC events*, the characteristic signature is the missing transverse momentum, P_{t}. The main CC selection requirements are summarized in the following, the cut values are collected in Table 3.9. In general, the level of background contamination increases and its removal becomes more difficult as the P_{t} cut decreases; in the latest analyses [74, 75], which assess CC reactions down to $P_{\mathrm{t}} = \mathcal{O}(10\,\mathrm{GeV})$, several additional conditions for background reduction have been imposed, which will not all be reproduced in detail here.

- The event vertex has to be reconstructed and has to be consistent with coming from an ep reaction. Whereas H1 uses only the tracking-chamber data for the vertex determination, ZEUS also determines an event vertex from the calorimeter timing information. The CC selection is based on this timing vertex for events with $\gamma_0 < 23°$, where γ_0 is the hadronic angle calculated according to (3.8), but with respect to the nominal IP.
- The overall transverse momentum, P_{t}, has to exceed some minimum value. ZEUS applies an additional cut on the value $(P_{\mathrm{t}})_{\mathrm{out}}$ calculated by excluding a cone around the forward beam hole from the sum in (3.3), thus reducing background from proton interactions with the beam pipe.

Table 3.9. Selection criteria for the ZEUS and H1 high-Q^2 CC analyses of the 1994–97 data [74, 75]. ZEUS applies two different sets of cuts for events with or without a tracking vertex, which are marked with (a) and (b), respectively. Note that the selection requirements are different from those in the previous H1 analysis [72] and in preliminary studies [428, 431–433, 435, 436]

	ZEUS	H1
Vertex position	$\|v_z\| < 50\,\text{cm}$ for (a) and (b)	$\|v_z\| < 35\,\text{cm}$
Missing P_t	(a) $P_t > 12\,\text{GeV}$ $(P_t)_{\text{out}} > 10\,\text{GeV}$ (b) $P_t > 14\,\text{GeV}$ $(P_t)_{\text{out}} > 12\,\text{GeV}$	$P_t > 12\,\text{GeV}$
P_t/E_t, $V_{\text{ap}}/V_{\text{p}}$	(a) $P_t/E_t > 0.55$ $(P_t \leq 20\,\text{GeV})$ $P_t/E_t > 0.40$ $(20 \leq P_t \leq 30\,\text{GeV})$ (b) $P_t/E_t > 0.60$ $(P_t \leq 30\,\text{GeV})$	$V_{\text{ap}}/V_{\text{p}} < 0.15$
Track-cal. matching	(a) $\Delta\phi < 2.0$ $(P_t \geq 20\,\text{GeV})$ $\Delta\phi < 1.0$ $(P_t < 20\,\text{GeV})$ (b) —	—
Electron veto	(a) Reject events with e candidates with $E'_e > 4\,\text{GeV}$ and matching track $(\theta_e < 164°)$ or $E'_e > 2\,\text{GeV}$ $(\theta_e > 164°)$ (b) —	Reject events with isolated track opposite to $\boldsymbol{P}_{t,\text{had}}$ direction $(\Delta\phi > 160°)$
Kinematic cuts	$Q^2_{\text{JB}} > 200\,\text{GeV}$ $y_{\text{JB}} < 0.9$	$0.03 < y_{\text{JB}} < 0.85$

- Further cuts are applied on transverse event characteristics to select reactions with the typical CC signature, i.e. missing transverse momentum balanced by a collimated hadronic system. These topology cuts are especially suited to reject events in which jets containing high-p_t neutrinos or muons (e.g. from charm decays) cause an overall P_t imbalance. ZEUS requires a lower bound on P_t/E_t and an upper bound on $\Delta\phi = |\phi_{\text{tracks}} - \phi_{\text{cal}}|$, i.e. demands that the energy flow be collimated along the direction of the net transverse momentum and that the reconstructed tracks and energy deposits in the calorimeter coincide in azimuth. H1 employs two variables characterizing the transverse energy flow:

$$V_{\text{p}} = \frac{\boldsymbol{P}_{\text{t,had}}}{(P_{\text{t}})_{\text{had}}} \cdot \sum_{\boldsymbol{P}_{\text{t,had}} \cdot \boldsymbol{P}_{\text{t},i} > 0} \boldsymbol{P}_{\text{t},i} \,,$$
$$V_{\text{ap}} = \frac{\boldsymbol{P}_{\text{t,had}}}{(P_{\text{t}})_{\text{had}}} \cdot \sum_{\boldsymbol{P}_{\text{t,had}} \cdot \boldsymbol{P}_{\text{t},i} < 0} \boldsymbol{P}_{\text{t},i} \,, \tag{3.26}$$

where $\boldsymbol{P}_{\text{t,had}} = \sum_i \boldsymbol{P}_{\text{t},i}$ is the net transverse momentum vector and the sums run over the same energy deposits as in (3.6)–(3.9); note that

$(P_t)_{had} = |\boldsymbol{P}_{t,had}|$. An upper bound on V_{ap}/V_p is imposed in the H1 CC selection, corresponding to constraining the ratio of the transverse energy flows parallel and antiparallel to the direction of the net transverse momentum (see also Sect. 3.5 and Fig. 3.12).

- For background rejection, tracking and timing information is used to eliminate beam–gas and beam–wall events, and electron identification provides a veto against NC events. Dedicated algorithms are applied to identify signatures of penetrating minimally ionizing particles not coming from the event vertex and thus to remove events faked or distorted by cosmic rays. Furthermore, H1 excludes events passing the final selection cuts for the analysis of reactions with a $\not{P}_t \ell j$ topology [88] (see Sect. 2.2.2).
- Kinematic regions suffering from bad resolutions, large background contaminations or uncertain acceptances are excluded by cuts on the kinematic variables.
- Finally, the remaining events are visually inspected, and events incompatible with the CC interpretation are removed from the sample.

The efficiency for the ZEUS CC selection is approximately 90% in the central y region ($0.1 < y < 0.9$) and degrades outside this region and at low x; the overall value for events with $Q^2 > 200\,\text{GeV}^2$ is $\approx 70\%$ [74]. In the case of H1, the trigger efficiency is the limiting factor at low P_t (see Table 3.3); for the study in [72], which required $P_t > 50\,\text{GeV}$, an overall selection efficiency of $\approx 80\%$ was quoted.

3.4.2 Background Studies

Both collaborations have made extensive studies of possible background contributions to their final data samples. In the 1996 publications [71, 72], various sources of background were investigated, and upper thresholds for residual contamination were quantified for NC reactions in the region $x_{DA} > 0.45$ and $y_{DA} > 0.25$ (ZEUS) and $Q^2 > 10^4\,\text{GeV}^2$ (H1). In the later analyses [73–75], these studies were extended to the full kinematic range under consideration.

Two types of background are relevant: events originating, completely or in part, from sources other than ep reactions (*non-ep background*), and ep reactions due to other interaction mechanisms than NC or CC DIS. The non-ep background (mainly collisions of beam particles with residual gas molecules in the beam pipe or with beam-line elements, cosmic rays, and overlays of detector signals from such sources with ep events) is greatly reduced by the vertex and timing cuts and is found to be negligible for NC. However, in the case of CC, a contamination of a few percent is left after offline event selection and has to be removed by visual inspection (see Sect. 3.4.1). In the final event samples, the level of non-ep background is estimated to be negligible, with a related uncertainty of at most 0.6% [73–75].

The investigation of background from ep reactions is mainly based on MC simulation of the following ep reaction types which potentially can contribute to the background (see Sect. 3.2 for a summary of the event generators used for these studies). Nonnegligible background contaminations estimated by MC studies are subtracted statistically from the data samples.

- Photoproduction (PHP) of jets with high transverse momentum, produced dominantly by the hard subprocesses $\gamma q \to qg$ and $\gamma g \to q\bar{q}$ (*direct PHP*) or by parton–parton scattering where one of the initial-state partons comes from a fluctuation of the photon into a virtual hadronic state (*resolved PHP*). The experimental signatures are (2+1) jets for direct PHP, and (2+1+1) jets for resolved PHP, where the additional jet is due to the photon remnant.
  A PHP event with high γ energy ($E_\gamma/E_e \gtrsim 0.7$ to fulfill the $(E - P_z)$ requirement) can be classified as an NC candidate if one of its jets is falsely identified as an electron, or if it contains an isolated high-p_t electron, e.g. from the decay of a charm or bottom meson.
  Owing to mismeasurement of hadronic energies or the presence of neutrinos or muons which escape full calorimetric detection, PHP reactions can exhibit a large transverse-momentum imbalance in the calorimeter. They thus contribute to the CC background, and in fact are the dominant contamination from ep reactions at low P_t.
  The scattered electrons in PHP events are detected by the luminosity monitors with an efficiency of typically $\mathcal{O}(10\%)$. This allows one to estimate the PHP background independently of MC simulations, albeit with limited statistical precision. H1 uses this method to cross-check the level of PHP background inferred from the MC (see Sect. 3.5, in particular Fig. 3.12).
- Prompt photon production (PPP) proceeds via the hard subprocess $\gamma q \to \gamma q$. Initial cross section measurements of PPP have recently been reported by the ZEUS collaboration [424] and are consistent with the theoretical predictions based on an NLO calculation. PPP is a potentially dangerous background to NC DIS since it is a source of isolated photons with high transverse momentum, which can be misidentified as electrons if they are outside the tracking volumes or produce matching tracks by conversion in the beam pipe or other inactive material between the IP and the sensitive tracking volumes. PPP does not contribute to CC background.
- The direct production of W or Z bosons yields events which can have high E_t and in addition can contain electrons or neutrinos from the leptonic decays $W \to \ell\nu$ or $Z \to \ell\ell$. Both the NC and the CC signatures can be faked, although not in the typical (1+1)-jet topology. Since W and Z production have small cross sections of $\lesssim 1\,\mathrm{pb}$ [83] and the decay leptons are typically emitted with high energies into the forward direction, they are potential sources of DIS background mainly at the highest Q^2.
- Pairs of charged leptons can be produced by the process $\gamma\gamma \to \ell\ell$, where the initial photons are radiated from the electron and proton. Reactions of this type, in which a lepton $\ell^\pm$, the scattered electron or the hadronic system has high p_t, contribute to both the NC and the CC background.

Table 3.10. Background contamination in the selected NC samples with $x_{\mathrm{DA}} > 0.45$ and $y_{\mathrm{DA}} > 0.25$ (ZEUS) and $Q^2 > 10^4\,\mathrm{GeV}^2$ (H1) (1996 analyses [71, 72]). The cross section estimates are results of high-statistics MC studies

	Background cross section (fb)	
	ZEUS (90% C.L.)	H1 (95% C.L.)
$\gamma p \to$ jets (PHP)	< 1.8	< 27
$\gamma p \to \gamma X$ (PPP)	0.28	< 6
W, Z production	$< 0.5^*$	< 25
$\gamma\gamma \to \ell\ell$	< 0.1	< 3

(∗) Does not contain hadronic decay modes of W and Z.

The NC background cross sections from the above sources as estimated for $x_{\mathrm{DA}} > 0.45$ and $y_{\mathrm{DA}} > 0.25$ (ZEUS) and for $Q^2 > 10^4\,\mathrm{GeV}^2$ (H1) are summarized in Table 3.10. Both collaborations expect less than 0.1 events background in these regions [431]. For the recent analysis at lower Q^2 [73], ZEUS quotes a maximum background contamination of 3% in a bin and 0.5% over the full phase space, dominantly from PHP. The corresponding numbers for H1 are 5% and 1%, respectively [75].

In the recent CC analysis by ZEUS [74], a background contamination of 1% at high Q^2 is found (mainly from W production), increasing to > 10% in the lowest Q^2 bin (where PHP and lepton pair production dominate the background). H1 finds a very similar level of background in the CC sample (2%, increasing to 10% at the lowest Q^2) [75].

In both the NC and the CC analyses the uncertainties related to background subtraction are below 1% for most bins and rise up to ≈ 5% only at the lowest Q^2 and high y (maximally 7% in the ZEUS measurement of the doubly differential CC cross section).

3.4.3 Acceptance and Migration Corrections

Even after background subtraction, the distributions of event variables in the selected samples cannot be directly used as measures for the corresponding differential cross sections for two reasons: (i) some signal reactions escape detection or are rejected by the selection cuts; (ii) the reconstruction of kinematic variables as described in Sect. 3.3.3 is not only limited in precision by the detector resolutions, but in addition suffers from systematic effects related both to physical peculiarities of the reactions (like ISR) and to measurement biases. As a result, the reconstructed kinematic variables exhibit systematic shifts with respect to the true values, and the distributions are therefore distorted owing to event migrations.

Different algorithms are used to correct for these effects. They can be categorized into *horizontal corrections* (HCs), applied to the variables whose distributions are investigated, and *vertical corrections* (VCs), applied to the distributions themselves.

Horizontal corrections are used to obtain unbiased estimates of variables which can be interpreted in terms of physical quantities, and to reduce migrations in the corresponding distributions. Usually, HCs are derived from MC studies by evaluating either the average measured value of a variable as a function of its true value (*measured-from-true*), or the average true value as a function of the measured one (*true-from-measured*). It is important to note that in many cases both approaches yield distinctly different results, the reason being that the true-from-measured method depends on the distribution of the true variable in the MC sample whereas the measured-from-true method doesn't. In particular, the true-from-measured method tends to induce a shift of the corrected variable into the direction of increasing differential cross sections and to depopulate the tails of distributions. True-from-measured HCs have been used e.g. by ZEUS to correct x and Q^2 in the 1994–96 NC analysis [71] and in the preliminary CC analyses [428, 432]. In contrast, H1 has applied measured-from-true HCs to the hadronic variables P_t and y_{JB} in the 1994–96 CC analyses [72]. In the more recent analyses, progress in detector understanding and reconstruction methods has allowed the determination of the kinematic variables without significant biases, whence horizontal corrections of the above-described type have no longer to be applied.

Acceptance losses and distortions of distributions due to event migrations are corrected by applying VCs to the event numbers in the single bins of a distribution. In the simplest case the correction factors are taken to be $N_{\mathrm{true}}/N_{\mathrm{meas}}$, where N_{true} is the number of MC events whose true variables fall into a bin, and N_{meas} is the number of MC events which pass the selection procedure and whose measured variables fall into the same bin (*bin-by-bin unfolding*). This procedure is justified if (i) the MC reproduces the data well and (ii) the bin sizes are large compared with the experimental resolutions and systematic shifts.[8] Both conditions are fulfilled for the HERA high-Q^2 analyses, and bin-by-bin corrections are being used throughout. Note that only after performing the VCs are the distributions independent of detector and analysis properties, and results of different experiments can be directly compared.

3.5 Summary of Analyses and Data Samples

Results of ZEUS and H1 high-Q^2 analyses have been reported in four stages: in February 1997 (analyses of the 1994–96 data) [71, 72], at the 1997 summer conferences (update including the first part of the 1997 data) [431, 439], in

[8] If these conditions do not hold, numerically demanding methods have to be used in order to guarantee unbiased results which are stable against small fluctuations of the input distributions [437, 438].

Table 3.11. Data sets used for the different high-Q^2 e^+p analyses of ZEUS and H1. The names are used for reference in this book only (see Chap. 4). In spring 1998 H1 reported NC results [433] both for a high-Q^2 data set (denoted HN3a) and for a set extending to lower Q^2 (HN3b). Note that small differences of the integrated luminosities as compared with those quoted in Table 1.1 are due to particular data quality requirements. The event numbers, N_{evt}, are after all selection cuts

NC	Experiment	$\mathcal{L}$ (pb^{-1})	Reference	Name	Kinematic cuts	N_{evt}
1994–96	ZEUS	20.1	[71]	ZN1	$Q^2 > 5000\,\mathrm{GeV}^2$	191
	H1	14.2	[72]	HN1	$Q^2 > 2500\,\mathrm{GeV}^2$	443
1994–	ZEUS	33.5	[431]	ZN2	$Q^2 > 5000\,\mathrm{GeV}^2$	326
July 1996	H1	23.7	[431, 439]	HN2	$Q^2 > 2500\,\mathrm{GeV}^2$	724
1994–97	ZEUS	47.7	[428]	ZN3	$Q^2 > 400\,\mathrm{GeV}^2$ $y_e < 0.95$	38 000
	H1	35.6	[433]	HN3a	$Q^2 > 2500\,\mathrm{GeV}^2$ $0.1 < y_e < 0.95$	1297
				HN3b	$Q^2 > 200\,\mathrm{GeV}^2$ $y_e < 0.9$ $0.005 < x < 0.65$	75 000
1994–97	ZEUS	47.7	[73]	ZN4	$Q^2 > 400\,\mathrm{GeV}^2$ $y_e < 0.95$	37 438
	H1	35.6	[75]	HN4	$Q^2 > 150\,\mathrm{GeV}^2$ $y_e < 0.9$	75 000
CC						
1994–96	H1	14.2	[72]	HC1	$P_{\mathrm{t}} > 50\,\mathrm{GeV}$	31
1994–	ZEUS	33.5	[431]	ZC2	$Q^2 > 1000\,\mathrm{GeV}^2$	445
July 1996	H1	23.7	[431, 439]	HC2	$Q^2 > 2500\,\mathrm{GeV}^2$	61
1994–97	ZEUS	47.7	[428]	ZC3	$Q^2 > 400\,\mathrm{GeV}^2$	869
	H1	35.6	[433]	HC3	$Q^2 > 2500\,\mathrm{GeV}^2$	100
1994–97	ZEUS	47.7	[74]	ZC4	$Q^2 > 200\,\mathrm{GeV}^2$ $y_{\mathrm{JB}} < 0.9$	1086
	H1	35.6	[75]	HC4	$Q^2 > 150\,\mathrm{GeV}^2$ $0.03 < y_{\mathrm{JB}} < 0.85$	700

spring 1998 (complete 1994–97 data) [428, 433] and in 1999 (final results for the 1994–97 data) [73–75]. Although the results of the intermediate two stages are marked preliminary and are partly superseded by the latest publications, we will include a selected subset of them in this book in order to demonstrate the development of the analyses and to discuss features of the data which are not documented in the final papers. In Table 3.11, symbolic names are assigned to the different analyses, and the properties of the corresponding data samples are summarized. The symbolic analysis names will be used in the following and also in Chap. 4 to refer to the corresponding results. For reasons of simplicity, Table 3.11 does not list the recent searches for physics beyond the SM [76–88] which partly use the same data sets as the cross section analyses, but usually with different kinematic cuts.

The analysis results, including the distributions of kinematic variables will be discussed in Chap. 4. We conclude this chapter by presenting distributions of a few of the most relevant auxiliary variables and compare them with the corresponding results of MC simulations. The distributions of the energy and polar angle of the scattered electron obtained in the ZEUS analysis ZN4 are shown in Fig. 3.7; the corresponding distributions resulting from the H1 analysis HN4 are reproduced in Fig. 3.8. For both experiments, good agreement between data and MC is found over the full spectrum of electron energies and angles. The dip in the ZEUS θ_e spectrum around $\theta_e \approx 0.65$ is caused by the fiducial-volume cuts applied in the electron identification algorithm (see Sect. 3.3.2). The few events at the highest $E'_e \gtrsim 200\,\text{GeV}$ are those which fall into the kinematic region of $Q^2 \gtrsim 2 \times 10^4\,\text{GeV}^2$, where an excess of events was reported (analyses ZN1 and HN1 [71, 72]).

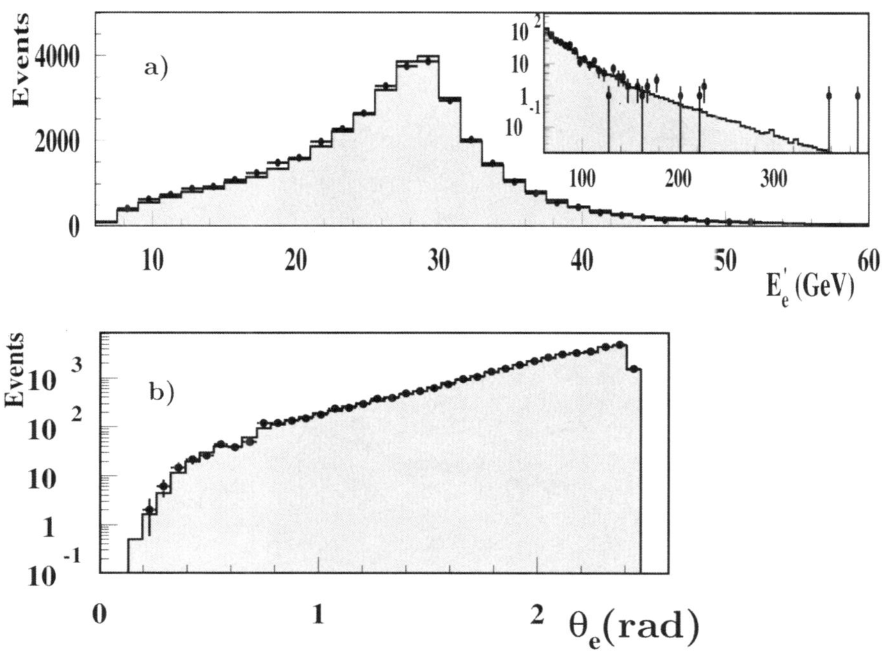

Fig. 3.7. Distributions of (**a**) the electron energy, E'_e, and (**b**) the electron scattering angle, θ_e, in ZEUS analysis ZN4. *Dots with error bars* represent the data; the *shaded histogram* is the MC prediction. The inset in (**a**) shows the high-E'_e tail of the distribution. The plots are reproduced from [73]

The global variables (see Sect. 3.3.1) are used for the reconstruction of kinematic variables and for calibration purposes (see Sects. 3.3.3 and 3.3.4), and their detailed understanding and correct simulation are essential for the reliability of physical results and the investigation of absolute energy scales and other systematic effects. A few selected distributions of global variables are shown and compared with the MC simulation in Fig. 3.9 ($E - P_z$ and γ_{had} from analysis ZN4) and Fig. 3.10 (hadronic variables from analysis HN4). Generally, good agreement is observed between data and MC. The H1 figures demonstrate particularly well the level of detailed understanding and precision which has been achieved in detector calibration and kinematic reconstruction: both the decomposition of y_{JB} into contributions from different sources and the control of the calorimeter noise subtraction shown in Fig. 3.10a, and the comparison of hadronic variables with their counterparts reconstructed from the energy and angle of the scattered electron (Figs. 3.10b–d) show that there are no residual systematic biases in the corrected hadronic variables, and that the resolutions of the hadronic and the electron measurements are accurately simulated in the MC.

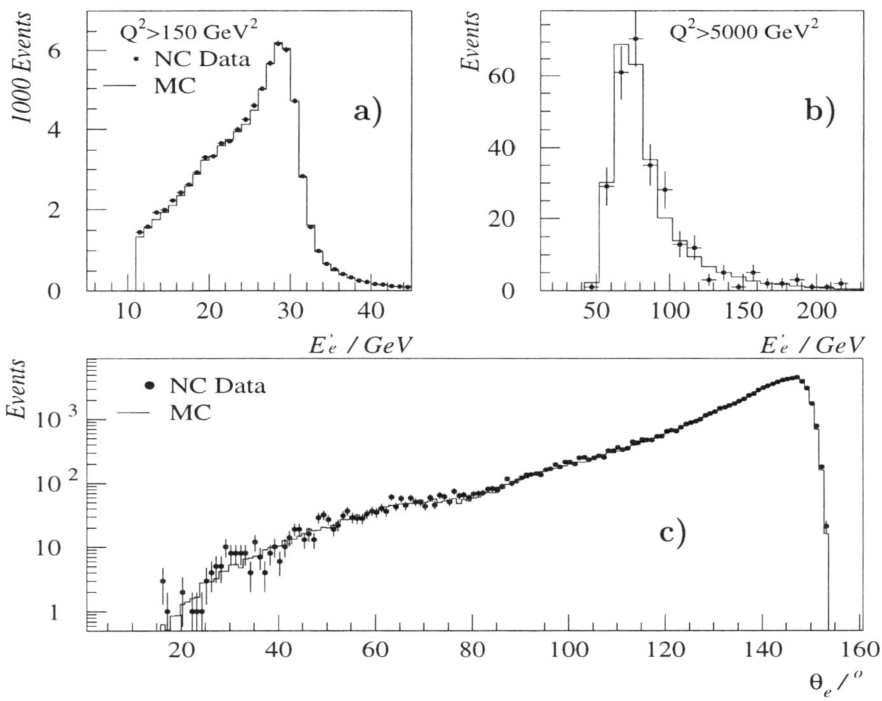

Fig. 3.8. Distributions of (**a**) the energy of the scattered electron, E_e', for $Q^2 > 200\,\mathrm{GeV}^2$, (**b**) of E_e' for $Q^2 > 5000\,\mathrm{GeV}^2$ and (**c**) of the electron scattering angle, θ_e in the H1 NC analysis HN4. *Dots with error bars* represent the data; the *histograms* are the MC predictions. The plots are reproduced from [75]

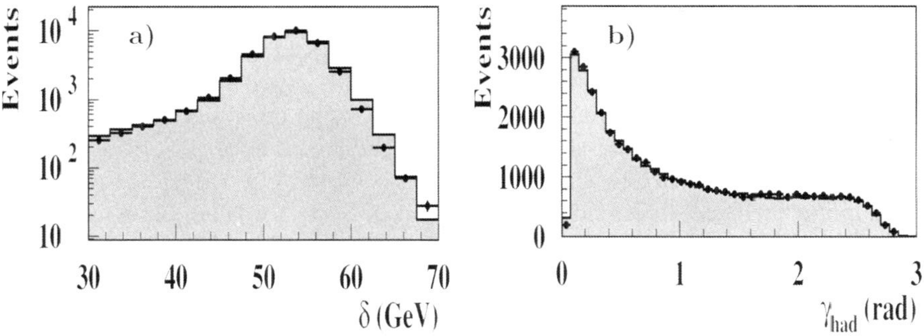

Fig. 3.9. Distributions of (**a**) $\delta = E - P_z$ and (**b**) the hadronic angle γ_{had} from the ZEUS analysis ZN4. The *dots with error bars* represent the data, the *shaded histograms* the results of the MC simulation. The plots are reproduced from [73]

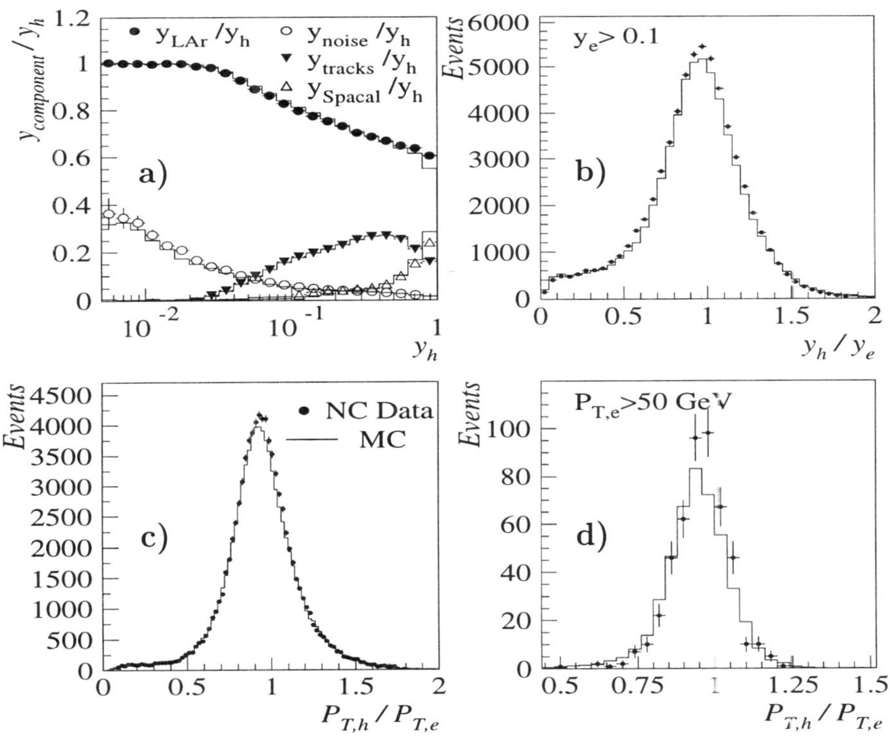

Fig. 3.10. MC–data comparison of distributions involving hadronic quantities in the H1 analysis HN4: (**a**) contributions to y_{JB} from LAC (*full dots*), from the rear calorimeter SPACAL (*open triangles*), from tracks (*full triangles*) and the subtracted contribution from calorimeter noise (*open dots*); (**b**) ratio y_{JB}/y_e; (**c**) ratio of the hadronic and the electron transverse momenta, $(P_{\mathrm{t}})_{\mathrm{had}}/P_{\mathrm{t}}'$; (**d**) $(P_{\mathrm{t}})_{\mathrm{had}}/P_{\mathrm{t}}'$ for $P_{\mathrm{t}}' > 50\,\mathrm{GeV}$. The *histograms* show the results of the MC simulation. The plots are reproduced from [75]

Finally, the distributions of some characteristic hadronic variables of the CC analyses ZC4 and HC4 are presented in Fig. 3.11 and Fig. 3.12, respectively. Again, very good agreement is observed between the distributions of the relevant hadronic quantities and the MC simulation. At the highest $P_{\rm t} \gtrsim 80\,{\rm GeV}$ in Fig. 3.11a, there is a slight indication of an excess of events over the MC expectation; the region of high Q^2 and x, where these events are located, will be discussed in more detail in Sect. 4.2. The distributions of $V_{\rm ap}/V_{\rm p}$ in Fig. 3.12 demonstrate not only that the simulation of event topologies and the reconstruction of hadronic quantities in CC events are well under control, but also that the background contributions from other ep reactions are correctly reproduced in the MC. Figure 3.12b furthermore shows that the photoproduction background to CC reactions (see Sect. 3.4.2) as determined from reactions in which the scattered electron is detected in the luminosity monitor is fully compatible with the corresponding MC simulation. These observations lend high credibility to the detector simulation since they involve predominantly events in the tails of the resolution functions, for which an accurate simulation is known to be difficult.

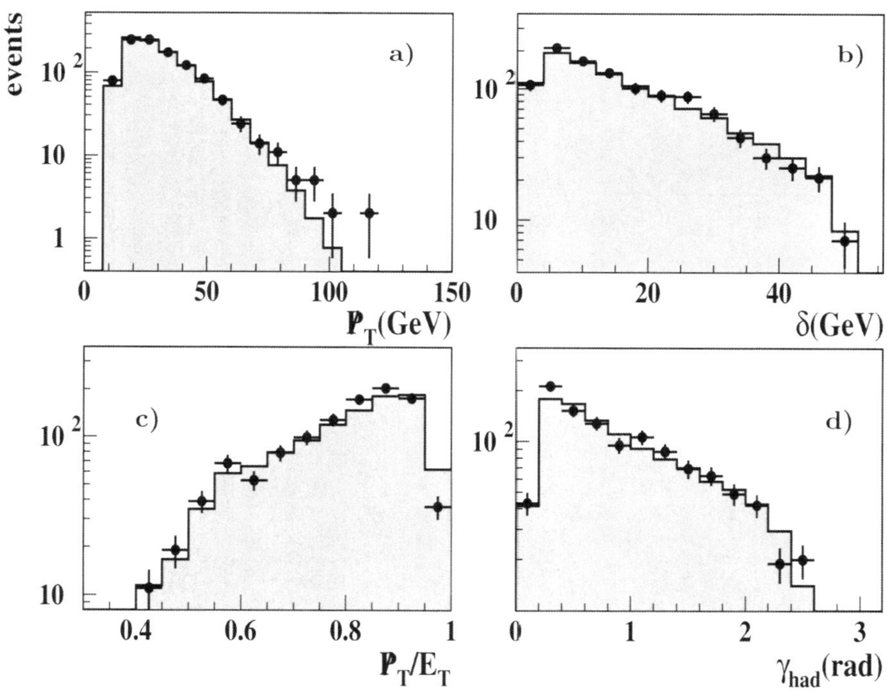

Fig. 3.11. Distributions of (**a**) the net transverse momentum, $(P_{\rm t})_{\rm had}$, (**b**) $\delta = E - P_z$, (**c**) $(P_{\rm t})_{\rm had}/E_{\rm t}$ and (**d**) the hadronic angle, $\gamma_{\rm had}$, from the ZEUS CC analysis ZC4. *Dots with error bars* represent the data; the *shaded histograms* are the MC predictions. The plots are reproduced from [74]

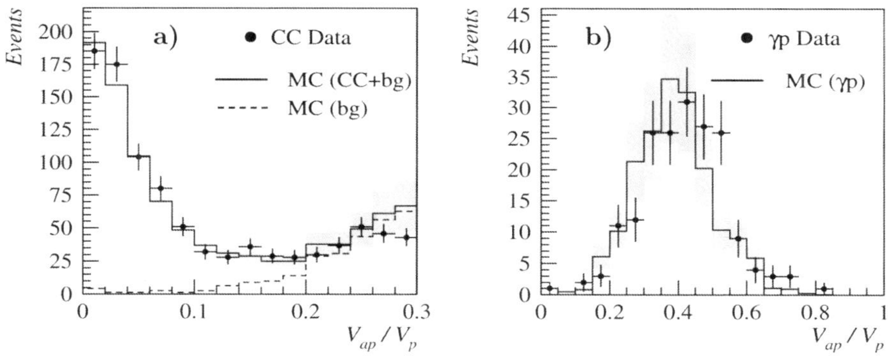

Fig. 3.12. Distributions of the ratio $V_{\mathrm{ap}}/V_{\mathrm{p}}$ (see (3.26)) from the H1 analysis HC4, (**a**) in the selected CC sample before applying the cut $V_{\mathrm{ap}}/V_{\mathrm{p}} < 0.15$, and (**b**) in photoproduction events with the detected electron passing the CC selection. The *dots with error bars* represent the data, the *full lines* the MC simulation and the *dashed line* in (**a**) the subtracted background determined from an MC simulation of photoproduction and NC events. The *shaded bands* indicate the uncertainty of the background subtraction. The plots are reproduced from [75]

4. Results from the ZEUS and H1 Experiments

The investigation of high-Q^2 DIS at HERA started out as early as 1992 with the very first data taken by ZEUS and H1. Reports on searches for resonant production of leptoquarks (LQs) [55, 60] were amongst the first physics publications of ZEUS and H1, followed soon by measurements of the e^-p CC cross section by H1 [52] and of $\mathrm{d}\sigma^{e^-p}/\mathrm{d}Q^2$ in NC and CC reactions with $Q^2 > 400\,\mathrm{GeV}^2$ by ZEUS [50]. These initial analyses were exploratory and statistics-limited, but already excluded gross deviations from the SM expectations. This picture consolidated when the first HERA e^+p data became available for analysis. Figure 4.1 shows the status of NC and CC cross section measurements as of mid-1996, when the 1994 data were analyzed [51, 53, 54] and preliminary results based on the 1995 data had been reported by ZEUS [57].

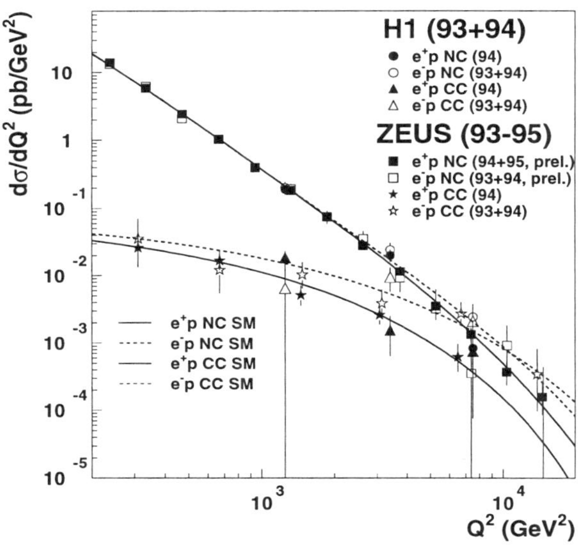

Fig. 4.1. Differential cross sections $\mathrm{d}\sigma/\mathrm{d}Q^2$ of NC and CC DIS as measured by ZEUS [51, 57] and H1 [53] by 1996. The *symbols with error bars* denote the experimental results; the *lines* are the corresponding SM predictions. Note that the region $Q^2 > 10^4\,\mathrm{GeV}^2$ is hardly covered by experimental data

Meanwhile, both experiments have increased their e^+p data samples by more than an order of magnitude as compared with the end of 1994. The high-Q^2 domain of NC and CC DIS has been investigated with significantly higher statistical precision and reduced systematic uncertainties. In February 1997, ZEUS and H1 announced the observation of an excess of NC events with $Q^2 \gtrsim 15\,000\,\mathrm{GeV}^2$ over the SM prediction [71, 72]. An update of these analyses including the first half of the 1997 data appeared in summer 1997 [431, 439], and although the new data did not confirm the previous observations, a clear indication of a discrepancy between the data and the SM expectation was still present. ZEUS and H1 also reported CC rates which exceeded the predictions. The preliminary results of the analyses including the full 1997 data sets became available in early 1998 [428, 433]. Recently, both ZEUS [73, 74] and H1 [75] have published final results on the high-Q^2 NC and CC cross sections. In the following, we will refer to the various analyses using the symbolic names defined in Table 3.11 in Sect. 3.5. In a few places, preliminary results derived from the e^-p data taken in 1998 and 1999 will also be quoted.

In this chapter, we will review the results on DIS in the Q^2 region above several hundred GeV^2 reported by ZEUS and H1 (Sects. 4.1 and 4.2), their interpretations in terms of the QCD and electroweak sectors of the SM (Sect. 4.3), and the searches for signals of physics beyond the SM in these data (Sect. 4.4). The impressive spectrum of cross section and structure function measurements at lower Q^2 has been covered in several detailed reviews [44–46] and will not be considered here. The basic questions to be discussed in the following are: (i) What quantities have been measured, and to what precision? (ii) Are the results consistent with the SM expectation,[1] or are there significant deviations from it? (iii) To what level of sensitivity can the HERA experiments exclude (or confirm) hypothesized deviations of the cross sections from the SM predictions? In some places it will be appropriate to discuss not only the latest, final results but also the intermediate analyses (ZN1, ZN2, ZC2, ZN3 and ZC3 from ZEUS; HN1, HC1, HN2, HC2, HN3a,b and HC3 from H1) in order to illustrate the historical development, the analysis techniques applied and the premises on which further analysis steps were built.

4.1 Neutral-Current Results

Of the HERA DIS data, the NC reaction channel offers higher precision than the CC channel for two reasons: the experimental advantage of assessing both the scattered electron and the hadronic system provides superior kinematic resolution, and the large cross section for Q^2 below a few $1000\,\mathrm{GeV}^2$ allows for statistically accurate measurements and investigations of systematic effects.

[1] Here and in the following, we use the terms *SM expectation* and *SM prediction* to refer to quantities which are determined by extrapolating lower-Q^2 measurements into the kinematic region of interest using the SM concepts (see Chap. 2).

In fact, the high-Q^2 excess reported in 1997 has gained a substantial part of its significance from this NC-specific precision.

In the following sections, two approaches are discussed which have been applied to analyze NC reactions and confront the results with the SM expectation described in Sect. 2.1: the comparison of distributions of reconstructed kinematic quantities with the corresponding predictions derived from the MC simulation (Sect. 4.1.1), and the evaluation of differential cross sections which can directly be compared with the theoretical expectations and between ZEUS and H1 (Sect. 4.1.2).

4.1.1 Investigation of Neutral-Current Event Distributions

The first high-Q^2 results presented in 1997, and parts of the later updates, were based on comparisons between data and MC simulations of the distributions of x, y and Q^2 in the regions of $Q^2 > 5000\,\mathrm{GeV}^2$ and $Q^2 > 2500\,\mathrm{GeV}^2$ for ZEUS and H1, respectively. No acceptance corrections or unfolding procedures were applied to calculate cross sections (see Sect. 3.4.3). Although these distributions therefore depend on the properties of the respective detectors and on details of the offline analysis algorithms, they are well suited to investigate the compatibility of the data with the SM prediction fed into the MC simulation. We first present the measured event distributions and the corresponding MC results. Thereafter, observed and expected event numbers in different kinematic regions are compared and used to quantify the probability that the apparent discrepancies are due to statistical fluctuations.

In this section we will concentrate on the data with $Q^2 > 2500\,\mathrm{GeV}^2$ and discuss the experimental evidence for the excess of events at the highest Q^2 and x which has triggered so much excitement. Note that some of the results presented in this section are "old", i.e. they do not include the full 1997 data set or they stem from intermediate status reports. This reflects the fact that the more recent publications do not quantify the compatibility of the data and the SM prediction and focus more on cross section measurements and the search for new phenomena than on event distributions.

The perhaps most intuitive way to comprehend these high-Q^2 data is to investigate their two-dimensional distribution in the (x, y) plane or, equivalently, in the (M, y) plane ($M = \sqrt{xs}$). These representations contain the full information on kinematic distributions in the data sample, and, although a direct comparison with theoretical predictions is difficult, allow one to scan for "anomalies" like events which cluster around one particular value of x and have a flat distribution in y (as would be expected for a scalar resonance, see Sect. 2.2.2).

The (x, y) event distribution obtained by ZEUS is shown in Fig. 4.2 for the analyses ZN2 and ZN3. The rectangle in the upper right corner of the plot denotes the "signal region" as defined in [71], i.e. the kinematic domain in which an excess of events was observed. Note that neither the rectangular shape of the signal region nor its exact boundaries are physics-motivated, and

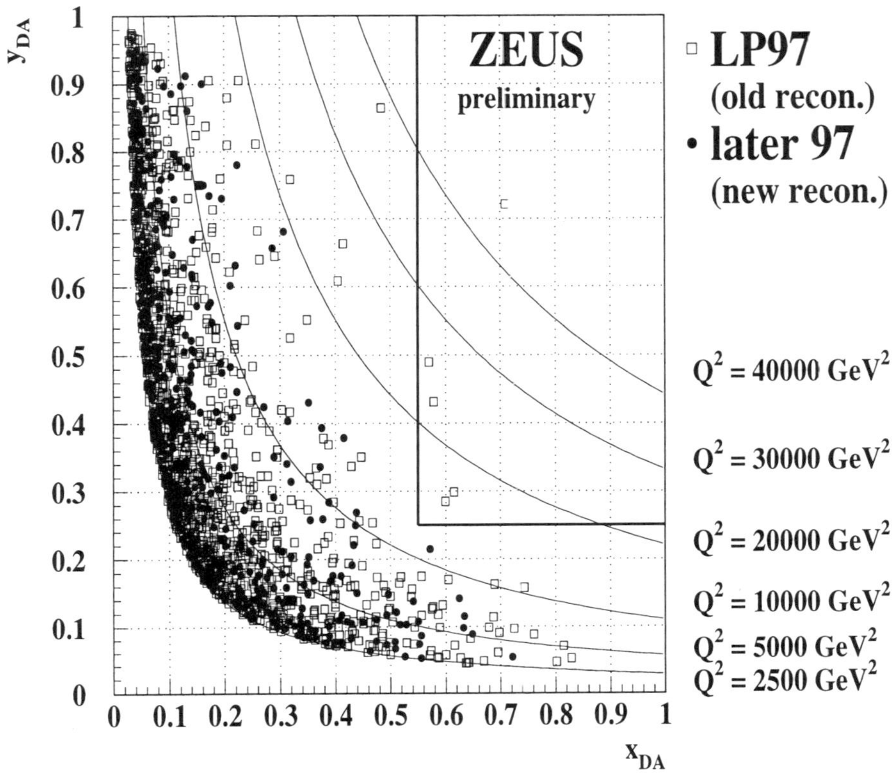

Fig. 4.2. Distributions of ZEUS events in the (x, y) plane. The *open squares* represent the result of analysis ZN2 (using the 1994–96 and the first half of the 1997 data). The *full dots* denote the events contained in the second half of the 1997 data (analysis ZN3). Note that the reconstruction methods for ZN3 have been refined with respect to ZN2, and that the distribution of the ZN2 subsample (*squares*) would be slightly modified if the ZN3 reconstruction was applied. The plot has been reproduced from [428]

that both are arbitrary to some extent. Only one further event occurred in this box in the 1997 data taking even though the integrated luminosity was more than doubled.

The corresponding H1 event distributions in the (M, y) plane are reproduced in Fig. 4.3. In order to ease the comparison of ZEUS and H1 results, a little conversion aid between x and M is provided in Table 4.1. The 1996 H1 data indeed exhibit a structure which resembles the resonance scenario mentioned above: there are six events with $y > 0.4$ lining up at $M \approx 205\,\text{GeV}$ (owing to refined reconstruction methods, the (M, y) pattern has slightly changed since the original analysis HN1, where seven events were clustered around $M = 200\,\text{GeV}$ [72]). It is this group of events which most significantly deviates from the SM prediction and suggests an interpretation in terms of

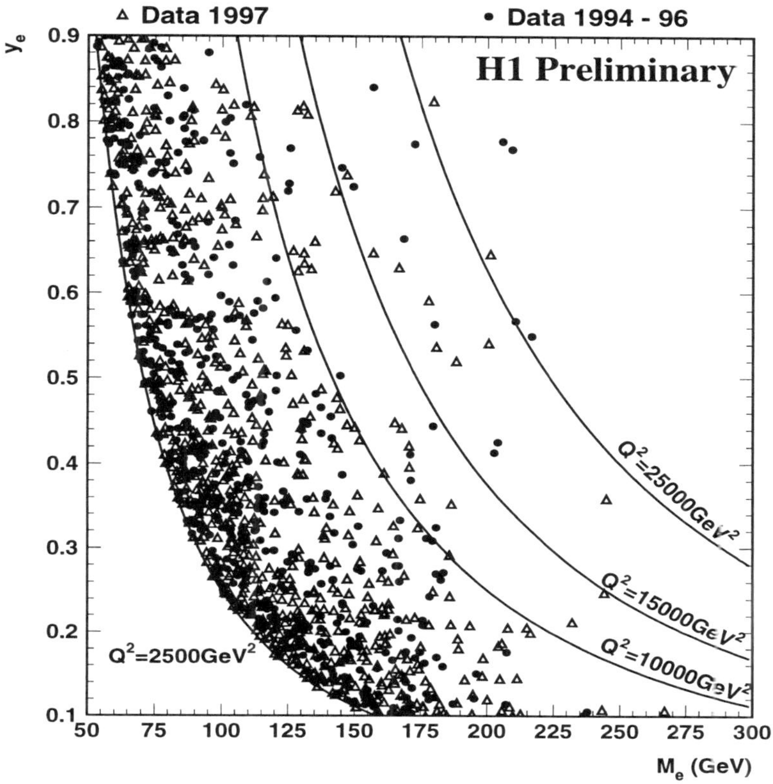

Fig. 4.3. Distributions of H1 events in the (M, y) plane. The *full dots* represent the 1994–96 data. The *open triangles* denote the events in the 1997 data. All variables were reconstructed using the HN3a algorithms, which induce slight shifts with respect to the earlier results (see [72, 431]). The plot has been reproduced from [433]

Table 4.1. Conversion between x and $M = \sqrt{xs}$ for $s = 90\,200$ GeV^2, corresponding to the beam energies used in 1994–97 (see Table 1.1)

x	0.20	0.25	0.30	0.35	0.40	0.45	0.50	0.55	0.60	0.65	0.70
M (GeV)	134.4	150.2	164.6	177.7	190.0	201.5	212.4	222.8	232.7	242.2	251.4

the resonant production of a new particle, an LQ or squark (see Sect. 2.2.2). However, the 1997 H1 data did not corroborate the particular pattern observed in the 1994–96 event sample.

In order to compare the data and the SM prediction on a more quantitative footing, the one-dimensional projections of the points in Figs. 4.2 and 4.3 have been superimposed on the corresponding expectations calculated from the MC simulation. Note that these distributions must not be interpreted as cross sections. However, differences between data and MC exceeding the

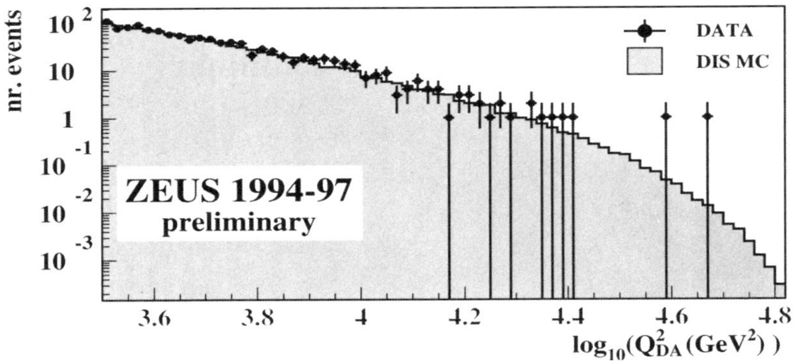

Fig. 4.4. Distributions of Q^2_{DA} for $Q^2_{\mathrm{DA}} \gtrsim 3000\,\mathrm{GeV}^2$ in the ZEUS e^+p data sample of 1994–97 (analysis ZN3). The *points with error bars* denote the data; the *shaded histogram* is the MC simulation. Note that the error bars are calculated as square roots of the event numbers and are only used for illustration but not in the significance analysis. The plot has been reproduced from [428]

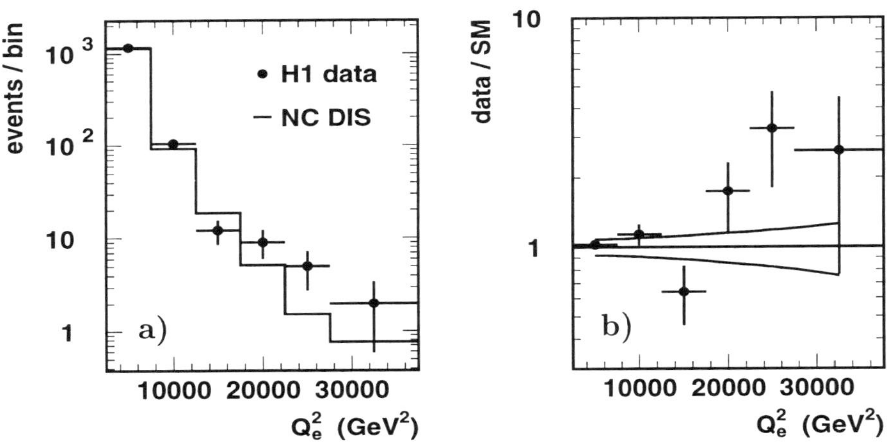

Fig. 4.5. (**a**) Distribution of Q_e^2 in the H1 e^+p data of 1994–97. The data are shown as *points with error bars*, the *histogram* represents the MC simulation. (**b**) Ratio data/simulation; the *lines* indicate the error band due to systematic uncertainties. The plots have been reproduced from [84]

level of systematic uncertainties related to the detector simulation would indicate that the underlying model(s) do not describe those data correctly. An inspection of the Q^2 distributions presented in Fig. 4.4 (ZEUS) and Fig. 4.5 (H1) actually reveals indications of an excess of events as compared with the expectation at the highest Q^2. Similarly, Fig. 4.6 demonstrates that the H1 data points are above prediction in the tails of the M distribution of events with high y. A discussion of the ZEUS spectrum of the invariant e-jet mass in NC DIS events can be found in Sect. 4.4.1.

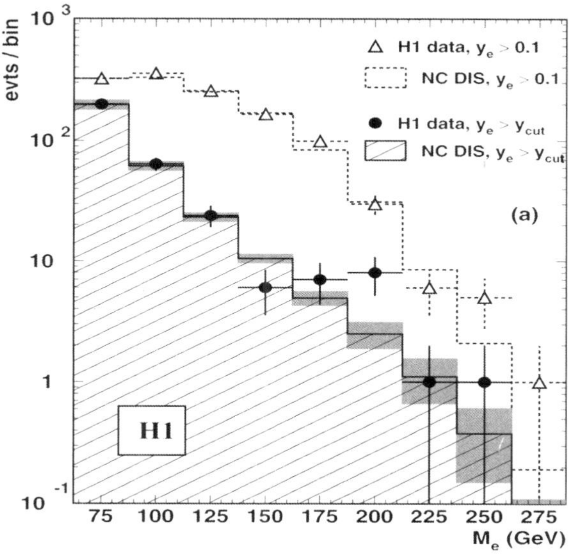

Fig. 4.6. Distribution of $M_e = \sqrt{x_e s}$ for H1 events with $y_e > 0.1$ (*open triangles with error bars*) and for events with $y_e > y_{\mathrm{cut}}$ (*full dots with error bars*), where y_{cut} is an M_e-dependent threshold which decreases with increasing M_e and has a value of $y_{\mathrm{cut}} = 0.4$ for $M_e = 200\,\mathrm{GeV}$. The *open* (*hatched*) *histogram* is the result of the MC simulation for $y > 0.1$ ($y > y_{\mathrm{cut}}$), with the *shaded band* indicating the systematic uncertainty for $y > y_{\mathrm{cut}}$. The plot has been reproduced from [84]

Before we turn to discuss the statistical significance of the observed discrepancies, we first want to recollect that the errors which are assigned to simulated distributions and event numbers, as indicated e.g. in Fig. 4.6, receive contributions from two main sources: (i) the uncertainty of the cross section prediction, which is dominated by the experimental errors on the input data for the PDF parameterizations and amounts to 3 – 8% in the kinematic regions of interest (see Sect. 2.1.6); (ii) experimental effects which cause systematic shifts of the kinematic variables. The MC statistical errors do not contribute significantly to the overall uncertainties. The most prominent example of an effect of the type (ii) is the error of the absolute calorimeter energy scale, which induces sizable uncertainties for kinematic variables reconstructed with the e method (see Table 3.7). In kinematic regions where the average differential cross section in adjacent bins differs by large factors, such shifts are amplified and lead to large uncertainties of the expected event numbers. This feature explains the shape of the error band in Fig. 4.5b, which rises from about 8% to 30% from the lowest to the highest Q^2, where migration effects are largest. ZEUS quotes an overall experimental systematic uncertainty of 6.5% for $y_{\mathrm{DA}} > 0.25$ and $x_{\mathrm{DA}} > 0.55$ (analysis ZN2 [431]). The systematic errors in the NC analyses ZN4 and HN4 are discussed in more detail in Sect. 4.1.2, those of the CC analyses ZC4 and HC4 in Sect. 4.2.2.

Both collaborations have reported observed and expected event numbers for various parts of the kinematic plane, and we want to concentrate on those which are directly comparable between ZEUS and H1. The available cross section measurements typically average over kinematic regions larger than appropriate for studying the significances of observed deviations between data and expectation and are therefore not used for this purpose here.

A selection of observed and expected event numbers in different kinematic regions and different analyses is given in Table 4.2. Before going into details of the statistical analysis and addressing the probabilities $\mathcal{P}$, a comment on the expected event numbers, $N_{\rm exp}$, seems appropriate. Obviously the ratio of the $N_{\rm exp}$ values from different analyses in a given kinematic region does not in all cases reflect the corresponding ratio of integrated luminosities. In particular, the ZEUS values of $N_{\rm exp}$ for the 1994–96 data are roughly twice as large as those of H1, whereas the luminosity ratio is about 1.4 (see Table 3.11). This mismatch is mainly due to different migration patterns induced by the kinematic reconstruction and correction algorithms (see Sects. 3.3.3 and 3.4.3). It is interesting to note that the later analyses exhibit the reverse tendency (H1 having higher $N_{\rm exp}$ per luminosity than ZEUS), demonstrating that both collaborations have substantially modified their analysis methods.

An inspection of Table 4.2 shows that significant deviations between data and expectation, if any, are of the type $N_{\rm obs} > N_{\rm exp}$, with a maximum discrepancy of $N_{\rm obs} - N_{\rm exp} \approx 10$ for $Q^2 > 15\,000\,{\rm GeV}^2$ (the excess did not increase on adding the 1997 data). Using an overall acceptance of about 80% and the integrated luminosities in Table 3.11, a very rough estimate for the apparent excess cross section of $\sigma_{\rm exc} \approx 0.4\,{\rm pb}$ for the 1994–96 data and of $\sigma_{\rm exc} \approx 0.15\,{\rm pb}$ for the 1994–97 data can be obtained.

The $\mathcal{P}$ values quoted in Table 4.2 are defined to be the probabilities to observe at least $N_{\rm obs}$ events if $N_{\rm exp} \pm \delta N_{\rm exp}$ are expected. It has been assumed that $N_{\rm obs}$ is distributed[2] according to a Poissonian with mean μ, and that the probability density of μ is given by a Gaussian with mean $N_{\rm exp}$ and width $\delta N_{\rm exp}$ which is normalized to unity in the region $\mu \geq 0$. Under these assumptions, $\mathcal{P}$ is given by

$$\begin{aligned}\mathcal{P} &= \mathcal{P}(N_{\rm obs} > N_{\rm exp}) \\ &= \frac{\int\limits_0^\infty {\rm d}\mu \, \exp\left[-(\mu - N_{\rm exp})^2/2(\delta N_{\rm exp})^2\right] \sum\limits_{n=N_{\rm obs}}^{\infty} \dfrac{{\rm e}^{-\mu}\,\mu^n}{n!}}{\int\limits_0^\infty {\rm d}\mu' \, \exp\left[-(\mu' - N_{\rm exp})^2/2(\delta N_{\rm exp})^2\right]} \, . \end{aligned} \tag{4.1}$$

[2] The concept of a distribution of $N_{\rm obs}$ needs some justification since there is only one such number per experiment and event category. However, one can consider the distribution of $N_{\rm obs}$ as obtained from a *sample of many identical hypothetical or simulated experiments*. Actually, this approach relates the probabilities discussed in this section to frequencies of observing a particular outcome in a series of equal experiments, and allows us to investigate heuristically questions of probability and significance using computer simulation techniques.

Table 4.2. Observed (N_{obs}) and expected (N_{exp}) NC event numbers in selected kinematic regions, for different data samples of both collaborations. The H1 data are subject to the additional cut $0.1 < y_e < 0.9$, which mainly affects the first two rows. The data sets denoted by "96", "LP97" and "all" comprise the 1994–96 data, the data available in summer 1997, and the full 1994–97 sample, respectively. The columns "ZEUS + H1" are calculated by adding the corresponding entries in the "ZEUS" and "H1" columns, and by combining the errors δN_{exp} in quadrature. The probabilities $\mathcal{P}$ are evaluated according to (4.1) (see text). The numbers are quoted from [71, 72] ("96"), [431] ("LP97") and [428, 433, 440] ("all")

Kinematic	**ZEUS**				**H1**				**ZEUS + H1**		
region	Data	N_{obs}	$N_{\mathrm{exp}} \pm \delta N_{\mathrm{exp}}$	$\mathcal{P}(\%)$	Data	N_{obs}	$N_{\mathrm{exp}} \pm \delta N_{\mathrm{exp}}$	$\mathcal{P}(\%)$	N_{obs}	$N_{\mathrm{exp}} \pm \delta N_{\mathrm{exp}}$	$\mathcal{P}(\%)$
$Q^2 > 2500\,\mathrm{GeV}^2$	all	1817	1792 ± 93	40.4	all	1297	1276 ± 98	42.0	3114	3068 ± 135	37.7
$Q^2 > 5000\,\mathrm{GeV}^2$	all	440	396 ± 24	8.3	all	322	336 ± 30	65.8	762	732 ± 38	26.2
$Q^2 > 10\,000\,\mathrm{GeV}^2$	all	66	60 ± 4.2	26.1	all	51	55 ± 6.4	66.8	117	115 ± 7.5	44.6
$Q^2 > 15\,000\,\mathrm{GeV}^2$	96	12	8.7 ± 0.7	17.0	96	12	4.7 ± 0.8	0.58	24	13.4 ± 1.0	0.75
	all	20	16.6 ± 1.2	24.0	all	22	14.8 ± 2.1	7.1	42	31.8 ± 4.1	8.7
$Q^2 > 20\,000\,\mathrm{GeV}^2$	96	5	2.8 ± 0.2	14.8	96	5	1.3 ± 0.3	1.4	10	4.1 ± 0.4	1.0
	all	8	5.7 ± 0.4	21.8	all	10	4.4 ± 0.7	2.0	18	10.1 ± 0.8	1.9
$Q^2 > 25\,000\,\mathrm{GeV}^2$	96	3	1.01 ± 0.09	8.2	96	3	0.51 ± 0.16	1.8	6	1.52 ± 0.18	0.53
	all	3	2.09 ± 0.13	34.8	all	6	1.58 ± 0.29	0.72	9	3.67 ± 0.32	1.4
$Q^2 > 35\,000\,\mathrm{GeV}^2$	96	2	0.15 ± 0.01	1.0	96	0	0.08 ± 0.04	100	2	0.23 ± 0.04	2.3
	all	2	0.29 ± 0.02	3.5							
$x > 0.55$ and	96	4	0.91 ± 0.08	1.4	96	1	0.45 ± 0.18	35.4	5	1.36 ± 0.20	1.4
$y > 0.25$	LP97	5	1.51 ± 0.13	2.0	LP97	1	0.75 ± 0.31	50.9	6	2.26 ± 0.33	3.1
$187.5\,\mathrm{GeV} < M$	96	2	1.75 ± 0.14	52.1	96	7	0.95 ± 0.18	0.10	9	2.70 ± 0.23	0.22
$< 212.5\,\mathrm{GeV}$	LP97	3	2.92 ± 0.24	55.6	all	8	3.01 ± 0.54	1.6	11	5.93 ± 0.59	4.4
and $y > 0.4$											

The probabilities $\mathcal{P}$ in Table 4.2 indicate the likelihood that the observed event numbers come from a statistical fluctuation of the SM expectation. $\mathcal{P}$ is at the level of a few percent at $Q^2 \gtrsim 20\,000\,\mathrm{GeV}^2$ and in the "signal regions" defined by ZEUS and H1 in [71, 72] (see the lower part of Table 4.2). All $\mathcal{P}$ values have increased significantly compared with the first analyses (ZN1, HN1), where values of $\mathcal{O}(1\%)$ and below were observed by both collaborations.

For $Q^2 \lesssim 15\,000\,\mathrm{GeV}^2$, excellent agreement between data and prediction is observed (the usual one-sigma confidence interval would correspond to $\mathcal{P}$ being in the range 15.9%–84.1%). No common tendency of observed event numbers to be below or above the expectation is visible in this range.

Although the $\mathcal{P}$ values are well-defined probabilities, it is not justified to take them as measures of the significance with which the data deviate from the SM prediction. The reason is that $\mathcal{P}$ depends strongly on the choice of the kinematic region for which it is determined, and is hence arbitrary to some extent.[3] Furthermore, no signal region is defined a priori (e.g. by a model predicting a specific LQ mass). Any deviation between the data and MC could therefore potentially be interpreted as a sign of new physics. Both ZEUS and H1 have estimated the probability that an excess of events with an associated value of $\mathcal{P}$ as low as those in Table 4.2 could occur *anywhere* in the kinematic region under investigation. They proceeded in the following steps:

- Choose a continuous parameter defining the signal region. In the case of ZEUS this was either x^*, determining the rectangle $x > x^*$, $y > 0.25$, or the lower bound on Q^2, $(Q^2)^*$. H1 applied a "sliding mass window technique" [64] with a mass parameter M^* defining a region $|M - M^*| < \Delta M/2$, $y > 0.4$. The analysis was repeated for different y cuts and values of ΔM.
- Calculate $\mathcal{P}$ as a function of x^*, $(Q^2)^*$ or M^* and record the minimum value, $\mathcal{P}_{\mathrm{min}}$. H1 scanned M^* in intervals of 1 GeV, determined the average $\overline{\mathcal{P}}$ over three consecutive steps at a time and recorded the minimum of $\overline{\mathcal{P}}$. The results from ZEUS (x^*, analysis ZN1) and H1 (analysis HN1) are shown in Fig. 4.7 and Fig. 4.8, respectively.
- Generate a large number of simulated event samples according to SM assumptions, each corresponding to the integrated luminosity of the experimental data ("Monte Carlo experiments", MCEs). Determine in which fraction of them $\mathcal{P}_{\mathrm{min}}$ is smaller than the value observed in the data.

ZEUS found that 7.2% (6.0%) of all MCEs yielded a minimum probability $\mathcal{P}_{\mathrm{min}}(x^*)$ ($\mathcal{P}_{\mathrm{min}}[(Q^2)^*]$) which was lower than the corresponding experimental result (analysis ZN1). In the M^* analysis of H1 with a mass window width of $\Delta M_e = 25\,\mathrm{GeV}$ a fraction of 0.9% of all MCEs was found in which $\mathcal{P}_{\mathrm{min}}(M^*)$ fell below the experimental value (analysis HN1). These fractions are interpreted as estimates of the probabilities that statistical fluctuations cause deviations from the SM expectation values somewhere in the kinematic

[3] A simple way to produce arbitrarily small values of $\mathcal{P}$ is to define the kinematic region to be a box around an observed event and letting its size go to zero.

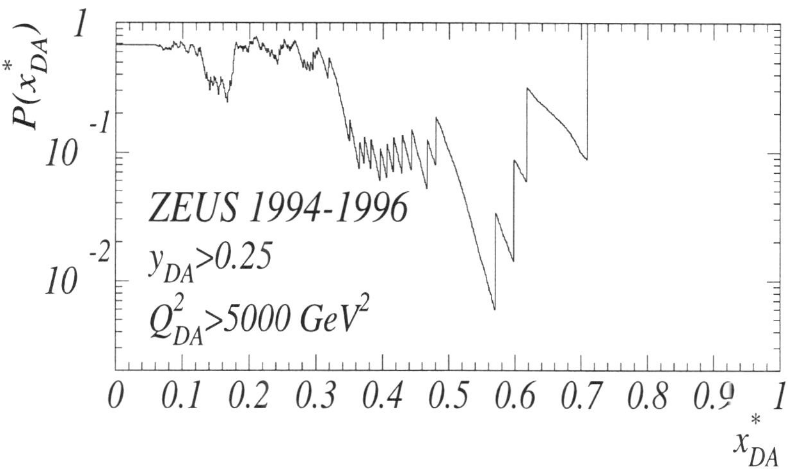

Fig. 4.7. Probability $\mathcal{P}$ calculated according to (4.1) for the kinematic region $x_{\mathrm{DA}} > x^*_{\mathrm{DA}}$, $y_{\mathrm{DA}} > 0.25$ and $Q^2_{\mathrm{DA}} > 5000\,\mathrm{GeV}^2$ in the ZEUS analysis ZN1. Note that $\mathcal{P}$ is unity by construction for very high x^*_{DA}, above which no events are observed. The plot has been reproduced from [71]

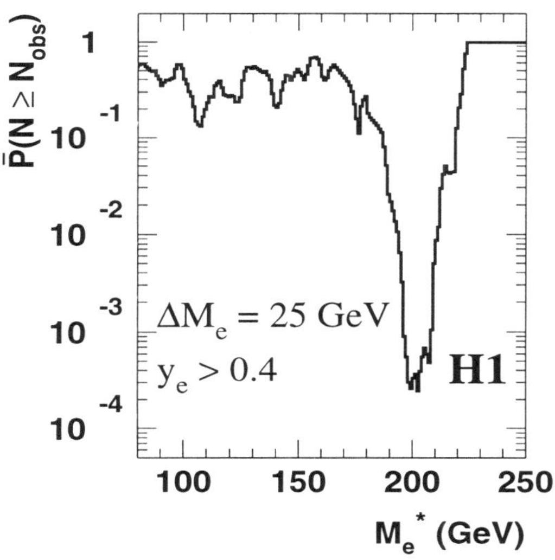

Fig. 4.8. Probability $\overline{\mathcal{P}}$ calculated according to (4.1) for the kinematic region $|M_e^* - M_e| < \Delta M_e/2$ and $y_e > 0.4$ in the H1 analysis HN1. Note that $\overline{\mathcal{P}}$ is unity by construction if no events are observed in the mass window. The plot has been reproduced from [72]

region under investigation which are at least as large as the observed ones. They should not, however, be taken as probabilities that "the SM is correct" since this statement would require us to specify the complete set of alternatives and assign probabilities to them. No update of these statistical analyses including the 1997 data is available. However, it seems safe to conclude from Table 4.2 that the new data have increased the probability that the observed excess of events at high x and Q^2 is due to a statistical fluctuation.

4.1.2 Differential Neutral-Current Cross Sections

The distributions of kinematic variables in the event samples selected by ZEUS and H1 allowed us to discuss the agreement of the data with the SM prediction in the previous section. However, the comparison of these data between ZEUS and H1,[4] as well as interpretations in terms of e.g. cross section predictions or QCD fits, requires one to apply the complete set of corrections (see Sect. 3.4.3) and to derive detector-independent quantities. In their more recent analyses, both experiments have therefore concentrated on determining cross sections, which are the canonical objects fulfilling these requirements and can directly be compared and also combined between both experiments.

For all results reported in the following sections, the cross sections have been derived using bin-to-bin corrections (see Sect. 3.4.3 for a discussion of unfolding methods) of the type

$$\sigma_{\text{meas}} = \frac{(N_{\text{obs}}/\mathcal{L})|_{\text{data}}}{(N_{\text{obs}}/\mathcal{L})|_{\text{MC}}}\,\sigma_{\text{MC}}\ , \tag{4.2}$$

where σ_{meas} is the cross section actually measured in a particular kinematic region. The double ratio in (4.2) is calculated from the experimental and simulated numbers of observed events in the relevant region, N_{obs}, and from the corresponding integrated luminosity values, $\mathcal{L}$, in the data and MC. The "true" cross section, σ_{MC}, is taken from the MC generator (HERACLES [148, 415] with DJANGO [416]) and includes all simulated radiative effects.[5] The measured cross section, σ_{meas}, can be extrapolated to kinematic regions outside the experimental acceptance by choosing σ_{MC} appropriately; ZEUS, for example, uses this method to determine cross sections covering the y range up to $y = 1$ although events with $y_e > 0.9$ are excluded from the event selection (see Table 3.8).

It is customary to define correction factors, C_{rad}, which transform σ_{meas} into so-called *radiatively corrected* or *Born* cross sections, σ_{Born}:

$$\sigma_{\text{Born}} = \sigma_{\text{meas}}\, C_{\text{rad}} = \sigma_{\text{meas}}\, \frac{\sigma_{\text{SM}}}{\sigma_{\text{MC}}}\ , \tag{4.3}$$

where σ_{SM} is calculated for the same kinematic region and using the same PDF set as σ_{MC}, but without electroweak radiative corrections; note that the cross section σ_{MC} cancels when inserting (4.2) into (4.3). The radiative corrections, C_{rad}, are typically in the range $0.85-1$ for NC differential cross sections. The resulting cross section, σ_{Born}, can directly be interpreted in terms of the LO formulae (2.4) or (2.12). It reflects the QCD effects and possibly residual electroweak radiative corrections included in σ_{SM}; in particular, σ_{Born} is sometimes referred to as the *LO (NLO) cross section* if LLA

[4] The ZEUS + H1 sums quoted in Table 4.2 are only given for providing a rough overview and are not used for the significance analysis in Sect. 4.1.1.

[5] H1 has verified σ_{MC} using the program HECTOR [441].

(NLLA) QCD corrections (see Sect. 2.1.2) are taken into account in the calculation of $\sigma_{\rm SM}$. The ZEUS and H1 cross sections presented in the following are radiatively corrected, using NLLA calculations of $\sigma_{\rm SM}$ and Q^2-independent electroweak parameters like $\alpha = \alpha(Q^2 = 0)$.

Equation (4.3) can also be applied to determine differential cross sections at given kinematic coordinates ξ, provided $\sigma_{\rm meas}$ is evaluated for a kinematic region including ξ and the appropriate differential SM cross section, evaluated at ξ, is used on the right-hand side. This procedure, which has been applied by ZEUS, assumes implicitly the validity of the SM cross section shape (i.e. the correctness of the PDFs) inside the kinematic region covered by $\sigma_{\rm meas}$. H1 follows a slightly different procedure by explicitly using the result of the H1 NLO QCD fit (see Sect. 4.3.1) to assign differential cross sections to kinematic coordinates ξ (*bin-centering corrections*).

Before discussing the ZEUS [73] and H1 [75] cross section results, we want to review the main contributions to the systematic uncertainties (all percentages given in the following quantify the corresponding modifications of the measured cross sections). Reflecting the different detector technologies and analysis methods, these sources of systematic error are distinctly different between both experiments.

For ZEUS, the main effects are due to:

- Variations of the reconstruction and simulation of the hadronic system, affecting the kinematic variables via $\gamma_{\rm had}$ (see Sect. 3.3.3). The dominant contribution ($< 3\%$ except at high Q^2 and x, where it increases to about 8%) comes from the differences between the ARIADNE [198] and LEPTO [196] implementations of fragmentation (see Sect. 2.1.7).
- The electron angle measurement, the electron identification efficiency and the variations of the electron isolation cut and of the fiducial cuts (Sect. 3.3.2), each of which causes an uncertainty of $\lesssim 2\%$.
- Variations of selection cuts and background simulation (mostly below 2%, rising to 3% at high y and to 8% at the highest Q^2).

Uncertainties related to the trigger, the calorimetric electron energy scale and the simulation of the vertex position distribution are estimated to be $\lesssim 1\%$. The 1.6% uncertainty on the luminosity measurement (see Sect. 3.1.4) affecting the overall normalization of the cross section results is not included in the systematic errors.

In the H1 analysis, systematic uncertainties are classified into uncorrelated and correlated ones, where the first type (indicated by "(u)" below) is assumed to be independent between bins, whereas the second type (marked with "(c)") modifies the cross sections by an overall factor. The main uncertainties arise from:

- the calorimetric electron energy scale, inducing $5\%(u)/3\%(c)$ error;
- a correlated uncertainty of $< 5\%(c)$ associated with the subtraction of photoproduction background;

- the hadronic energy scales (calorimeters, tracking), yielding a correlated contribution of $\lesssim 4\%(c)$;
- the procedure for noise subtraction in the LAC, leading to a correlated contribution which is largest at low y, up to $15\%(c)$ at $Q^2 \leq 2500\,\text{GeV}$ and $x = 0.65$;
- the electron identification efficiency, yielding an uncertainty of $2\%(u)$, rising to $4\%(u)$ at large y and low Q^2;
- the electron angle measurement, contributing $\lesssim 2\%(u)/1\%(c)$.

Uncertainties related to trigger efficiencies, track–cluster matching in the electron identification process and QED radiative corrections are estimated to be $\lesssim 1\%$ each. Like ZEUS, H1 does not include the luminosity uncertainty ($1.6\%(c)$) in the systematic errors. For the cross section presentation below, correlated and uncorrelated systematic uncertainties are combined in quadrature with the statistical errors to obtain "total errors".[6]

In the remainder of this section, we will present the ZEUS [73] and H1 [75] measurements of the NC DIS differential cross sections $\mathrm{d}\sigma/\mathrm{d}Q^2$, $\mathrm{d}\sigma/\mathrm{d}x$ and $\mathrm{d}\sigma/\mathrm{d}y$, and of the doubly differential cross sections $\mathrm{d}^2\sigma/(\mathrm{d}Q^2\,\mathrm{d}x)$, which are expressed in terms of the reduced cross section $\tilde{\sigma}^{e^+p\to e^+X}$ (2.46). These measurements are based on the complete 1994–97 e^+p data sets and cover the Q^2 range from a few hundred GeV^2 to several $10^4\,\text{GeV}^2$.

The results for $\mathrm{d}\sigma/\mathrm{d}Q^2$ from ZEUS and H1 are compared with the SM prediction in Figs. 4.9 and 4.10, respectively. The inner error bars indicate statistical uncertainties, the outer ones statistical and systematic uncertainties combined in quadrature (total errors). Whereas the measurement uncertainties are statistics-dominated for $Q^2 \gtrsim 5000\,\text{GeV}^2$, the systematic effects take over at low Q^2. Within the errors, excellent agreement with the SM cross section is observed for $Q^2 \lesssim 20\,000\,\text{GeV}^2$, over more than five orders of magnitude of $\mathrm{d}\sigma/\mathrm{d}Q^2$. A moderate excess of the measured over the predicted values is present in both experiments' data at higher Q^2. Note that the cross section definitions and the SM curves are slightly different in the two cases: ZEUS quotes $\mathrm{d}\sigma/\mathrm{d}Q^2$ for the full y range and uses SM predictions calculated with the PDF set CTEQ4D [185], whereas H1 restricts $\mathrm{d}\sigma/\mathrm{d}Q^2$ to the region $y < 0.9$ and takes the SM prediction from a QCD fit to fixed-target data and to H1 structure function and cross section data, including the high-Q^2 results from [75] (see Sect. 4.3.1).

A direct comparison between the $\mathrm{d}\sigma/\mathrm{d}Q^2$ results of ZEUS and H1 is presented in Fig. 4.11. The NC points shown there are identical to those in Figs. 4.9 and 4.10, and in particular differ in the y range which they represent. An LO calculation shows that the ratio $(\mathrm{d}\sigma/\mathrm{d}Q^2)/[\mathrm{d}\sigma/\mathrm{d}Q^2(y<0.9)]$ is about

[6] Since both the uncorrelated and the correlated uncertainties are themselves obtained by combining the respective contributions in quadrature, no distinction is actually made between both types in this procedure. However, correlated and uncorrelated systematic errors are treated differently in the χ^2 calculation for the H1 QCD fit (see Sect. 4.3.1).

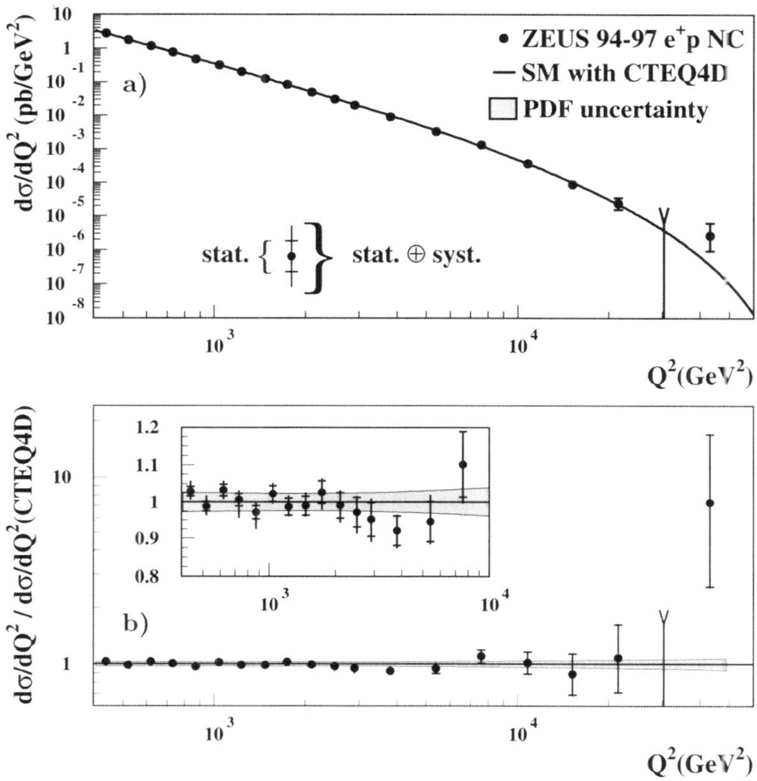

Fig. 4.9. ZEUS results from analysis ZN4: (**a**) NC cross section $\mathrm{d}\sigma/\mathrm{d}Q^2$ (*dots with error bars*), compared with the SM prediction (*line*); (**b**) ratio of measured to predicted cross sections (*dots*). The *error bars marked with a downward-pointing arrow* indicate the 1σ upper limits in a bin with zero observed events. The *shaded band* shows the PDF uncertainty. The plots have been reproduced from [73]

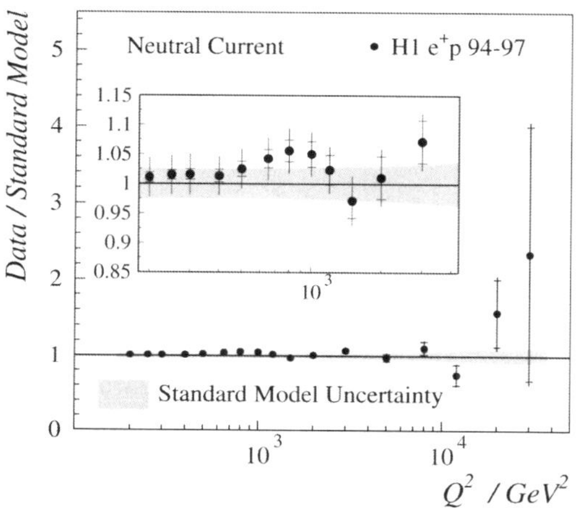

Fig. 4.10. H1 results for the ratio of measured to predicted NC cross sections $\mathrm{d}\sigma/\mathrm{d}Q^2$ in the region $y < 0.9$ (*dots with error bars*, analysis HN4). The PDF uncertainty (see Sect. 2.1.6) is indicated by the *shaded band*. The *inset* is a magnification of the region $Q^2 \lesssim 6000\,\mathrm{GeV}^2$. The plots have been reproduced from [75]

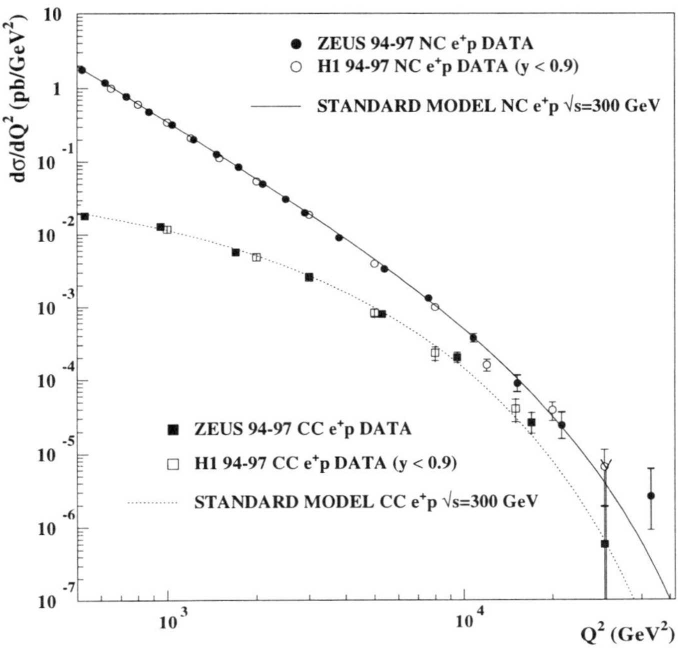

Fig. 4.11. Comparison of the ZEUS and H1 results for $\mathrm{d}\sigma/\mathrm{d}Q^2$, for NC (ZEUS: analysis ZN4 (*filled dots*), H1: analysis HN4 (*open dots*)) and for CC scattering (ZEUS: analysis ZC4 (*filled squares*), H1: analysis HC4 (*open squares*)). The SM prediction for NC (CC) scattering is indicated by the *full* (*dotted*) *line*. The ZEUS results and the SM curves represent the full y range; the H1 points are for $y < 0.9$. The offset of the H1 points induced by this difference corresponds roughly to half of their total uncertainties, except for CC at low Q^2, where it is smaller (see text)

1.02 at $Q^2 = 200\,\mathrm{GeV}$, rising approximately linearly with Q^2 to about 1.3 at $Q^2 = 30\,000\,\mathrm{GeV}^2$; in the full Q^2 range, the offsets of the H1 points with respect to the SM expectation in Fig. 4.11 are roughly given by half of their total uncertainties. Unaffected by this offset, very good agreement between the ZEUS and H1 data is observed, extending to $Q^2 > 20\,000\,\mathrm{GeV}^2$. Note that the Q^2 coverage has been significantly extended as compared with the pre-1996 status documented in Fig. 4.1. We will come back to Fig. 4.11 in the discussion of CC cross sections (Sect. 4.2.2) and of electroweak aspects (Sect. 4.3.2).

Figure 4.12 presents the ZEUS results on $\mathrm{d}\sigma/\mathrm{d}x$ above three different lower thresholds of Q^2 (corresponding to lower limits on y for fixed x). Again, good agreement is found between the ZEUS data and the SM expectation, demonstrating that the x shape of the PDFs determined from the fixed-target data describes well the ep data for Q^2 values up to and beyond $10\,000\,\mathrm{GeV}^2$, after a QCD evolution over two decades in Q^2. No obvious discrepancy between the data and SM prediction is present, not even in the two highest x

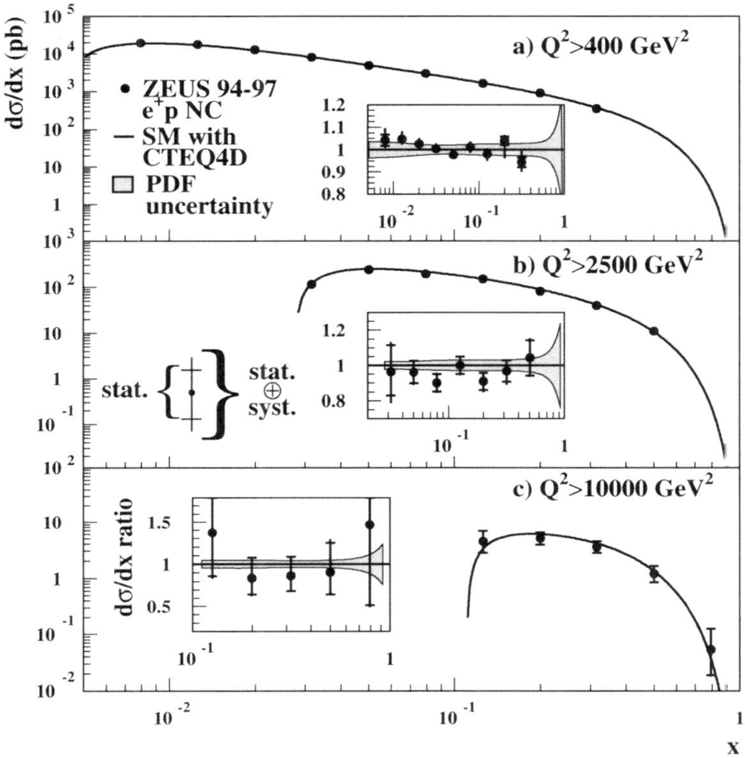

Fig. 4.12. ZEUS results from analysis ZN4: differential NC cross section $d\sigma/dx$ (*dots with error bars*) for (**a**) $Q^2 > 400\,\mathrm{GeV}^2$, (**b**) $Q^2 > 2500\,\mathrm{GeV}^2$ and (**c**) $Q^2 > 10\,000\,\mathrm{GeV}^2$. The *lines* represent the SM prediction. The *insets* show the ratio of measured to predicted cross sections, with the *shaded bands* indicating the PDF uncertainty (see Sect. 2.1.6). The plots have been reproduced from [73]

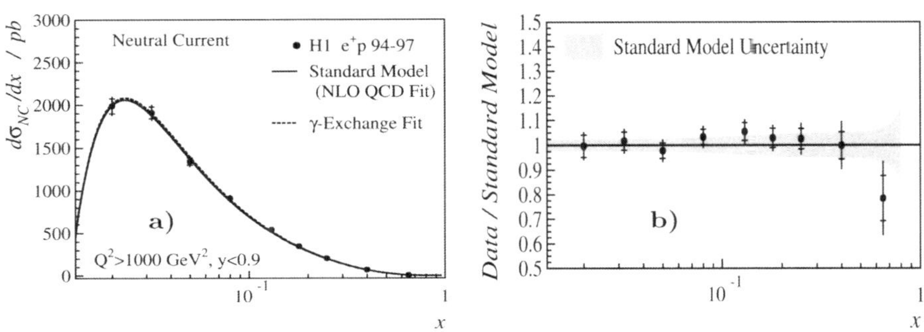

Fig. 4.13. H1 results from analysis HN4: (**a**) differential NC cross section $d\sigma/dx$ in the region $Q^2 > 1000\,\mathrm{GeV}^2$ and $y < 0.9$ (*dots with error bars*), compared with the SM prediction (*full line*) and a fit based on γ exchange only (*dashed line*, see Sect. 4.3.2); (**b**) ratio of measured to predicted cross sections (*dots*) and uncertainty of SM prediction (*shaded band*). The plots have been reproduced from [75]

bins for $Q^2 > 10\,000\,\mathrm{GeV}^2$. The same is true for the H1 results on $\mathrm{d}\sigma/\mathrm{d}x$ for $Q^2 > 1000\,\mathrm{GeV}^2$ presented in Fig. 4.13. Note that these statements are not incompatible with the evidence for event excesses discussed in Sect. 4.1.1 and summarized in Table 4.2, since none of the bins in Figs. 4.12 and 4.13 corresponds to one of the kinematic regions specified there; however, this observation demonstrates how sensitive statements on the HERA high-Q^2 excess are to the choice of phase space for which they are made. The difference between the solid and the dashed lines in Fig. 4.13 indicates the contribution of Z exchange to the cross section, which becomes sizable at Q^2 values exceeding the squares of the W and Z masses.

The ZEUS results on $\mathrm{d}\sigma/\mathrm{d}y$ are shown in Fig. 4.14, again for three different lower thresholds of Q^2 (corresponding to lower limits on x for fixed y). The good agreement between the ZEUS data and the SM expectation completes the suite of SM–data comparisons of singly differential cross sections

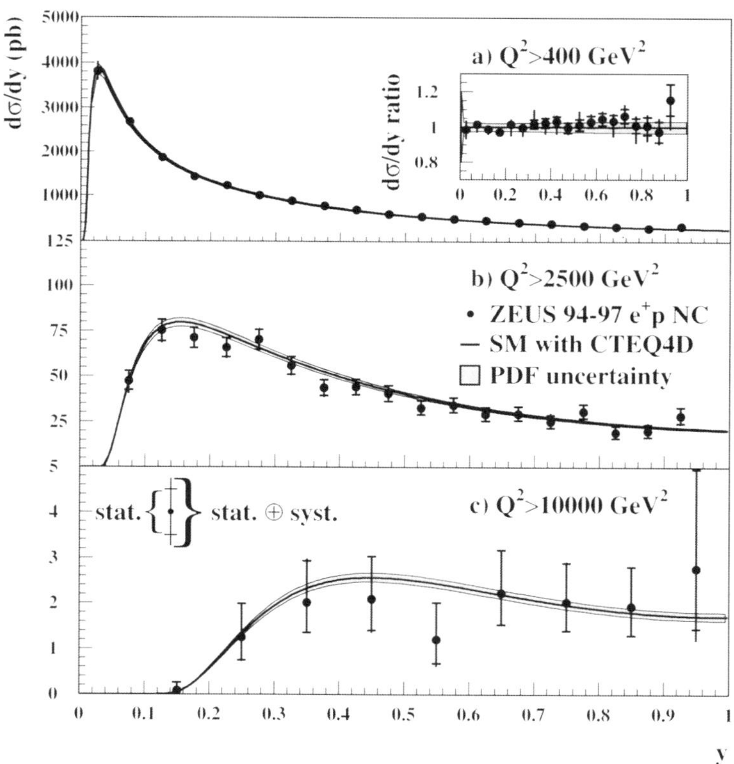

Fig. 4.14. ZEUS results from analysis ZN4: differential NC cross section $\mathrm{d}\sigma/\mathrm{d}y$ (*dots with error bars*) for (**a**) $Q^2 > 400\,\mathrm{GeV}^2$, (**b**) $Q^2 > 2500\,\mathrm{GeV}^2$ and (**c**) $Q^2 > 10\,000\,\mathrm{GeV}^2$. The *lines* represent the SM prediction, with the *shaded bands* indicating the PDF uncertainty (see Sect. 2.1.6). The *inset* shows the ratio of measured to predicted cross sections. The plots have been reproduced from [73]

confirming the SM. Since the y dependence of the NC DIS cross section (2.4) is dominated by the $[1 \pm (1-y)^2]$ terms which originate from the electroweak helicity structure of lepton–quark scattering, this result also corroborates the basic paradigm that DIS is mediated by electroweak interactions of point-like fermions. An analysis of the reduced NC cross section (2.46) in terms of $(1-y)^2$ has been performed by H1 and is discussed in Sect. 4.3.2.

We want to conclude this section by investigating whether the impressive agreement between measured and predicted singly differential cross sections described above also holds for doubly differential cross section. H1 [75] has reported results[7] on the reduced cross section, $\tilde{\sigma}^{e^+p\to e^+X}$, determined from $\mathrm{d}^2\sigma/(\mathrm{d}x\,\mathrm{d}Q^2)$ according to (2.46).

These measurements are particularly important since they (i) allow us to detect possible deviations from the SM predictions which might be hidden by the integration implicit in the singly differential cross sections, and (ii) constitute the central high-Q^2 input for QCD analyses which quantify the agreement of the data with the evolution equations and serve to disentangle the single PDFs. To leading order, the reduced NC cross section is given by $\mathcal{F}_2 - f(y)\mathcal{F}_3$, where $f(y) = [1-(1-y)^2]/[1+(1-y)^2] < 1$ induces the only dependence on the center-of-mass energy. Since the $\mathcal{F}_3$ term vanishes asymptotically for $Q^2 \to 0$, $\tilde{\sigma}^{e^+p\to e^+X}$ is appropriate for direct comparisons of HERA cross sections with fixed-target data recorded at lower energies.

The H1 results for $\tilde{\sigma}^{e^+p\to e^+X}$ are shown as functions of x for different fixed values of Q^2 in Fig. 4.15. The price to pay for the amount of detail contained in the 130 data points is an increase of the statistical errors, which dominate the overall uncertainties at Q^2 values of a few $100\,\mathrm{GeV}^2$ and above. The two sets of almost indistinguishable curves in Fig. 4.15 indicate the results of two different QCD fits performed by H1 (see Sect. 4.3.1), of which one (dashed lines) is restricted to fixed-target data and H1 structure function measurements at $Q^2 < 150\,\mathrm{GeV}^2$, while the other (solid lines) in addition uses the H1 high-Q^2 data (i.e. the data shown in Fig. 4.15 and the reduced CC cross sections discussed in Sect. 4.2.2). The fact that the results of both QCD fits are almost identical over the full kinematic range displayed in Fig. 4.15 confirms that the high-Q^2 data agree with the predictions derived from lower-energy data. In Fig. 4.16, $\tilde{\sigma}^{e^+p\to e^+X}$ is presented as a function of Q^2 for fixed values of x above 0.08. Also shown are fixed-target measurements at lower Q^2. The QCD-induced scaling violation, leading to a decrease of $\tilde{\sigma}^{e^+p\to e^+X}$ with rising Q^2 at high x and to an increase at low x, is clearly visible. The H1 QCD fit (solid line) describes both the fixed-target and the H1 data well; in this case, however, a slight excess is seen in the two highest Q^2 bins at $x = 0.4$. The turnover of the SM prediction at $Q^2 \gtrsim 10^4\,\mathrm{GeV}^2$ is caused by the destructive interference of Z and γ exchange amplitudes, i.e. by the negative $\mathcal{F}_3$ term. Correspondingly, the H1 fit assuming pure photon exchange (dashed line) deviates significantly from the SM prediction in this region.

[7] Corresponding results from ZEUS are not yet available.

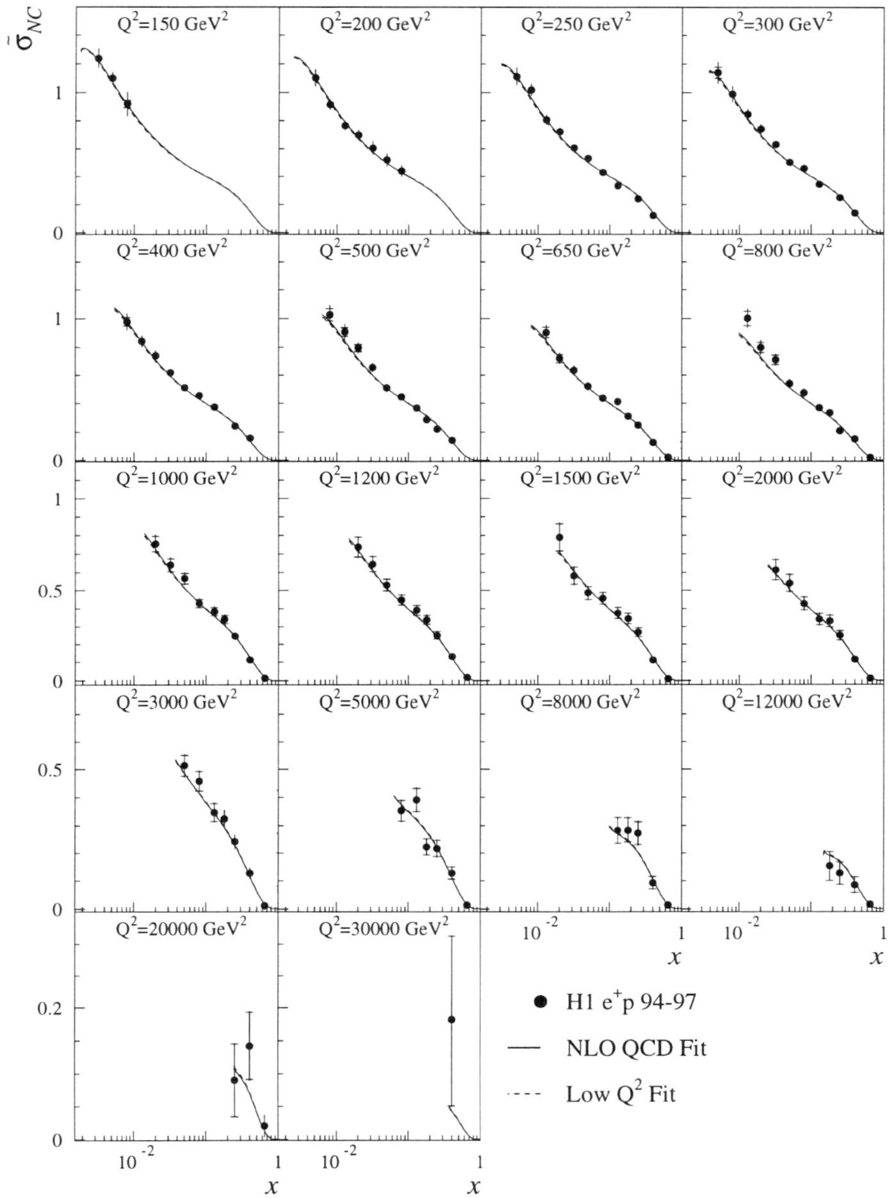

Fig. 4.15. H1 results from analysis HN4: reduced cross section $\tilde{\sigma}^{e^+p\to e^+X}$ as a function of x for different fixed values of Q^2 (*dots with error bars*), compared with results of two different QCD fits (see Sect. 4.3.1), one using fixed-target data and lower-Q^2 H1 measurements as input (*dashed lines*), the other employing in addition the cross section data shown in this plot and the H1 CC results discussed in Sect. 4.2.2 (*full lines*). The *inner* (*outer*) *error bars* indicate the statistical (total) uncertainties. The plot has been reproduced from [75]

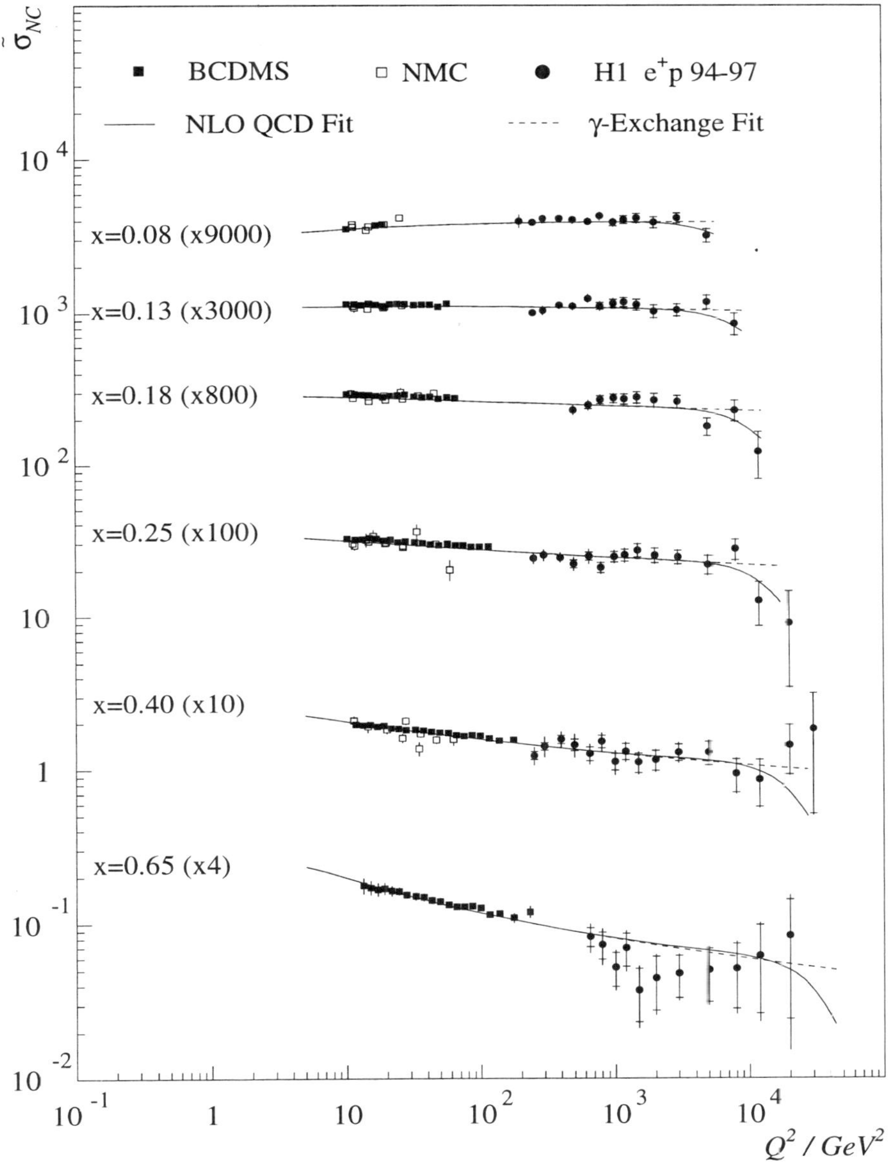

Fig. 4.16. Reduced cross section $\tilde{\sigma}^{e^+p\to e^+X}$ as a function of Q^2 for different fixed values of x. The *filled circles* are the same H1 data as in Fig. 4.15 (analysis HN4). The *filled and open squares* at lower Q^2 show the corresponding measurements of the fixed-target experiments as indicated in the plot header. The *solid curves* are the result of the H1 QCD fit (see Sect. 4.3.1); the *dashed lines* represent a fit including γ exchange only (see Sect. 4.3.2). For presentational clarity, all reduced cross sections have been multiplied with the factors indicated in parantheses. The plot has been reproduced from [75]

4.2 Charged-Current Results

The investigation of CC *ep* reactions at HERA suffers from several disadvantages as compared with the NC case: the total CC cross section is much smaller (see Sect. 2.1.6), restricting the statistical accuracy and impeding systematic studies; the kinematic reconstruction is less precise and subject to larger systematic uncertainties (Sect. 3.3.3); there are manifold potential background reactions which are difficult to separate from the signal sample (Sect. 3.4.2); and last but not least, the uncertainties of the SM cross section prediction are large (Sect. 2.1.6), reducing the significance of comparisons between experimental data and the SM expectation.

Despite these difficulties, a variety of CC results exists. ZEUS and H1 have reported CC event distributions [74, 84, 431, 433] and differential cross sections [74, 75], which will be presented in the following sections. The consistency of the CC results with the electroweak sector of the SM (which enters e.g. via the W mass in the propagator) will be discussed in Sect. 4.3.2.

4.2.1 Investigation of Charged-Current Event Distributions

For the same reasons as in Sect. 4.1, we start inspecting the CC data by having a look at the two-dimensional distributions of the high-Q^2 event samples in the kinematic plane. Both collaborations reconstruct the kinematic variables according to the Jacquet–Blondel method (see Sect. 3.3.3). The (x, Q^2) distribution of CC events has been reported by ZEUS in [74] (analysis ZC4) and is reproduced in Fig. 4.17. Note that the distribution is distinctly different from the NC case, where the event density is by far largest at lowest Q^2. No unexpected patterns are present in the event distribution of Fig. 4.17, nor in the corresponding H1 (M, y) distribution [84] (not shown). However, we must be aware of the fact that any fine-grain structures would be washed out by the limited resolution.

Figure 4.17 is well suited to discuss the impact of CC event topologies on the quality of cross section measurements in different kinematic regions. As can be understood from (3.23), the lower bound on $P_{\mathrm{T}} = (P_{\mathrm{t}})_{\mathrm{had}}$ applied in the event selection predominantly excludes events at low Q^2. The systematic uncertainty due to background contaminations is largest close to this bound. The pole-like instability of (3.23) at $y = 1$ causes a deterioration of the kinematic resolution at high y. At low y, on the other hand, event reconstruction is difficult since the hadronic system emerges at small angles in the forward direction (see Fig. 1.2) and tends to be at the edge of tracking acceptance and to overlap with the proton remnant jet, inducing measurement difficulties both for P_{T} and for the vertex coordinates. ZEUS uses the calorimeter timing information to determine the vertex position of such events, which are indicated as filled symbols in Fig. 4.17 (see Sects. 3.3.3 and 3.4.1).

In order to study the compatibility of experimental data and the SM predictions, the distributions of Q^2 are compared with the SM expectations

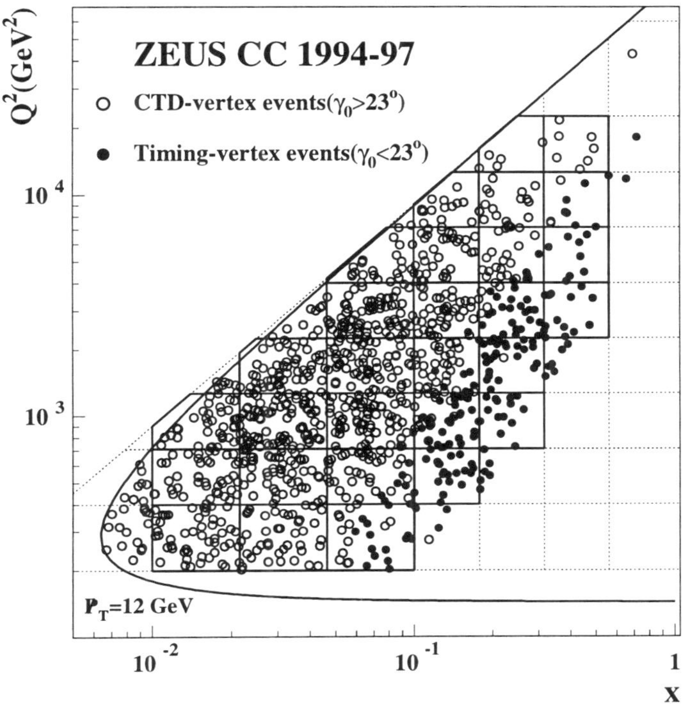

Fig. 4.17. Distribution of ZEUS e^+p CC events in the (x, Q^2) plane (analysis ZC4). Events reconstructed using the CTD vertex are indicated by *open symbols*, those for which the timing vertex was used by *filled dots* (see Sect. 3.4.1). The kinematic region selected by the cut $P_t > 12\,\text{GeV}$ is indicated by the *solid line*. The plot has been reproduced from [74]

in Figs. 4.18 and 4.19 for ZEUS and H1, respectively. The x distribution reported by ZEUS is shown in Fig. 4.20. Note that the ZEUS Q^2 and x distributions only contain a subset of the data included in Fig. 4.17. The data and SM expectation are in good agreement for $Q^2 \lesssim 10\,000\,\text{GeV}^2$ and for $x \lesssim 0.35$. In the high-end tails of the Q^2 and x distributions the data tend to exceed the predictions. However, the significance of these deviations is small owing to the large systematic errors of the predictions, which are dominated by the absolute calibration of the hadronic calorimeter energy scales and by the uncertainty of the d-quark distribution in the relevant kinematic region.

In intermediate reports prior to the final publications, ZEUS (analyses ZC2 and ZC3) and H1 (HC2 and HC3) have compared observed and expected CC event numbers above different thresholds of Q^2 and x. These numbers and the corresponding probabilities, $\mathcal{P}$, calculated according to (4.1), are summarized in Table 4.3. The most unlikely feature of the data is the presence of the few events with $Q^2 > 20\,000\,\text{GeV}^2$. However, all the probabilities are of $\mathcal{O}(10\%)$ or larger, indicating that the excess in the data is not particularly

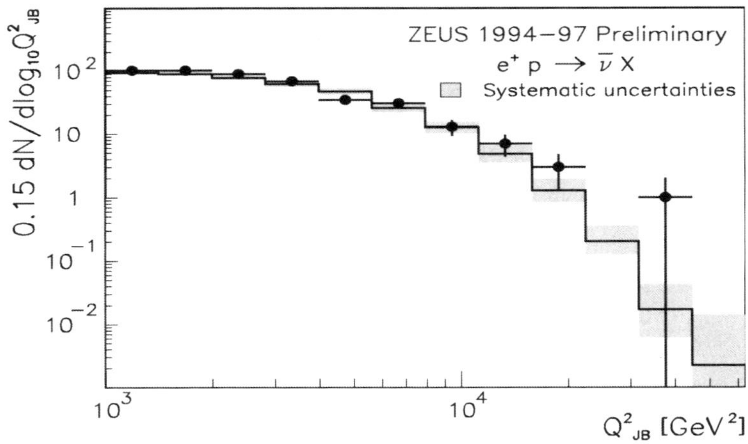

Fig. 4.18. Distribution of Q^2 in the ZEUS e^+p CC sample (analysis ZC2). Q^2 has been reconstructed using the Jacquet–Blondel method and corrected to account for reconstruction biases. The *dots with error bars* represent the data; the *histogram* is the SM expectation. The *shaded areas* indicate the systematic errors of the expectation due to PDF and energy scale uncertainties. The plot has been reproduced from [431]

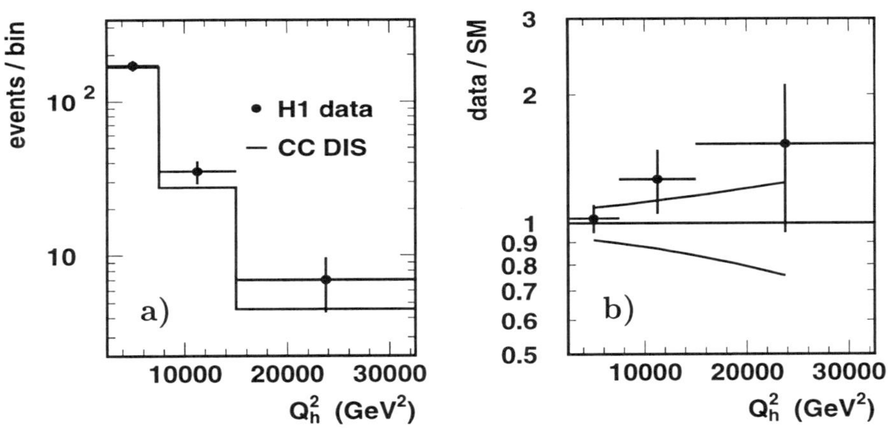

Fig. 4.19. Distribution of Q^2 in the high-Q^2 H1 e^+p CC sample with $y < 0.9$; Q^2 and y have been reconstructed using the Jacquet–Blondel method. In (**a**), the *dots with error bars* represent the data, the *histogram* is the SM expectation; (**b**) shows the ratio measured/expected (*dots*), with the systematic uncertainty indicated by the *band marked by the lines*. The plots have been reproduced from [84]

significant. Since the different correction methods applied by ZEUS and H1 in the kinematic reconstruction (see Sect. 3.4.3) induce large differences in the numbers of expected events per integrated luminosity (for example, this ratio differs by a factor of 1.6 for $Q^2 > 15\,000\,\mathrm{GeV}^2$), no attempt has been made to evaluate combined numbers as in Table 4.2.

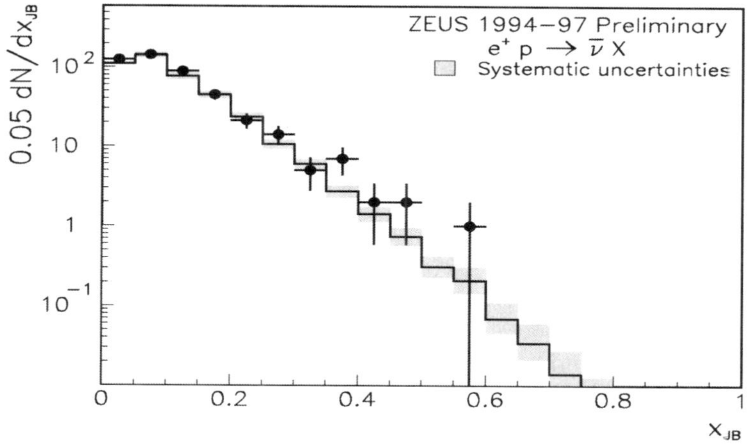

Fig. 4.20. Distribution of x in the ZEUS e^+p CC sample (analysis ZC2); x has been reconstructed using the Jacquet–Blondel method and corrected to account for reconstruction biases. The *dots with error bars* represent the data; the *histogram* is the SM expectation. The *shaded areas* indicate the systematic errors of the expectation due to PDF and energy scale uncertainties. The plot has been reproduced from [431]

Table 4.3. Numbers of observed (N_{obs}) and expected (N_{exp}) CC events above different thresholds of Q^2 and x. The errors δN_{exp} are dominated by the uncertainties of the absolute energy scale and of the d-quark distribution. For each kinematic region, only the latest available number is quoted. The probabilities $\mathcal{P}$ have been calculated according to (4.1)

Kinematic region	Experiment	Data set	N_{obs}	$N_{\text{exp}} \pm \delta N_{\text{exp}}$	$\mathcal{P}$ (%)
$Q^2 >$ 1000 GeV^2	ZEUS	ZC3	586	600 ±52	59.7
$Q^2 >$ 2500 GeV^2	H1	HC3	100	95.3 ±16.7	40.6
$Q^2 >$ 5000 GeV^2	H1	HC2	43	34.7 ±6.9	19.2
$Q^2 >$ 7500 GeV^2	H1	HC3	41	27.6 ±8.4	10.1
$Q^2 >$ 10 000 GeV^2	ZEUS	ZC3	22	$17.0\ ^{+5.7}_{-5.2}$	25.2
	H1	HC2	13	8.3 ±3.1	16.2
$Q^2 >$ 15 000 GeV^2	ZEUS	ZC3	8	$3.9\ ^{+1.9}_{-1.6}$	10.7
	H1	HC3	9	5.1 ±2.8	17.5
$Q^2 >$ 20 000 GeV^2	ZEUS	ZC3	3	$0.97^{+0.65}_{-0.47}$	11.6
	H1	HC2	4	1.21±0.64	5.9
$Q^2 >$ 30 000 GeV^2	ZEUS	ZC3	1	$0.06^{+0.08}_{-0.04}$	8.6
$x > 0.1$	ZEUS	ZC2	186	167 ±18	20.0
$x > 0.2$	ZEUS	ZC2	52	45.5 ±6.8	25.9
$x > 0.3$	ZEUS	ZC2	17	11.5 ±2.2	11.3
$x > 0.4$	ZEUS	ZC2	5	2.8 ±0.7	16.8
$x > 0.5$	ZEUS	ZC2	1	0.65±0.22	46.6

4.2.2 Differential Charged-Current Cross Sections

Both ZEUS [50, 51] and H1 [52–54] have investigated the cross sections of the CC reactions $e^-p \to \nu X$ and $e^+p \to \bar{\nu} X$ in the data taken in 1993–95. These measurements were severely statistics-limited, but they allowed the collaborations to determine the total cross sections with a relative precision of less than 20% (30%) for e^+p (e^-p) reactions, and to measure differential cross sections with respect to x, y and Q^2 ($\mathrm{d}\sigma/\mathrm{d}Q^2$ is shown in Fig. 4.1). The total cross sections and $\mathrm{d}\sigma/\mathrm{d}Q^2$ were also used to derive the W propagator mass and, for the first time, to establish experimentally a finite propagator mass in CC DIS [52]. All results were found to be in reasonable agreement with the SM predictions and with direct measurements of the W mass.

We will in the following review the updated CC cross section measurements obtained by ZEUS [74] and H1 [75] from analyses of the complete sets of e^+p data taken in 1994–97. The determination of the W propagator mass and other electroweak aspects will be discussed in Sect. 4.3.2. The CC cross sections discussed below have been determined following the same procedure described in Sect. 4.1.2 for the NC case; in particular, they are radiatively corrected, with $\mathcal{C}_{\mathrm{rad}}$ ranging from 0.97 to about 1.1 at high y and Q^2 [74, 75].

The dominant contributions to the systematic uncertainties of the CC cross section measurements by ZEUS are due to:

- the uncertainty of the hadronic calorimeter energy scale, inducing a cross section error which is mostly below 10% but increases to about 30% (25%) in the highest Q^2 (x) bins;
- variations of the selection cuts, the largest effect being an 8% uncertainty in the lowest Q^2 bins caused by the P_{T} cut;
- differences between the fragmentation simulation in ARIADNE [198] and LEPTO [196] (see Sect. 2.1.7), leading to uncertainties of up to 8% at the lowest and highest Q^2, and at low x and high y;
- the photoproduction background subtraction, which causes cross section errors of less than 1% except for $Q^2 < 400\,\mathrm{GeV}^2$, where up to 7% are reached.

The systematic effects related to the determination of the trigger efficiency, to the PDF choice for the MC simulation and for the calculation of σ_{SM} in (4.3), to the effect of $\mathcal{F}_{\mathrm{L}}$, and to the radiative corrections are estimated to be less than 3.5% each.

In the case of H1, the largest systematic uncertainties[8] are due to:

- the uncertainty of the hadronic calorimeter energy scale, contributing up to 8%(c) at the largest Q^2;
- a variation of the cut on $V_{\mathrm{ap}}/V_{\mathrm{p}}$ (see Sect. 3.5), which yields uncertainties of up to 12%(c) at low x and Q^2;

[8] As in Sect. 4.1.2, "(c)" and "(u)" denote correlated and uncorrelated systematic errors, respectively.

- the trigger efficiency, inducing a 3 – 8%(u) error;
- the vertex-finding efficiency, which is uncertain by 2%(u) for most of the kinematic region, but up to 5%(u) at low y;
- the photoproduction background subtraction, leading to cross section uncertainties of up to 4%(c) at low Q^2 and x;
- the subtraction of calorimeter noise (up to 4%(c)).

The contributions from radiative corrections and from background identification efficiencies are estimated to be $< 3\%(u)$ each.

The ZEUS measurement of $\mathrm{d}\sigma/\mathrm{d}Q^2$ is shown in Fig. 4.21, the corresponding H1 result for the ratio of measured and predicted cross sections is presented in Fig. 4.22. The SM predictions have been evaluated by ZEUS and H1 using the CTEQ4D PDF set and the H1 NLO QCD fit, respectively. Within their statistical and systematic uncertainties, the data agree with the SM predictions in the whole Q^2 range. The slight enhancement of the data over the expectation at $Q^2 \gtrsim 10\,000\,\mathrm{GeV}^2$ is not significant (confirming the same conclusion drawn from the event numbers in Table 4.3). Nevertheless, the result of the ZEUS QCD fit (dash-dotted line in Fig. 4.21b) indicates that the data suggest a larger $\mathrm{d}\sigma/\mathrm{d}Q^2$ in that region than inferred from the CTEQ4D PDF set. The data are compatible with an increase by as much as a factor of two for $Q^2 \gtrsim 20\,000\,\mathrm{GeV}^2$.

The ZEUS and H1 results for $\mathrm{d}\sigma/\mathrm{d}Q^2$ are directly compared with each other in Fig. 4.11. As in the NC case, the H1 points are evaluated for $y < 0.9$ and are therefore offset with respect to the SM curve. An LO calculation shows that $(\mathrm{d}\sigma/\mathrm{d}Q^2)/[\mathrm{d}\sigma/\mathrm{d}Q^2(y<0.9)]$ rises about linearly from 1.02 at $Q^2 = 300\,\mathrm{GeV}^2$ to 1.15 at $Q^2 = 15\,000\,\mathrm{GeV}^2$, implying that the offset is somewhat less than half of the total error of the H1 points, except at low Q^2, where it is smaller. Both sets of measurements are in good agreement also at $Q^2 > 10\,000\,\mathrm{GeV}$, where both collaborations measure a cross section exceeding the SM prediction by about the same amount.

In Fig. 4.23 the ZEUS results for $\mathrm{d}\sigma/\mathrm{d}x$ in the region $Q^2 > 200\,\mathrm{GeV}^2$ are shown; the corresponding H1 results for $Q^2 > 1000\,\mathrm{GeV}^2$ are reproduced in Fig. 4.24. The different behavior of the SM curve for $x \lesssim 0.03$ is induced by the different Q^2 restrictions. In the range $x \gtrsim 0.4$, the ZEUS measurements exhibit a moderately significant cross section excess over the CTEQ4D-based SM expectation. The ZEUS QCD fit describes the ZEUS data better than the CTEQ4D PDFs in that region, in spite of the fact that the ZEUS data have not been used as input to the fit (see Sect. 4.3.1). Figure 4.23b demonstrates that, like the ZEUS fit, also the recent MRST PDF set [141] and the approach by Bodek and Yang [126] predict a $\mathrm{d}\sigma/\mathrm{d}x$ which is larger than the CTEQ4D prediction in the high-x range; in all cases, this increase is caused by a modification of the d valence density. Obviously, no hypothesis beyond the SM is needed to explain the deviation of the high-x CC data from the CTEQ4D prediction. It is interesting to examine these results in the light of the suggested PDF modifications discussed in Sects. 2.1.6 and 2.2.1.

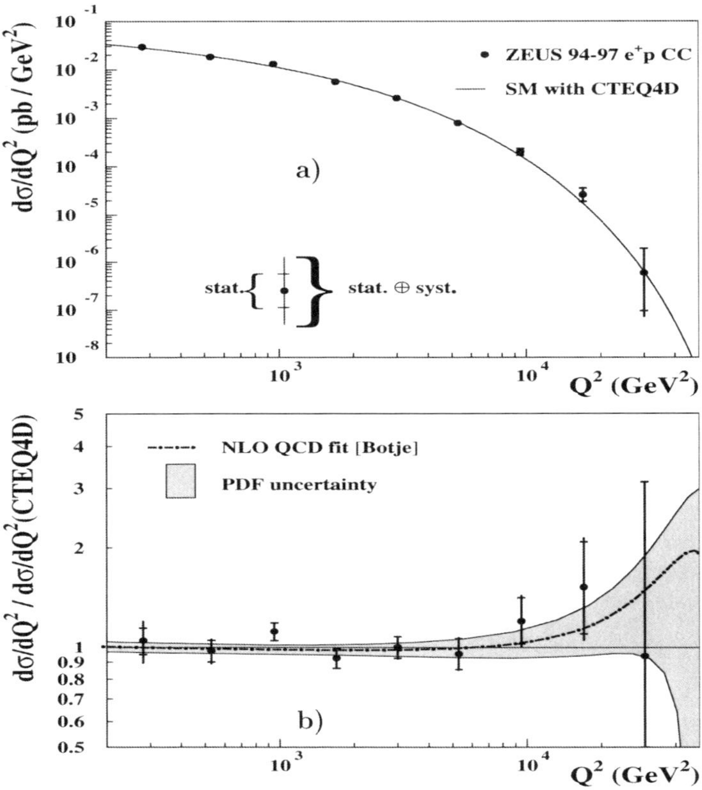

Fig. 4.21. ZEUS results from analysis ZC4: (**a**) CC cross section $\mathrm{d}\sigma/\mathrm{d}Q^2$ (*dots with error bars*), compared with the SM prediction (*line*); (**b**) ratio of measured to predicted cross sections (*dots*). The *dash-dotted line* is the result of the ZEUS QCD fit (see Sect. 4.3.1); the *shaded band* indicates the associated PDF uncertainty (see Sect. 2.1.6). The plots have been reproduced from [74]

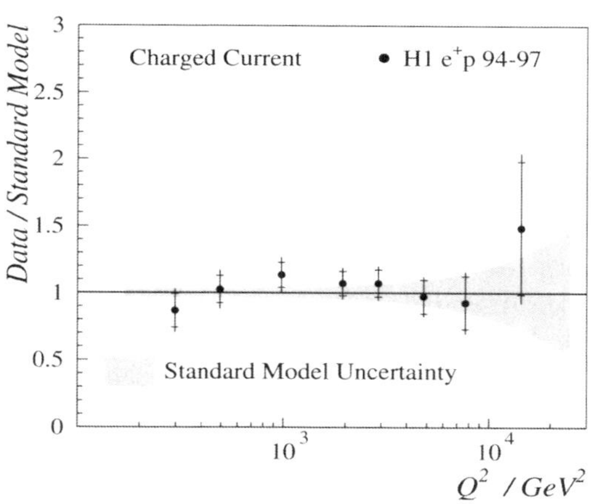

Fig. 4.22. H1 results for the ratio of measured to predicted CC cross sections $\mathrm{d}\sigma/\mathrm{d}Q^2$ in the region $y < 0.9$ (*dots with error bars*, analysis HC4). The SM prediction is taken from the H1 NLO QCD fit, the PDF uncertainty (see Sect. 2.1.6) is indicated by the *shaded band*. The plot has been reproduced from [75]

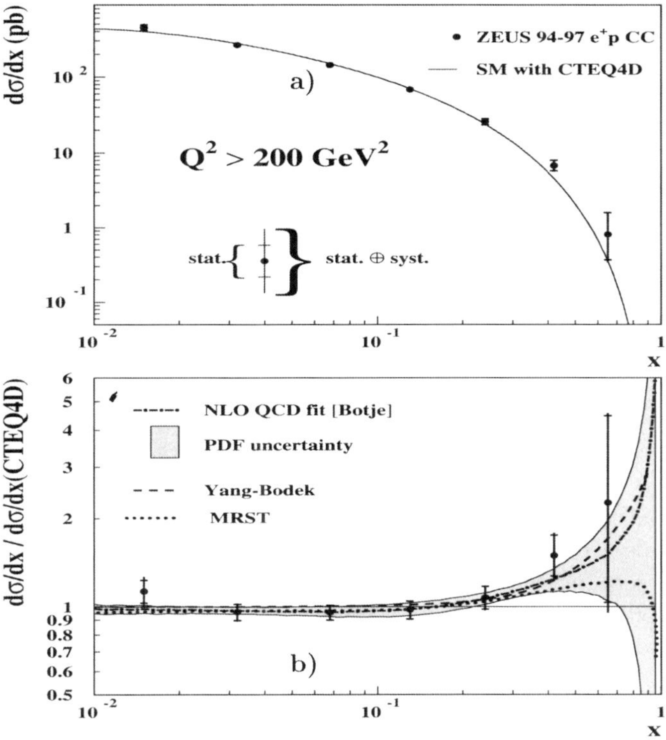

Fig. 4.23. ZEUS results from analysis ZC4: (**a**) CC cross section $\mathrm{d}\sigma/\mathrm{d}x$ in the region $Q^2 > 200\,\mathrm{GeV}^2$ (*dots with error bars*), compared with the the SM prediction (*line*); (**b**) ratio of measured to predicted cross sections (*dots*). The *dash-dotted line* is the result of the ZEUS QCD fit (see Sect. 4.3.1); the *shaded band* indicates the associated PDF uncertainty (see Sect. 2.1.6). The *dashed* and the *dotted lines* represent the cross sections calculated from the MRST [141] and the Bodek–Yang [126] PDF sets, respectively. The plots have been reproduced from [74]

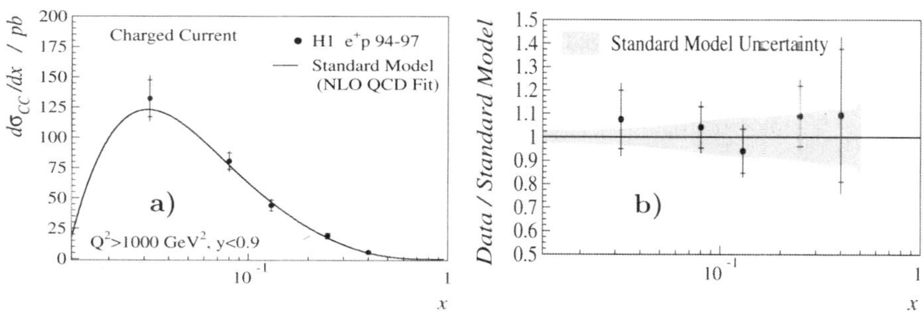

Fig. 4.24. H1 results from analysis HC4: (**a**) CC cross section $\mathrm{d}\sigma/\mathrm{d}x$ in the region $Q^2 > 1000\,\mathrm{GeV}^2$ and $y < 0.9$ (*dots with error bars*), compared with the SM prediction as calculated from the H1 NLO QCD fit (*full line*); (**b**) ratio of measured to predicted cross sections (*dots*) and uncertainty of SM prediction (*shaded band*). The plots have been reproduced from [75]

Both the assumption of a nonvanishing d/u ratio at $x \to 1$ and the presence of intrinsic charm would affect the CC cross section most prominently at large x. An increase of $\mathrm{d}\sigma/\mathrm{d}x$ by 20 – 100% at $x \approx 0.5$ may be expected (see Figs. 2.13 and 2.17), accompanied by a rise of $\mathrm{d}\sigma/\mathrm{d}Q^2$ by as much as 30% at $Q^2 \gtrsim 20\,000\,\mathrm{GeV}^2$. The relative precision of the points in Fig. 4.21 is $\mathcal{O}(10-15\%)$ for $Q^2 \lesssim 10\,000\,\mathrm{GeV}^2$ and rises to above 30% for larger Q^2; the total error of $\mathrm{d}\sigma/\mathrm{d}x$ rises from about 10% at low x to $\mathcal{O}(20-30\%)$ at $x \approx 0.4$. These uncertainties are mostly statistics-dominated. We conclude that the current precision of the differential CC cross section measurements is roughly of the size of the expected effects but by far not yet sufficient to distinguish different hypotheses.

Figure 4.25 presents the ZEUS result for the differential cross section $\mathrm{d}\sigma/\mathrm{d}y$ for $Q^2 > 200\,\mathrm{GeV}^2$. Again, good agreement with the SM prediction is observed, confirming that the quark and antiquark terms in (2.19) contribute in the right proportion. The comparison with the ZEUS fit shows that the modification of the d quark density discussed above affects $\mathrm{d}\sigma/\mathrm{d}y$ most strongly at small y, below the current experimental acceptance.

In the rest of this section we will discuss the doubly differential cross section $\mathrm{d}^2\sigma/(\mathrm{d}x\,\mathrm{d}Q^2)$, which will be presented in terms of the reduced cross section $\tilde{\sigma}^{e^+p\to\bar{\nu}_e X}$ defined in (2.47). The ZEUS and H1 results for $\tilde{\sigma}^{e^+p\to\bar{\nu}_e X}$ are shown in Figs. 4.26 and 4.27 as functions of Q^2 for fixed values of x and as functions of x for fixed values of Q^2, respectively. In Fig. 4.26 (Fig. 4.27), the H1 points from [75] have been shifted to the x (Q^2) values of the ZEUS points using the x and Q^2 dependence of the reduced cross sections resulting from the ZEUS NLO QCD fit. Within the experimental uncertainties, the

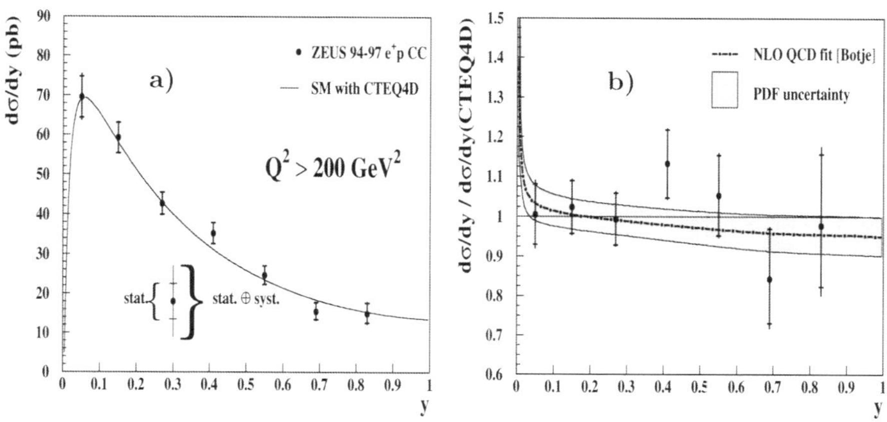

Fig. 4.25. ZEUS results from analysis ZC4: (**a**) CC cross section $\mathrm{d}\sigma/\mathrm{d}y$ (*dots with error bars*), compared with the SM prediction (*line*); (**b**) ratio of measured to predicted cross sections (*dots*). The *dash-dotted line* is the result of the ZEUS QCD fit (see Sect. 4.3.1); the *shaded band* indicates the associated PDF uncertainty (see Sect. 2.1.6). The plots have been reproduced from [74]

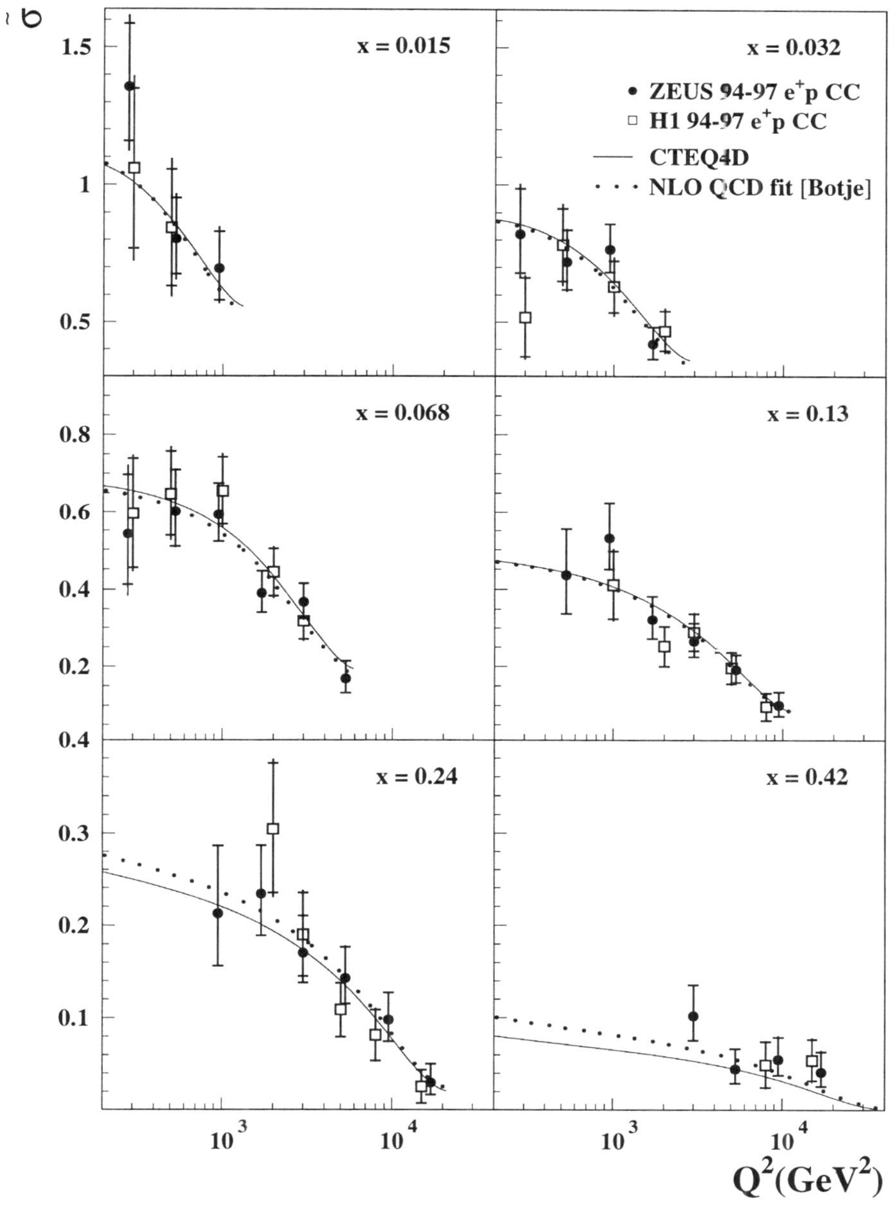

Fig. 4.26. Reduced CC cross sections $\tilde{\sigma}^{e^+p\to\bar{\nu}_eX}$ as functions of Q^2 at fixed values of x as measured by ZEUS (*filled dots*, analysis ZC4) and H1 (*open squares*, analysis HC4). The *full lines* indicate the SM prediction calculated from the CTEQ4D PDF set; the *dotted line* is the result of the ZEUS QCD fit (see Sect. 4 3.1). The plot is based on the x binning of the ZEUS analysis [74]; the H1 points from [75] have been shifted to the x values of the ZEUS points using the x dependence of the reduced cross sections resulting from the ZEUS NLO QCD fit

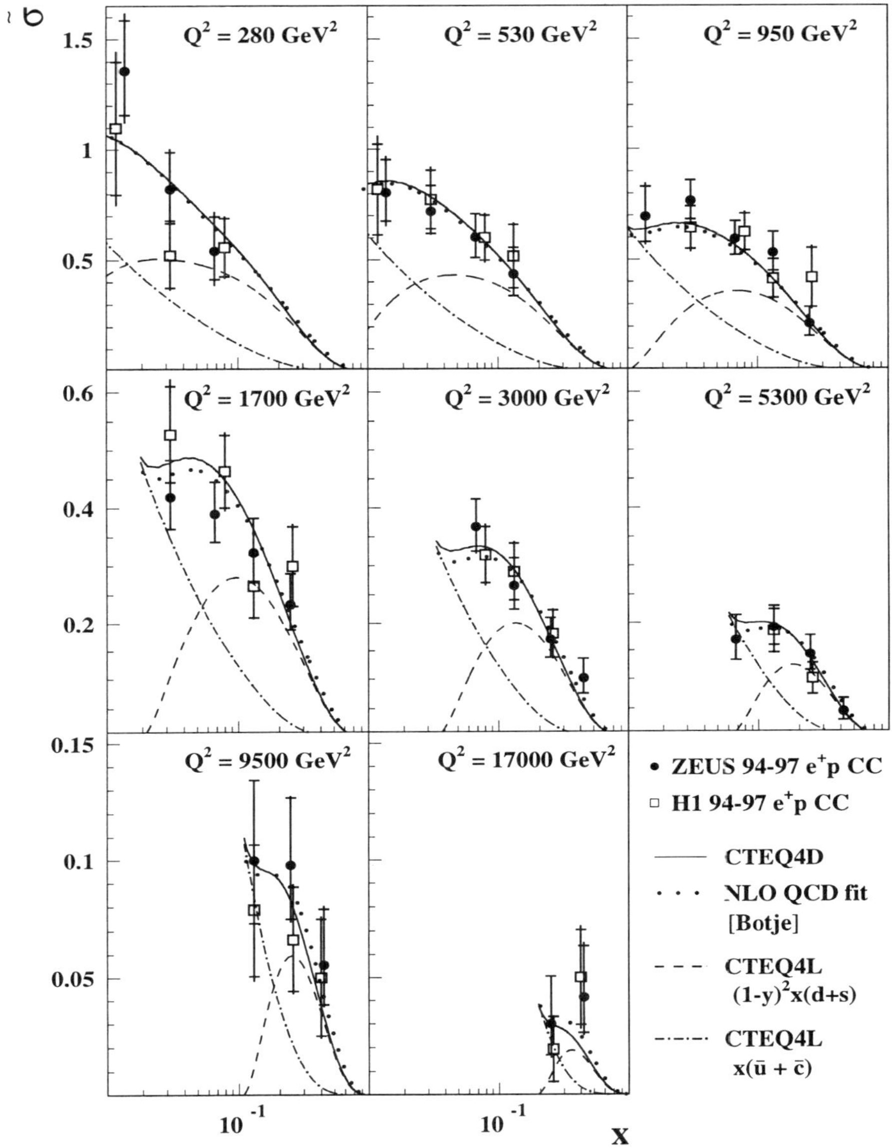

Fig. 4.27. Reduced CC cross sections $\tilde{\sigma}^{e^+p\to\bar{\nu}_e X}$ as functions of x at fixed values of Q^2 as measured by ZEUS (*filled dots*, analysis ZC4) and H1 (*open squares*, analysis HC4). The *full lines* indicate the SM prediction calculated from the CTEQ4D PDF set; the *dotted line* is the result of the ZEUS QCD fit (see Sect. 4.3.1). The *dashed* and *dash-dotted lines* represent the LO contributions to the reduced cross sections from quarks and antiquarks, respectively. The plot is based on the Q^2 binning of the ZEUS analysis [74]; the H1 points from [75] have been shifted to the Q^2 values of the ZEUS points using the Q^2 dependence of the reduced cross sections resulting from the ZEUS NLO QCD fit

data are consistent with each other and with the SM prediction. Owing to the large statistical errors, no preference for the CTEQ4D-based prediction (solid lines) or the ZEUS QCD fit (dotted lines) is obvious.

In the above discussion, we have related singly differential CC cross sections to the PDFs and to the helicity structure of the electroweak electron–(anti)quark interaction which induces the y terms in (2.12). The influence of these effects on CC reactions can most effectively be studied using the doubly differential measurements. The LO contributions to $\tilde{\sigma}^{e^+p\to\bar{\nu}_e X}$ coming from scattering on quarks and antiquarks are indicated in Fig. 4.27 by the dashed and dash-dotted lines, respectively. The following observations can be made. (i) At high x, the cross section is dominated by the quark contribution, which is to good approximation given by the d valence density in this region. Modifications of the SM prediction for high-x e^+p CC cross sections hence always involve changes of $d_{\mathrm{v}}(x,Q^2)$. (ii) The x values kinematically allowed for a given Q^2 range from $x_{\mathrm{min}} = Q^2/s$ to 1, corresponding to y decreasing from unity to zero. Owing to the $(1-y)^2$ factor in (2.19), the cross section at x_{min} is completely due the antiquark contribution. The shapes of the quark and antiquark curves in Fig. 4.27 are therefore markedly Q^2-dependent and do not directly reflect the PDFs. (iii) Assuming that the Q^2 dependence of the PDFs is governed by the QCD evolution equations, measurements of $\tilde{\sigma}^{e^+p\to\bar{\nu}_e X}$ can in principle be used to separate quark and antiquark PDFs. Furthermore, e^-p CC scattering is sensitive to u-type quark and d-type antiquark distributions, complementary to the e^+p case. Future high-statistics e^+p and e^-p data sets are therefore expected to be of central importance for precision PDF analyses in the high-x region.

4.3 QCD Analyses and Tests of the Electroweak Theory

All the results we have discussed in this chapter are interrelated by the basic quantities and principles which govern DIS in the SM: the PDFs and their Q^2 evolution in the QCD sector, and the $\mathrm{SU(2)_L \times U(1)_Y}$ gauge theory including its parameters in the electroweak domain. In the previous sections, we have commented on characteristic properties of the data and details of the analyses related to different aspects of both sectors of the SM. In general, very good agreement between the SM predictions and data was found. We now want to reinforce some of these statements and to describe interpretative analyses which use both NC and CC data or investigate derived quantities such as the W and Z propagator masses.

In Sect. 4.3.1, the QCD fits performed by M. Botje (ZEUS) and by the H1 collaboration will be discussed. These fits, to which we referred repeatedly in the previous sections, use different subsets of the HERA data and, in addition, fixed-target and other lower-energy data as input. They thus establish a test of the QCD concepts in a wide kinematic range and quantify their compatibility with the HERA data. These QCD fits explicitly use the DIS cross sections (2.4) and (2.12) and hence rely on the underlying lepton–parton cross sections given by electroweak theory. It is not the physics goal of the HERA experiments to compete with the impressive diversity and precision of the electroweak measurements at LEP, SLC and the Tevatron (see [149] for a recent compilation). However, the HERA data allow for the first-ever tests of some SM predictions at space-like virtualities of $\mathcal{O}(M^2_{W,Z})$, which extend considerably the experimentally verified region of validity of the SM. These tests will be summarized in Sect. 4.3.2.

4.3.1 The ZEUS and H1 QCD Fits

The ZEUS [187, 442] and H1 [75] collaborations have performed QCD fits to the HERA and to lower-energy data, with the objective to test the compatibility of their data with the pQCD predictions and to estimate the uncertainties of these predictions. The ZEUS fit includes fixed-target results and the H1 and ZEUS data taken prior to 1996. H1 produces two variants of the fit, one restricted to fixed-target data and H1 DIS measurements with $Q^2 < 150\,\mathrm{GeV}^2$ ("low-Q^2 fit"), the other using in addition the full set of new H1 NC and CC results ("H1 QCD fit"). The latter fit is also repeated neglecting Z exchange in NC reactions, i.e. with $P_Z = 0$ in (2.7) and (2.8) ("H1 γ-exchange fit").

Both collaborations apply fitting techniques with similar specifications (H1 characteristics are given in parentheses if different from ZEUS):

- *Input parameterizations.* Functions of the type (2.42)[9] are used for u_v, d_v, g and $q_s = \bar{u} + \bar{d} + \bar{s} + \bar{c}$ at $Q_0^2 = 4\,\mathrm{GeV}^2$; the difference $\Delta = \bar{d} - \bar{u}$ is supplied as a further input function with independent parameters determined in the fit (is taken from [141] in the case of H1). The strange-quark PDF is taken to be $s = \bar{s} = 0.2 q_s$ ($s = \bar{s} = 0.5\bar{u}$). The sum rules (2.43)–(2.45) are imposed to constrain the parameters at Q_0^2.
- *QCD evolution.* The pQCD evolution of the PDFs is performed using the QCDNUM package [443] by M. Botje. The heavy-quark densities are generated according to the VFNS (see Sect. 2.1.2), with the charm evolution setting in at Q_0^2 ($c = \bar{c} = 0.02 q_s$ at Q_0^2) and the b threshold at $Q^2 < 25\,\mathrm{GeV}^2$. The strong coupling constant is set to $\alpha_\mathrm{s}(M_Z^2) = 0.118$.
- *Input data.* The input data sets used for the fits are summarized in Table 4.4 (see Sect. 2.1.4 for a discussion of these data). Corrections are applied for nuclear and target-mass effects (see [75, 187] and references therein). For the ZEUS fit, the input data are constrained to $x > 0.001$ ($x > 0.01$ for $\mathcal{F}_3$ data), $Q^2 > 3\,\mathrm{GeV}^2$ ($Q^2 > 1\,\mathrm{GeV}^2$ for $\mathcal{F}_2^d/\mathcal{F}_2^p$ data) and $W^2 > 7\,\mathrm{GeV}^2$; higher-twist terms are parameterized as a fourth-order polynomial in x with the coefficients being determined in the fit. In the case of H1, the cuts $x < 0.7$, $Q^2 > 10\,\mathrm{GeV}^2$ and $W^2 > 20\,\mathrm{GeV}^2$ are applied and higher-twist terms (which are expected to be small owing to the W cut, see Sect. 2.1.2) are neglected.

[9] The functions $P(x; c_p, \ldots)$ in (2.42) are chosen differently by ZEUS and H1.

Table 4.4. Data samples used as input for the ZEUS and H1 QCD fits. The three left columns indicate the experiments, the reactions and the reference from which the results were taken (D.Y. means Drell–Yan lepton pair production). The numbers of data points used in the fits, N_{pnt}, and the respective contributions to χ^2 as calculated from the statistical and point-to-point uncorrelated systematic uncertainties are shown in the columns denoted χ^2/N_{pnt}; the numbers for the H1 low-Q^2 fit are given in parentheses. The scale factor applied to the data is reported only for those cases where it deviates from unity and is determined from the fit. The overall values of χ^2 and the numbers of degrees of freedom (i.e. N_{pnt} minus the numbers of fit parameters) are summarized in the last line

Experiment	Reaction	Ref.	ZEUS fit χ^2/N_{pnt}	ZEUS fit Scale	H1 fits χ^2/N_{pnt}	H1 fits Scale
ZEUS	e^+p NC	[3]	235/147		—	—
H1	e^+p NC	[9]	98/150		65(67)/ 77	1.02(1.01)
	e^+p NC	[75]	—	—	114 /130	0.98
	e^+p CC	[75]	—	—	18 / 25	0.98
SLAC	ep NC	[156]	61/ 56	0.99	—	—
	ed NC	[156]	52/ 57	0.98	—	—
BCDMS	μp NC	[152]	155/177	0.98	104(102)/139	0.96(0.97)
	μd NC	[153]	159/159	1.00	112(111)/133	0.98(0.98)
NMC	μp NC	[154]	250/233		143(143)/ 90	0.98(0.99)
	μd NC	[154]	190/233		126(125)/ 90	0.98(0.99)
	$\mathcal{F}_2^p/\mathcal{F}_2^d$	[190]	189/205		—	—
E665	μp NC	[155]	53/ 41	1.02	—	—
	μd NC	[155]	45/ 41	1.00	—	—
CCFR	$\overset{(-)}{\nu}$Fe CC	[166]	33/ 68	1.01	—	—
E866	pp, pd D.Y.	[174]	17/ 11		—	—
Sum			1537/1557		683(548)/671(516)	

- *Fit method.* The free parameters are determined from a χ^2 minimization taking into account the statistical and uncorrelated systematic uncertainties of the data, as well as normalization offsets constrained by the reported luminosity measurement errors. Systematic checks are performed by varying input assumptions such as $\alpha_s(M_Z^2)$, the renormalization and factorization scales, the heavy-quark thresholds, the strange-sea normalization, the nuclear corrections and the kinematic cuts applied to the input data.

All three fits yield a value of χ^2 per degree of freedom close to unity, demonstrating that the data coming from different reactions covering a large range of characteristic energy scales are all well described by global PDFs conforming to the pQCD evolution equations. The fact that the H1 fit also describes the new high-Q^2 data well[10] (overall $\chi^2/N_{\mathrm{pnt}} = 132/145$), and that it is almost indistinguishable from the low-Q^2 fit (see Fig. 4.15), indicates

[10] The slight discrepancy between the H1 fit and the data observed in the highest x bin of Fig. 4.16 may be explained by higher-twist terms, which are neglected in the H1 fits, but prove to be important also at $W^2 > 20\,\mathrm{GeV}^2$ for the high-x fixed-target data in the ZEUS fit [187].

that this statement is also true for the H1 high-Q^2 data. Moreover, we have seen in Sects. 4.1.2 and 4.2.2 that the ZEUS fit agrees perfectly with the ZEUS high-Q^2 data and partly describes them better than e.g. the CTEQ4D PDF set although these data have not been used for the fit.

Given the good agreement of the ZEUS and H1 QCD fits with the data, it is interesting to investigate whether the interpretation of the data in the light of the fit results yields new information on the PDFs, in particular at high x, where they are still controversial (see [192] and Sect. 2.1.6). Actually, the HERA e^+p NC and CC data sets provide complementary sensitivity: the u distribution produces the largest contribution to the NC cross section at high x, whereas the CC data are dominated by d in this region. H1 has exploited this fact to extract xu and xd as functions of Q^2 for $x = 0.25$ and $x = 0.4$. Figure 4.28 shows experimental data and fit results for the structure function terms $\Phi_{\rm NC}$ and $\Phi_{\rm CC}$, which are related to the corresponding reduced cross sections by

$$\begin{aligned} \Phi_{\rm NC} &= \left(1 + (1-y)^2\right) \tilde{\sigma}^{e^+p\to e^+X} \left(1 + \delta_{\rm rad}^{\rm NC}\right) , \\ \Phi_{\rm CC} &= \tilde{\sigma}^{e^+p\to\bar{\nu}_e X} \left(1 + \delta_{\rm rad}^{\rm CC}\right) . \end{aligned} \tag{4.4}$$

The $\delta_{\rm rad}$ terms symbolize weak-interaction-induced radiative corrections that are not included in the reduced cross sections reported by H1 and are negligible in the current context. Up to corrections of $\mathcal{O}(10\%)$ or below, $\Phi_{\rm NC}$ and $\Phi_{\rm CC}$ are given by their respective u and d contributions, $\Phi_{\rm NC}^u$ and $\Phi_{\rm CC}^d$ (dashed lines in Fig. 4.28). The results for xu and xd are finally obtained by multiplying $\Phi_{\rm NC}$ and $\Phi_{\rm CC}$ with $(xu/\Phi_{\rm NC})|_{\rm fit}$ and $(xd/\Phi_{\rm CC})|_{\rm fit}$, respectively, where the ratios are taken from the low-Q^2 fit in order to avoid circular arguments. Uncertainties of 50% are assigned to the other PDFs which are subtracted by this procedure.[11] In contrast to previous experiments, this determination of the high-x PDFs does not require one to combine data of reactions with different initial states, thus reducing considerably the systematic and theoretical uncertainties. The results, although statistics-limited, are consistent with the H1 low-Q^2 fit, which does not use CC data as input (see Table 4.4). More data will be needed to perform high-precision measurements at even higher x, which may serve to pin down the valence distributions in the limit $x \to 1$.

The ZEUS QCD fit demonstrates the variety of scenarios for the d/u ratio at high x which are still consistent with the fixed-target and early HERA data. Owing to the sparse experimental input and the large associated uncertainties, the x dependence of this ratio is determined to a larger extent by the functional form chosen for the valence PDFs than by the exact values of the corresponding parameters. Three typical cases are shown in Fig. 4.29. The ZEUS fit and CTEQ4M both use PDFs of the type (2.42), but ZEUS finds $b_{u_{\rm v}} > b_{d_{\rm v}}$ (corresponding to a diverging d/u at $x = 1$), whereas CTEQ4M

[11] Note, however, that the presence of intrinsic charm or other nonstandard contributions to the PDFs would not be taken correctly into account by this procedure.

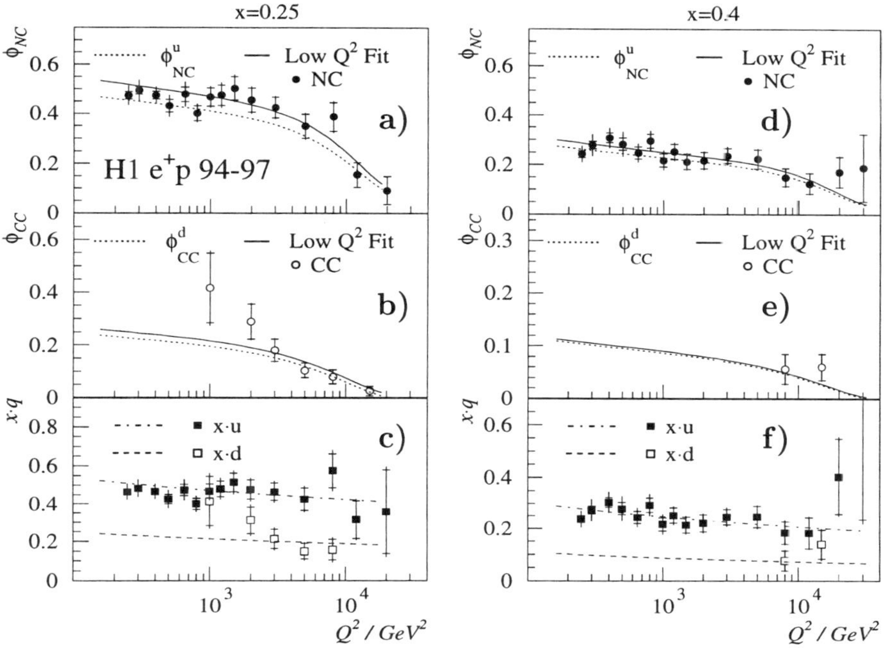

Fig. 4.28. Determination of the u and d quark densities from the reduced NC and CC cross sections measured by H1 (analyses HN4 and HC4), as functions of Q^2 for **(a–c)** $x = 0.25$ and **(d–f)** $x = 0.4$. The structure function sums Φ_{NC} and Φ_{CC} (see text) are shown in **(a,d)** and in **(b,e)**, respectively. The *filled dots* (*open squares*) represent the measured values of Φ_{NC} (Φ_{CC}), the *solid lines* are the results of the H1 low-Q^2 fit (see Sect. 4.3.1) and the *dashed lines* indicate the contributions of the dominant valence quark flavors. In **(c,f)** the extracted quark densities xu (*filled squares*) and xd (*open squares*) are compared with the prediction of the H1 low-Q^2 QCD fit (*dash-dotted* and *dashed lines* for xu and xd, respectively). The plots have been reproduced from [75]

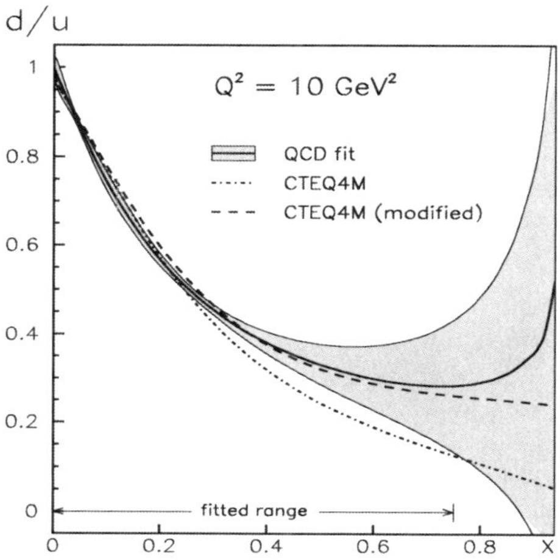

Fig. 4.29. NLO PDF ratio d/u as resulting from different analyses: the *solid line with error band* shows the result of the ZEUS QCD fit [187], the *dash-dotted line* represents the PDF set CTEQ4M [185] and the *dashed line* indicates the modification of CTEQ4M according to Bodek and Yang [126]. Note that the region $x \gtrsim 0.75$, where the error band diverges, is not covered by experimental data included in the QCD fits. The plot has been reproduced from [187]

(like all "standard" PDF parameterizations) has $b_{u_v} < b_{d_v}$ (i.e. $d/u \to 0$). The Bodek–Yang variant [126] (dashed line in Fig. 4.29) enforces $d/u = 0.2$ at $x = 1$ by construction. All three possibilities are compatible with each other within the error band of the ZEUS fit, indicating again that precise high-x HERA data are needed to distinguish between these scenarios. An integrated luminosity of $250\,\mathrm{pb}^{-1}$ per experiment has been estimated [192] to be sufficient to discriminate models with and without the nuclear corrections for deuteron targets, which induce a nonzero d/u ratio at $x \to 1$ (see Sect. 2.1.6).

4.3.2 Sensitivity to Electroweak Aspects

As discussed in Sect. 2.1.1, the electroweak sector of the SM enters *ep* DIS via the lepton–quark scattering amplitude. In principle, measurements of *ep* DIS cross sections carry three types of electroweak information: the weak coupling constants (2.3) and (2.11),[12] the propagator masses of the W and Z bosons, and the helicity structure of the lepton–quark interaction inducing the y terms multiplying the structure functions in (2.4) and (2.12). Sensitivity to the effects caused by the weak force, such as Z exchange and its interference with photon-induced amplitudes, is mainly provided by NC reactions at $Q^2 \gtrsim \mathcal{O}(M_{W,Z}^2)$ and by the CC channel. Since both have low cross sections, the corresponding investigations are still statistics-limited and currently result in consistency checks of the SM in a new kinematic domain, rather than contributing to the precision measurement of its parameters. A detailed account of electroweak measurements at HERA which will become possible with future high-statistics data can be found in [43].

The helicity structure has been discussed by H1 [75] using the results for Φ_{NC} and Φ_{CC} for x values between 0.08 and 0.25 (see Fig. 4.30). In this kinematic region, the QCD scaling violation is small, implying that the dependence of $\Phi_{\mathrm{NC,CC}}$ on $(1-y)^2$ at fixed x is predominantly given by the y terms in (2.19) and (2.4), i.e. by the scattering-angle distributions in the electron–quark center-of-mass system. In the CC case, the $\bar{u}$ and $\bar{c}$ contributions cause a flat "pedestal"; the d and s quarks yield the term rising linearly with $(1-y)^2$. Figure 4.30a does not only support the predicted y dependences (in agreement with the ZEUS CC results on $\mathrm{d}\sigma/\mathrm{d}y$, see Fig. 4.25) but also demonstrates that the CC cross section is increasingly dominated by the quark contributions as x rises. For NC, we have $\Phi_{\mathrm{NC}} \approx \left[1 + (1-y)^2\right] x\mathcal{F}_2$ at low Q^2 (i.e. at high $(1-y)^2$). At larger Q^2, this simple behavior is modified owing to the $\mathcal{F}_3$ terms induced by the Z exchange amplitude. As in the CC case, the data are in good agreement with the prediction, whereas they significantly deviate from the γ-exchange fit (dashed lines in Fig. 4.30b). Both the NC and the CC data sets hence confirm the SM helicity structure of lepton–quark interactions. In particular, no indication of scattering processes

[12] The electromagnetic coupling constant, α, is known to extremely high precision (see Table 2.1) and is not a subject of investigation in the ZEUS and H1 analyses.

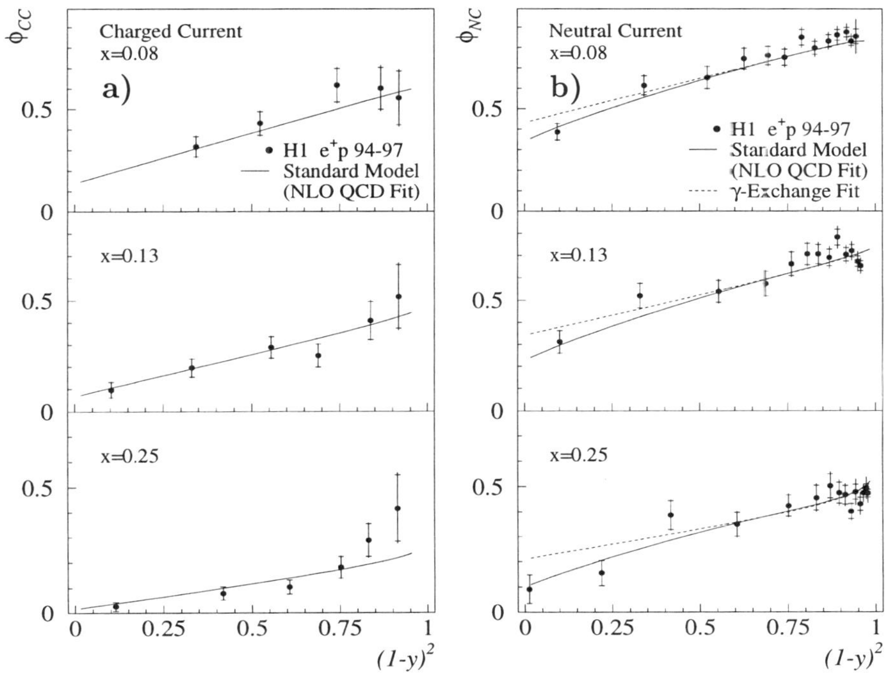

Fig. 4.30. H1 results (**a**) for the structure function sum Φ_{CC} and (**b**) for Φ_{NC} as functions of $(1-y)^2$ at fixed values of x (analyses HC4 and HN4). The *filled circles* show the experimental results, the *solid lines* indicate the SM prediction according to the H1 NLO QCD fit and the *dashed lines* are the result of the H1 QCD fit including γ-exchange only (see Sect. 4.3.1). The plots have been reproduced from [75]

involving scalar constituents of the proton or other than the SM currents is observed.

The difference between the photon-only scenario and the SM prediction in Fig. 4.30 is a consequence of the cross section contributions proportional to the term P_Z in (2.7) and (2.8) and therefore quantifies the sensitivity of the data to M_Z. ZEUS has investigated the agreement of the measured NC cross sections and the SM prediction for different values of the Z propagator mass, while keeping the couplings fixed. In Fig. 4.31a, the ratio of measured and predicted cross sections $\mathrm{d}\sigma/\mathrm{d}Q^2$ is shown for $M_Z = 40$ and $91\,\mathrm{GeV}$ and for $P_Z = 0$, i.e. $M_Z \to \infty$. The cross sections $\mathrm{d}\sigma/\mathrm{d}x$ obtained in the region $Q^2 > 10\,000\,\mathrm{GeV}^2$ for the same three M_Z values are compared with the ZEUS data in Fig. 4.31b. The corresponding result from H1 is shown in Fig. 4.32. Whereas $M_Z = 40\,\mathrm{GeV}$ is clearly disfavored by the data at $Q^2 \lesssim 10\,000\,\mathrm{GeV}^2$, the photon-only scenario, $P_Z = 0$, is incompatible with the ZEUS and the H1 data at $Q^2 > 10\,000\,\mathrm{GeV}^2$. ZEUS quotes a χ^2 probability of $p_{\chi^2} < 1.4\%$ for $P_Z = 0$, whereas $p_{\chi^2} = 41\%$ is obtained for the SM configuration. The

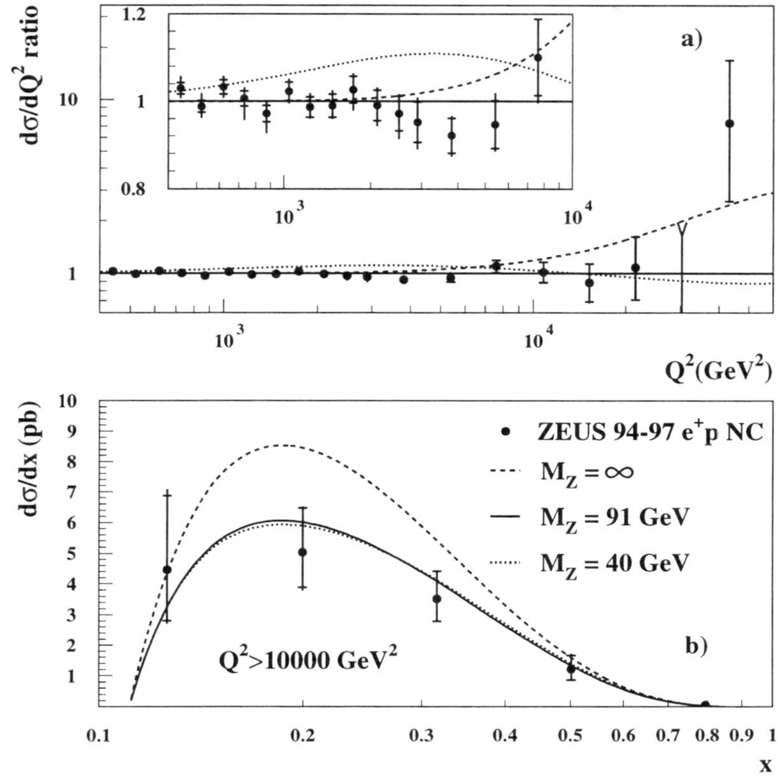

Fig. 4.31. Sensitivity of the ZEUS high-Q^2 NC data to the Z propagator: (**a**) ratio of measured to predicted differential cross sections $\mathrm{d}\sigma/\mathrm{d}Q^2$ and (**b**) differential NC cross section $\mathrm{d}\sigma/\mathrm{d}x$ for $Q^2 > 10\,000\,\mathrm{GeV}^2$. The *filled circles* denote the experimental results and the *dashed* and *dotted lines* indicate the expectations for the hypothetical cases $M_Z \to \infty$ and $M_Z = 40\,\mathrm{GeV}$, respectively. The plots have been reproduced from [73]

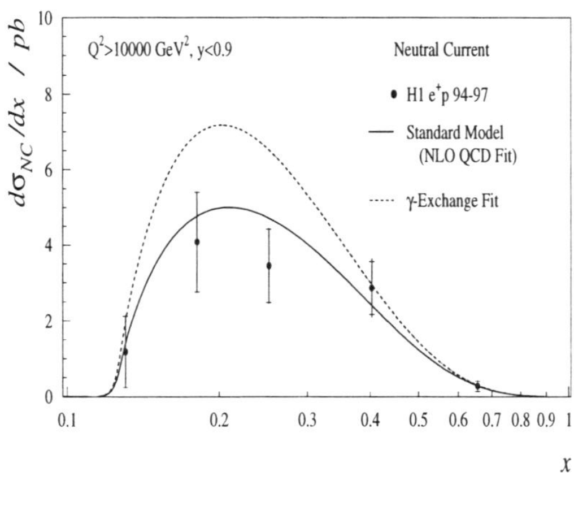

Fig. 4.32. H1 results for the differential NC cross section $\mathrm{d}\sigma/\mathrm{d}x$ in the region $Q^2 > 10\,000\,\mathrm{GeV}^2$ and $y < 0.9$. The *filled circles* show the experimental results, the *solid line* is the result of the H1 NLO QCD fit and the *dashed line* indicates the H1 QCD fit including γ-exchange only (see Sect. 4.3.2). The plot has been reproduced from [75]

HERA high-Q^2 data thus for the first time allow the direct observation of the Z contributions to unpolarized NC DIS cross sections.[13]

Finally, we turn to the CC cross sections, which depend quadratically on the W propagator and therefore allow a determination of M_W with much higher accuracy than M_Z in the NC case. Already in 1994, H1 established a finite W propagator mass [52], for the first time in the history of CC DIS (see [445] and references therein). The precision of the M_W determinations from HERA data improved continuously in the following years [50, 51, 53] and is now defined by the CC data discussed in Sect. 4.2.2. The sensitivity of these cross section measurements to M_W is demonstrated in Fig. 4.33. From χ^2 fits of the SM prediction to $\mathrm{d}\sigma/\mathrm{d}Q^2$ (ZEUS [74]) or $\mathrm{d}^2\sigma/(\mathrm{d}x\,\mathrm{d}Q^2)$ (H1 [75]), with M_W in (2.19) treated as a free parameter and G_F fixed at the PDG value (see Table 2.1), the W propagator mass is inferred to be

$$M_W = 81.4^{+2.7}_{-2.6}\,(\text{stat.}) \pm 2.0\,(\text{syst.})\,^{+3.3}_{-3.0}\,(\text{PDF})\ \text{GeV}\ \ (\text{ZEUS})\,, \qquad (4.5)$$

$$M_W = 80.9 \pm 3.3\,(\text{stat.}) \pm 1.7\,(\text{syst.}) \pm 3.7\,(\text{theo.})\ \text{GeV}\ \ (\text{H1})\,, \qquad (4.6)$$

where the "PDF" and "theo." errors reflect variations of the assumptions used for the underlying QCD fits and uncertainties of the input data for these fits. The precision of the HERA fit results for M_W is inferior to the direct measurements at LEP and the Tevatron, but their good agreement with the current world average of $M_W = (80.371 \pm 0.03)$ GeV (see Table 2.1) confirms the validity of the SM paradigms in the kinematic region of space-like W's. Also, since M_W enters (2.19) only in the combination M_W^2/Q^2, and Q^2, in turn, is directly affected by the hadronic calorimeter energy scale (see Table 3.7), these results constitute sensitive tests of the calorimeter calibration procedures in both experiments.

ZEUS has extended the M_W fit by treating G_F as a second free parameter, varied independently of M_W. Whereas M_W is mainly sensitive to the shape of $\mathrm{d}\sigma/\mathrm{d}Q^2$, G_F is determined by the total CC cross section (and the potential precision of its measurement is therefore limited by the luminosity uncertainty). The central result of the two-dimensional fit and the 70% C.L. χ^2 contour are shown in Fig. 4.34, together with the current world-average PDG values [108] of M_W and G_F (dotted lines). Very good agreement with these values is observed. In particular, allowing G_F to vary only marginally shifts the M_W value obtained in the one-dimensional fit (filled circle with horizontal error bar in Fig. 4.34). Note that this is actually the only significant test of the SM prediction for the size of weak couplings (see (2.18)) which is possible with the current HERA data; the weak contribution to the NC cross section is currently at the edge of measurability (see discussion above), implying that the errors on the weak NC couplings (2.3) in a corresponding analysis would be of $\mathcal{O}(100\%)$.

[13] The first experimental evidence for the presence of the Z propagator in NC DIS was obtained from the observation of parity violation in the scattering of polarized electrons on deuterons at $Q^2 = \mathcal{O}(1\,\mathrm{GeV}^2)$ [444].

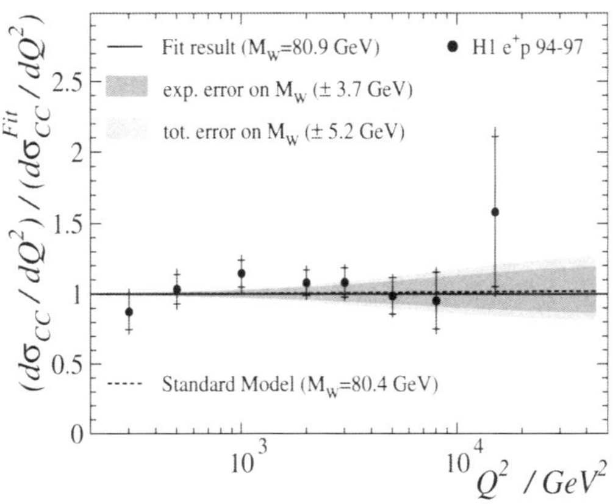

Fig. 4.33. H1 results for the W propagator mass (analysis HC4). The *filled circles* show the ratio of the measured differential CC cross section $\mathrm{d}\sigma/\mathrm{d}Q^2$ to the result of the propagator mass fit (see text). The *dark* (*light*) *shaded bands* indicate the variations corresponding to the experimental (total) mass errors. The plot has been reproduced from [75]

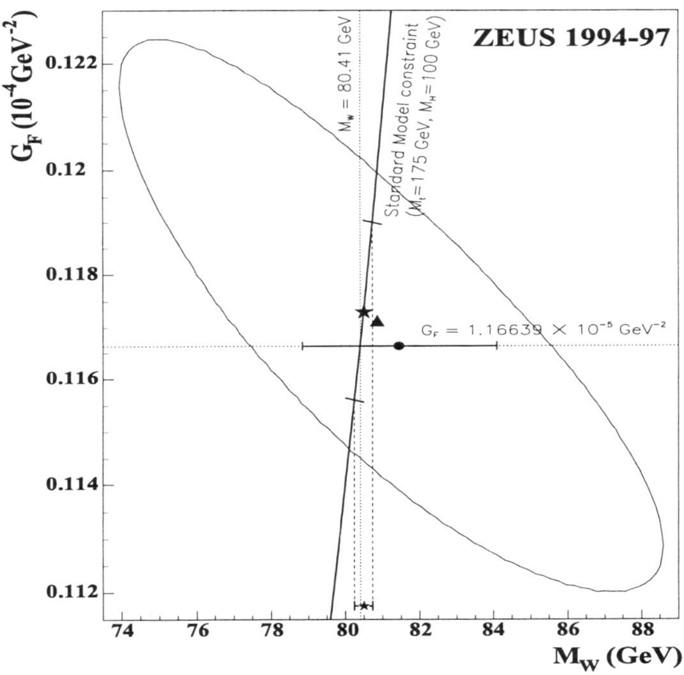

Fig. 4.34. Results of the ZEUS fits to determine the Fermi constant, G_F, and the W propagator mass, M_W, from the differential CC cross section $\mathrm{d}\sigma/\mathrm{d}Q^2$. The χ^2 minimum of the two-parameter fit is shown by the *filled triangle*; the corresponding 70% C.L. region is given by the *ellipse*. The *full line* is the SM constraint (4.7); the *large star* and the *solid bars* indicate the χ^2 minimum and the 1σ confidence interval along this line. The *small star with error bar* symbolizes the result on M_W obtained by imposing the SM constraint (see text). The *filled circle with error bar* shows the result of the M_W determination in a one-parameter fit with G_F fixed at the PDG value (*horizontal dotted line*). The *vertical dotted line* indicates the world average of direct W mass measurements [108]. All errors and confidence regions reflect statistical uncertainties only. The plot has been reproduced from [74]

Even though M_W and G_F have been treated as independent parameters in the ZEUS fit, they are actually related to each other by the SM formula

$$G_{\mathrm{F}} = \frac{\pi\alpha}{\sqrt{2}} \frac{M_Z^2}{M_W^2(M_Z^2 - M_W^2)} \frac{1}{1-\Delta r} , \tag{4.7}$$

which in addition depends on M_Z, on the fine-structure constant, α, and on the radiative correction term Δr. In the tree-level approximation, $\Delta r = 0$, and (4.7) can be directly derived from (2.18) and the relation $\cos\theta_{\mathrm{W}} = M_W/M_Z$. The dominant leading-order contributions to Δr are given by $\Delta r_0 = 1 - \alpha(0)/\alpha_{\overline{\mathrm{MS}}}(M_Z^2) = 0.0664$ and by a term quadratic in the top mass, $0.00952\,(m_t/174.3\,\mathrm{GeV})^2$ [109]. The constraint imposed by (4.7) is shown in Fig. 4.34, together with the 1σ confidence region along the resulting contour. Usually, (4.7) is applied to infer M_W from α, G_F and M_Z, i.e. from the three most precisely measured quantities which fix the electroweak SM couplings and gauge boson masses. ZEUS, in contrast, investigated the precision with which M_W can be determined from the ZEUS data if the SM and the world-average values of M_Z and α are assumed to be correct, but other measurements of G_F are ignored. M_W is then obtained by projecting the SM-constrained confidence interval in Fig. 4.34 onto the M_W axis, resulting in

$$M_W = 80.5^{+0.24}_{-0.25}\,(\mathrm{stat.})\,^{+0.13}_{-0.16}\,(\mathrm{syst.})\,\pm 0.31\,(\mathrm{PDF})^{+0.03}_{-0.06}\,(\mathrm{masses})\ \mathrm{GeV}\,. \tag{4.8}$$

The error labeled "masses" is due to variations of M_Z and m_t within their experimental uncertainties and of the Higgs mass in the range 100 – 220 GeV. Note that the SM constraint forces the resulting M_W to lie close to the PDG value, even if the error ellipse in Fig. 4.34 were significantly displaced from the SM point. The M_W value in (4.8) should therefore not be interpreted as a measurement; rather, its agreement with the world average is a successful check of the consistency of the ZEUS CC data with the SM prediction.

4.4 Searches for New Physics

In Sect. 2.2.2, we have discussed different hypothesized scenarios which could modify the DIS cross section at high x and Q^2. In the presence of an excess of events in this kinematic region which seems to be consistent with clustering around a certain value of x, it is of course highly desirable to test these scenarios against the data and to exploit all experimental signatures which might help to differentiate between hypotheses beyond the SM, scenarios involving modifications of the PDFs and the "simple" explanation assigning the excess events to a statistical fluctuation.

We will in the following review the HERA results of searches for two prominent scenarios which might modify the high-Q^2 DIS cross sections and have been proposed by many authors as possible explanations of the

HERA excess events: the resonant production of LQs or squarks (Sect. 4.4.1), and contact interactions as the effective "low-energy" phenomenology of new physics happening at mass scales far beyond the HERA center-of-mass energy (Sect. 4.4.2). With the exception of alternative, non-LQ decay channels of R_P-violating squarks, we will not consider search analyses in which all or most high-Q^2 DIS reactions, specifically those in the excess region, are excluded by the respective event selections. This applies in particular to the searches for excited fermions [60, 66–68, 70, 80, 87], for lepton flavor violation [59, 64, 84] and for pair production of supersymmetric particles [62, 82], although some of these are related to the high-Q^2 DIS sector either by providing reaction signatures which might fulfill the DIS selection criteria (like excited electrons, see Sect. 2.2.2), or by having hypothesized physical origins (such as lepton-flavor violating LQs) closely related to those discussed in the following.

Since search analyses require a detailed understanding of both the SM "backgrounds" and the phenomenology of the postulated signal, they are often impeded by experimental or theoretical difficulties which do not affect other studies at the same level. Consequently, many results presented in this section have been published only recently, or are preliminary or even incomplete. Still missing are, for example, a detailed simulation of pQCD processes in resonant LQ production including matrix-element calculations of high-p_t QCD splittings, and a systematic search for all possible decay modes of R_P-violating squarks, with a mapping of the corresponding exclusion limits to the SUSY parameter space. On the other hand, some preliminary results are already available for the 1998–99 e^-p data, which are particularly helpful to constrain scenarios to which the e^+p data are less sensitive. All in all, we will see that the existing results already allow us to draw meaningful conclusions, both on the possible nature of the high-Q^2 excess and on strategies to solve remaining questions in investigations of future HERA and other data.

4.4.1 Electron–Quark Resonances

The experimental searches for electron–quark resonances performed by the HERA experiments use the LQ classification presented in Sect. 2.2.2 (Table 2.3). The initial quark flavor, the spin of the resonance and its branching ratios $\beta_e = \mathrm{BR}(\mathrm{LQ} \to eq)$ and $\beta_\nu = \mathrm{BR}(\mathrm{LQ} \to \nu q')$ (if the latter channel is used for the analysis) have to be specified in order to properly simulate the experimental signatures. The search for squarks in the LQ-like decay modes[14] differs from the LQ case mainly because squarks are always scalar and have a reduced spectrum of couplings to quarks, and since they have different decay properties from LQs. For squarks, β_e is a variable number depending on the SUSY parameters, and β_ν is either zero ($\tilde{u}$-type squarks) or variable ($\tilde{d}$-type squarks); for "conventional" LQs, the branching ratios are given by

[14] In the following, we will use "LQ" generically for any electron–quark resonance decaying to eq or $\nu q'$.

the $SU(2)_L \times U(1)_Y$ symmetry, fulfill $\beta_e + \beta_\nu = 1$ and are either zero, half or unity.

In the following, we will first discuss the search strategies for LQ signals in the NC and CC channels and present the relevant experimental data. After an excursion dealing with the effects of radiative corrections on the different mass reconstruction methods and with the question whether the ZEUS and H1 excess data, if interpreted as resonance signals, are consistent with each other, we will present the LQ limits in terms of cross sections, coupling constants and branching ratios. Finally, searches for R_p-violating squarks in non-LQ decay channels will be briefly summarized, which also involve the events with $\not{P}_t \ell j$ signature observed by H1.

Experimental Data and Search Analyses

The basic input for the search analyses is the two-dimensional (x, y) or (M, y) distribution of NC and CC events. As discussed in Sect. 2.2.2, the signature of an LQ would be an excess of events over the SM prediction which cluster around a given value of M or x corresponding to the mass of the resonance, and which would be distributed flat in y or like $(1-y)^2$ for scalar or vector LQs, respectively. In Fig. 4.35, the signals simulated by H1 for scalar LQs

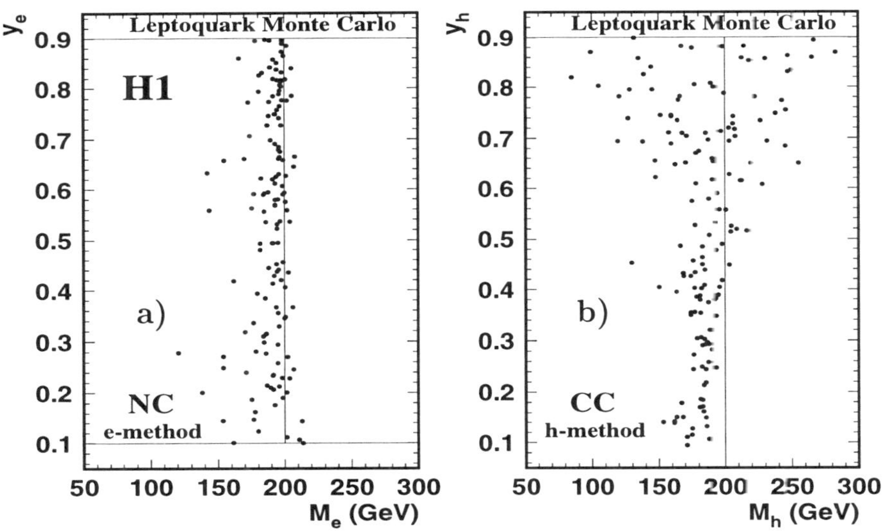

Fig. 4.35. Event distributions in the (M, y) plane as expected in the H1 analysis for the production of a hypothetical scalar LQ with $F = 0$ and mass $M_{LQ} = 200\,\text{GeV}$ which decays (**a**) in the NC channel ($LQ \to eq$) and (**b**) in the CC channel ($LQ \to \nu q'$). The *dots* indicate the reconstructed kinematic variables for a simulated LQ event sample corresponding to the data luminosity, with an LQ coupling of $\lambda = 0.05$ and a branching ratio of one in the respective decay channel. The plots have been reproduced from [84]

with a mass $M_{\mathrm{LQ}} = 200\,\mathrm{GeV}$ and a coupling[15] $\lambda = 0.05$ are shown for the decay modes $\mathrm{LQ} \to eq$ and $\mathrm{LQ} \to \nu q'$, which will be called "NC-like" and "CC-like" in the following.

The two-dimensional event distributions of the NC channel have already been discussed in Sect. 4.1.1 (see Figs. 4.2 and 4.3). Indeed, a pattern resembling a resonance signal is observed in the high-Q^2 region of the H1 data (see Fig. 4.6). The significance of this excess has been investigated in Sect. 4.1.1; see in particular Table 4.2. Before describing the analysis strategies applied to quantify the excess in terms of cross sections and LQ couplings, we want to present the corresponding ZEUS data, and also the experimental evidence available for the CC channel. Whereas H1 applies the electron (Jacquet–Blondel) method to reconstruct the mass $M_e = \sqrt{x_e s}$ ($M_h = \sqrt{x_{\mathrm{JB}} s}$) in the NC (CC) channel (see Sect. 3.3.3), ZEUS uses the invariant mass $M_{\ell \mathrm{j}}$ ($\ell = e, \nu$) calculated from the measured energies and directions of the scattered lepton and the jet with highest transverse momentum according to

$$M_{\ell \mathrm{j}} = \sqrt{E'_\ell E_{\mathrm{j}} \, (1 - \cos \xi)} \tag{4.9}$$

for all resonance searches. In (4.9) E_{j} is the jet energy and ξ is the angle between the lepton and jet. The CC lepton variables are determined from the Jacquet–Blondel method. The mass resolution for the NC channel improves for growing M_{LQ} and is about 4% at $M_{\mathrm{LQ}} = 200\,\mathrm{GeV}$, with a bias induced by QED and QCD radiation of about -2% for H1 and -1% for ZEUS. For the CC channel, H1 quotes a mass resolution of $\mathcal{O}(10\%)$ on average, with a bias of -6%. As can be seen in Fig. 4.35, the CC mass reconstruction is not only less precise on average, but in addition deteriorates at high y, i.e. in the region of the main sensitivity to LQ production. For this reason, ZEUS and H1 currently do not use the CC data to set limits on the production cross section of hypothesized LQ species having both NC- and CC-like decay channels.

For the limit-setting procedure, ZEUS uses the "constrained-jet" mass reconstruction method,

$$M_{\mathrm{CJ}} = \sqrt{2E_e(E + P_z)} \,, \tag{4.10}$$

where $E + P_z$ is calculated analogously to (3.4) from the final-state lepton and all jets with pseudorapidity $\eta_{\mathrm{j}} < 3$ and a transverse momentum above $15\,\mathrm{GeV}$. In the MC simulation of LQ production, M_{CJ} improves the mass resolution at high y by a factor of about two compared with $M_{\ell \mathrm{j}}$. However, (4.10) is based on energy–momentum conservation constraints and is sensitive to the non-LQ part of the hadronic final state. It is therefore only applied in cases where the final state is assumed to be exactly known, i.e. for the comparison of the data with specific models of eq resonance production as implemented in the MC simulation.

[15] In Sect. 2.2.2, the symbol g was used for the Yukawa couplings, in accordance with the conventions generally used. In this section, we will instead use λ in order to comply with the plots produced by ZEUS and H1.

In Fig. 4.36a,b the ZEUS M_{ej} distribution is compared with the SM expectation [76]. A clear excess in the data is observed at masses $M_{ej} \gtrsim 210\,\text{GeV}$, where 49 events are observed while 24.7 ± 5.6 events are expected. This excess also contains those five high-Q^2 events from the 1994–96 data sets which were reported in [71] (see Table 4.2); their average M_{ej} value is about 7 GeV below the mass reconstructed originally using the double-angle method.[16] Although the ZEUS M_{ej} spectrum resembles the expected signal of eq resonance production, this hypothesis is not supported by the distribution of $\cos\theta^* = 1 - 2y$ (1.6) in the events with $M_{ej} > 210\,\text{GeV}$, which is shown in Fig. 4.36c: the events cluster at $\cos\theta^* \approx 1$, as expected for NC DIS. No significant fraction of the excess is located at $\cos\theta^* < 0.6$, i.e. at $y > 0.2$. ZEUS therefore concludes that the data are not indicative of LQ production.

Some of the LQ species in Table 2.3 have CC-like decay modes. In addition, R_P-violating d-type squarks can decay into a quark and a neutrino. Results on searches for such resonances are available from H1 [84] and, preliminarily, from ZEUS [77]. The relevant mass spectra are compared with the SM predictions in Figs. 4.37a and 4.38. No obvious sign of a resonance structure is visible. Whereas the H1 data agree well with the SM prediction, a slight event excess is present in the ZEUS data at masses above 200 GeV; this corresponds to the excess at high x discussed in Sect. 4.2. The distribution of $\cos\theta^*$ in the ZEUS high-mass region, shown in Fig. 4.37b, reveals that the event excess lacks a significant component which is distributed uniformly in y, and hence does not support the presence of a scalar LQ. Note that, owing to the absence of the photon propagator, the y distribution of CC DIS events at fixed x is much more "LQ-like" than in the NC case and is in particular very similar to that expected for a vector LQ. The PDF modification suggested by Bodek and Yang [126] (see Sects. 2.1.6 and 4.2) leaves the shape of the expected $\cos\theta^*$ distribution mostly unmodified and hence does not affect the conclusions with respect to the LQ search.

Both collaborations derive upper limits on the product $\sigma_{\text{LQ}}\beta$ of the production cross section for LQs and their branching ratio into the decay channel under study. The results are based on comparisons of the expected and observed event numbers in M_{LQ}-dependent search windows which are optimized for maximal sensitivity using samples of simulated DIS and LQ events. The MC simulation of LQs includes QED as well as QCD radiation in the leading-log approximation. In the ZEUS analysis, only the resonant s-channel contribution to the LQ cross section is taken into account, and the search is restricted to the region where the narrow-width approximation (NWA) is valid (see Sect. 2.2.2). H1 considers the full LQ-induced cross section, thereby gaining sensitivity for LQs with masses around and even beyond the HERA center-of-mass energy. For a given M_{LQ}, the search window is defined by a mass interval, which reflects the mass resolution and the reconstruction bias,

[16] We will come back to differences between the mass reconstruction methods and their sensitivity to QCD radiation effects in the next subsection.

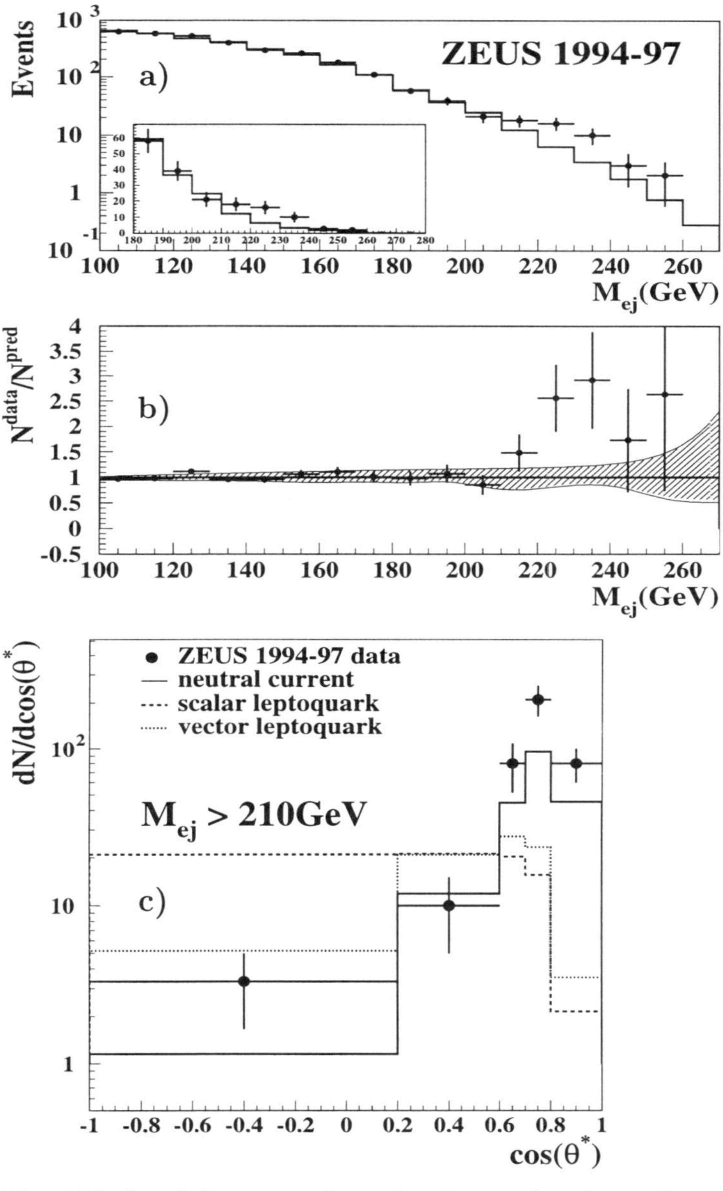

Fig. 4.36. Search for resonantly produced states decaying to electron and jet in the 1994–97 ZEUS NC data. (**a**) Distribution of the invariant electron–jet mass, M_{ej} (*filled circles*), compared with the SM prediction derived from the CTEQ4 PDF set [185] (*solid line*). The *inset* shows the region $M_{ej} > 180\,\mathrm{GeV}$ on a linear scale. (**b**) Ratio of measured to expected M_{ej} distributions, with the uncertainty of the prediction indicated by the *hatched band*. (**c**) Distribution of $\cos\theta^*$ (see text) for $M_{ej} > 210\,\mathrm{GeV}$; the *dashed* (*dotted*) *lines* indicate the expectations for a scalar (vector) LQ. The plots have been reproduced from [76]

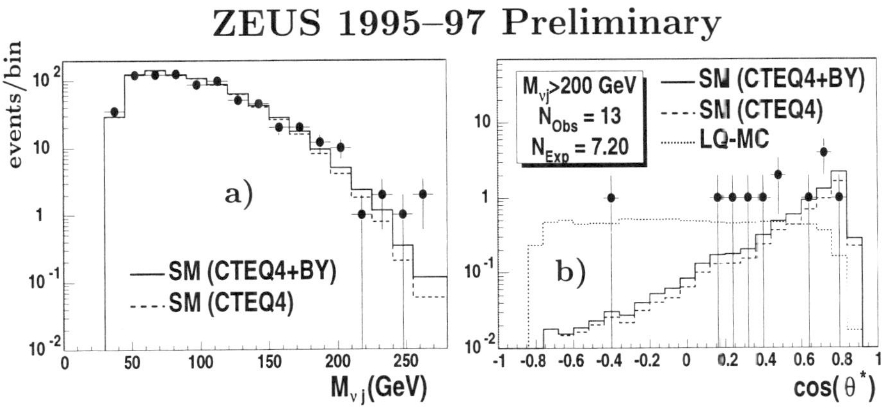

Fig. 4.37. Search for resonantly produced states decaying to neutrino and jet in the 1995–97 ZEUS CC data. (**a**) Distribution of the invariant neutrino–jet mass, $M_{\nu \mathrm{j}}$ (*filled circles*), compared with the SM prediction derived from the CTEQ4 PDF set [185] with (*solid line*) and without (*dashed line*) the modification suggested by Bodek and Yang [126]. (**b**) Distribution of $\cos\theta^*$ (see text) for $M_{\nu \mathrm{j}} > 200\,\mathrm{GeV}$; the *dotted line* indicates the expectation for a scalar LQ, normalized to 8 events in the relevant mass range. The plots have been reproduced from [77]

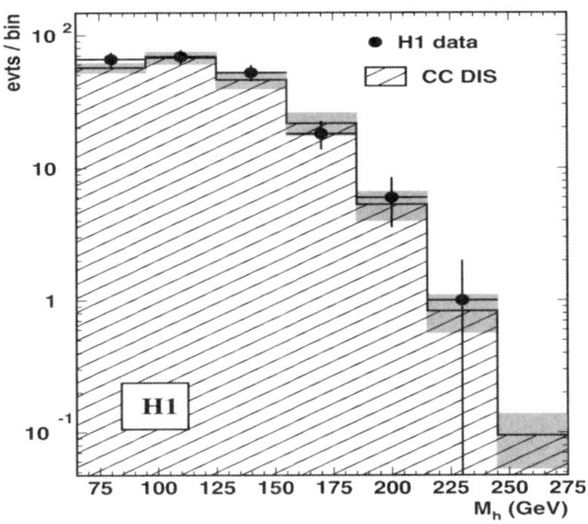

Fig. 4.38. Distribution of the mass $M_h = \sqrt{x_{\mathrm{JB}} s}$ (see (3.24)) in the H1 CC sample (1994–97 data). The *filled circles* represent the H1 data; the *histogram* is the SM expectation derived from the MRST PDF set [141], with the overall uncertainty indicated by the *shaded area*. The plot has been reproduced from [84]

and by a lower y cut, y_{cut}, which removes the bulk of DIS reactions. In the NC channel, the width of the mass window is typically 20 GeV. The y cut decreases with increasing M_{LQ}, e.g. from 0.45 to 0.05 (from 0.25 to 0.05) for the mass range 150 – 290 GeV in the ZEUS search for scalar (vector) LQs. The lower bound of the mass window and y_{cut} for the H1 analysis are shown as functions of M_{LQ} in Fig. 4.39a,b. At high masses, the inclusion of the non-resonant contributions distorts the mass shape of the signal, necessitating a widening of the mass window accompanied by an increase of y_{cut} to keep the

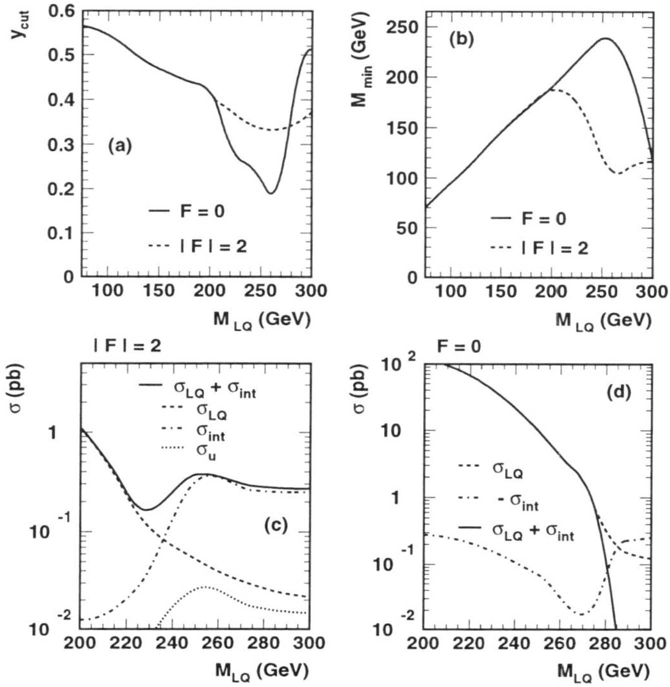

Fig. 4.39. H1 LQ search: lower cuts on (**a**) y and (**b**) on the reconstructed mass, $\sqrt{x_e s}$, in the searches for $F=0$ (*solid line*) and $|F|=2$ (*dashed line*) scalar LQs; (**c**) contributions to the cross section induced by $S_{0,\mathrm{R}}$ ($|F|=2$) LQs with $\lambda = 0.5$ in the kinematic range selected by these cuts: u-channel contribution (σ_u, *dotted line*), interference with SM (σ_{int}, *dash-dotted line*), sum of s-channel and u-channel contributions (σ_{LQ}, *dashed line*), and total difference from the SM cross section ($\sigma_{\mathrm{LQ}} + \sigma_{\mathrm{int}}$, *solid line*); (**d**) as in (**c**), but for $S_{1/2,\mathrm{L}}$ ($F=0$) LQs. In (**d**), σ_u is too small to be shown and σ_{int} is negative. The plots have been reproduced from [84]

DIS background low. The efficiency for accepting LQ-induced events rises with M_{LQ} and ranges from 20% to about 50% for $M_{\mathrm{LQ}} = 75-250\,\mathrm{GeV}$ in the H1 analysis.

For LQs coupling to first-generation (anti)quarks, the s-channel, u-channel and SM $\times$ LQ interference contributions to the e^+p NC cross section in the H1 search windows are shown as functions of M_{LQ} in Figs. 4.39c,d for $|F|=2$ and $F=0$, respectively. The interference terms are important for masses $\gtrsim 220\,\mathrm{GeV}$ ($\gtrsim 280\,\mathrm{GeV}$) for $|F|=2$ ($F=0$), i.e. for LQs produced from antiquarks (quarks). The u-channel contribution is always small and completely negligible for $F=0$. Figure 4.39d demonstrates that the approximation used in the ZEUS analysis (s-channel only) is valid almost up to the kinematic threshold for LQs which can be produced from valence quarks; we will report ZEUS results only for these channels, i.e. for $F=0$ in the case of e^+p and for $|F|=2$ in the case of e^-p reactions.

Radiative Effects, Mass Reconstruction, and the Compatibility of the ZEUS and H1 Data with a Single Narrow Resonance

The question whether the data of ZEUS and H1 are compatible with each other and with the interpretation as a single narrow eq resonance has been discussed since anomalous event rates were reported for the very first time. Most of the arguments focused on the excess events of analyses HN1 and ZN1, which cluster around mass values between 200 and 230 GeV. Although the resonance interpretation is not corroborated by the 1997 data and these events are no longer in the center of public attention, it is instructive to investigate the impact of QED and QCD radiative effects on the mass values reconstructed with the different methods described above, and to reexamine the compatibility of the ZEUS and H1 data in the light of the resulting estimates of mass reconstruction biases.

As discussed in Sect. 2.2.2, QED and QCD radiation affects the observed resonance shape and has different consequences for different reconstruction methods. The impact of ISR on the kinematic reconstruction has been addressed in Sect. 3.3.3; it induces an average difference of about 4% between x_{DA} and x_e, corresponding to a mass difference of about 4 GeV at $M_{\mathrm{LQ}} = 200\,\mathrm{GeV}$. For the DA method, ZEUS [431] estimated a probability of 2% to overestimate the x of a given event by more than 5% owing to ISR. A direct check of the impact of ISR was performed by Bassler and Bernardi from H1, who applied the ω method (see Sect. 3.3.3) to the ZEUS "signal events" described in [71] and found an average reduction of M by 10 GeV, corresponding to a reduction of the average x by almost 10% [429]. However, this result depends strongly on the averaging procedures applied.

QCD radiation modifies several aspects of the simple QPM-type event kinematics that have to be presupposed in order to identify the reconstructed value of x with M_{LQ}^2/s in the case of resonant LQ production. The most important of these effects are [269]: (i) owing to initial-state QCD radiation, the incoming quark acquires space-like virtuality, V_q, and nonzero transverse momentum, p_{t}^q (see Sect. 2.1.7); (ii) the decay quark of the LQ carries time-like virtuality, M_q, corresponding to the invariant mass of the jet evolving from the quark. The impact of these effects has been studied by simulating in a simplified model the calculation of kinematic variables in events with an invariant mass of the initial-state electron–quark system of $M_{\mathrm{LQ}} = 200\,\mathrm{GeV}$ and with $p_{\mathrm{t}}^q = 3.6\,\mathrm{GeV}$, $V_q = 14.3\,\mathrm{GeV}$ and $M_q = 32\,\mathrm{GeV}$ (these numbers have been quoted in [269] as the respective expected average values). Neither QED radiation nor detector simulation are included. The result, shown in Fig. 4.40, indicates that the main effect comes from the virtuality of the final-state quark. The M values reconstructed by the e and the DA methods differ by typically 10 – 15 GeV in the y region of interest, they increase with y, and in addition a nonnegligible spread is caused by averaging over the dependence of M on the azimuthal degree of freedom induced by nonzero p_{t}^q. For the electron–jet method (4.9), the bias is smaller, depends only weakly

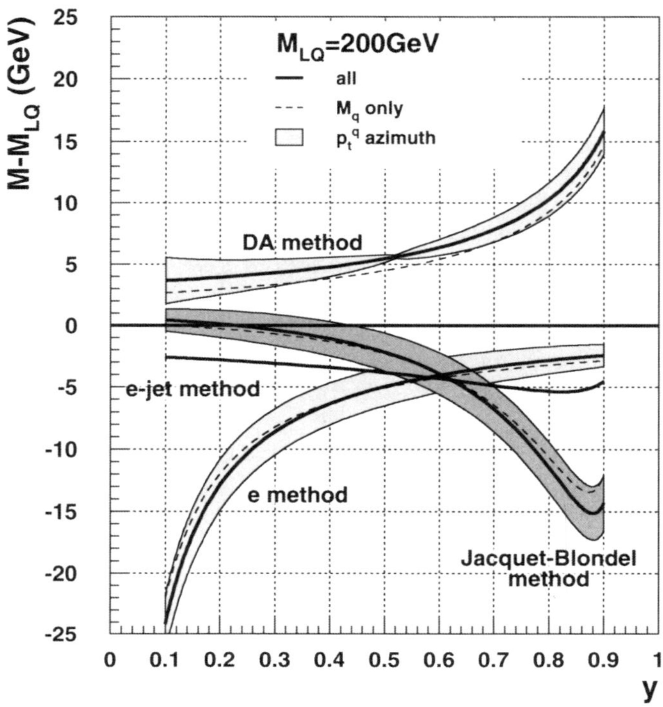

Fig. 4.40. Reconstruction of M for production of LQs with mass $M_{\mathrm{LQ}} = 200\,\mathrm{GeV}$. The *solid lines* indicate the y dependence of $M - M_{\mathrm{LQ}}$ for events with an initial-state quark carrying transverse momentum $p_{\mathrm{t}}^{q} = 3.6\,\mathrm{GeV}$ and space-like virtuality $V_q = 14.3\,\mathrm{GeV}$, and with a time-like virtuality of the LQ decay quark of $M_q = 32\,\mathrm{GeV}$ (numbers taken from [269]). Also shown are the results obtained with $p_{\mathrm{t}}^{q} = 0$, $V_q = 0$ and $M_q = 32\,\mathrm{GeV}$ (*dashed lines*). The *shaded areas* show the R.M.S. variation of M obtained by averaging over the azimuthal degree of freedom induced by nonzero p_{t}^{q}. The *curves belonging to the light shaded areas* are for the mass reconstruction in the NC channel, i.e. for the DA method (*top*) and the e method (*bottom*). Also shown is the result obtained for the invariant electron–jet mass, $M_{e\mathrm{j}}$ (4.9); the azimuthal dependence of $M_{e\mathrm{j}}$ is negligible. The *dark shaded area* shows the mass reconstruction bias in the CC channel (Jacquet–Blondel method). QED radiation, experimental resolutions and detector acceptances are not taken into account

on y and has no azimuthal dependence. The constrained jet reconstruction (4.10) yields a bias of less than 1 GeV in the whole y range and no azimuthal dependence (not shown). The M_e and $M_{e\mathrm{j}}$ curves in Fig. 4.40 can be compared with Fig. 4.35a and with the average resolution quoted in the previous subsection.[17] In general, the tendencies of radiation-induced resolution biases roughly agree between [76, 84] and our simple model, confirming that

[17] Note that we compare quantities derived from average sizes of QCD effects (Fig. 4.40) with results of truncated Gaussian fits to the resolution functions as quoted in [76, 84].

the dominant radiative QCD effects are included in the MC simulations. One should note, however, that p_t^q, V_q and M_q are broadly distributed (see [269]), and that Fig. 4.40 only gives a rough estimate of the average size of the impact of QCD effects on the kinematic reconstruction of LQ events.

For the Jacquet–Blondel method, i.e. in the CC channel, the reconstruction bias is negative and increases with rising y. However, in this case the experimental uncertainties related to the hadronic energy measurement exceed the QCD effects by far (see Fig. 4.35b).

Having discussed the mass reconstruction biases, we return to the question whether the ZEUS and H1 data are compatible with a single narrow resonance. As can be seen from Figs. 4.7 and 4.8 and also from Table 4.2, the excess events occur at different x in both experiments: in ZEUS, the most significant excess is observed for $x_{\mathrm{DA}} \gtrsim 0.55$ (implying $M_{\mathrm{DA}} \gtrsim 223\,\mathrm{GeV}$), whereas the H1 events cluster around $x_e = 0.44 - 0.47$ (equivalent to a mass $M_e = 200 - 205\,\mathrm{GeV}$). The corresponding x difference can be compared with the experimental resolutions of better than 10% (see Table 3.7), and it has been argued [446] on the basis of these numbers that the ZEUS and H1 events are no more compatible with each other than with the SM. However, several additional aspects have to be taken into account to reach a sound conclusion:

- The SM expectation in the "signal region" is above unity for both experiments, and even if a resonance signal was present it would be contaminated with SM background and, depending on the production mechanism, possibly also with nonresonant SM × LQ interference contributions. The mean and R.M.S. of the M values of the signal events can thus not be taken as estimates of a hypothetical resonance mass and of the experimental resolution.
- QED and QCD radiative effects would induce an observed mass difference of roughly $10 - 20\,\mathrm{GeV}$, depending on the y range of the events and on the interplay of QED and QCD corrections (which has not been studied here).
- The latest H1 search for production of eq resonances [84] finds the strongest indication for such a resonance (i.e. the weakest exclusion limits) for an LQ mass of about 210 GeV (see next subsection), 5% above the average mass quoted originally [72]. H1 assigns 6 GeV of the 10 GeV average shift to the new electron energy calibration and 4 GeV to the difference between M_e and M_{LQ} [433]. The latter number roughly agrees with the M_e shift indicated in Fig. 4.40 for the relevant y range.

Combining the above-described effects (which all tend to reduce the apparent discrepancy between the average M values of the ZEUS and H1 "event clusters"), it seems premature to exclude the possibility that these data actually contain events coming from a single narrow resonance. However, this statement only refers to the properties of the ZEUS and H1 events and addresses neither the question whether the data actually bear indications of LQ production nor whether such an interpretation could be reconciled with the stringent Tevatron exclusion limits (see Sect. 2.2.2).

Limits on Cross Sections and Couplings

Although the HERA data discussed in the previous subsections exhibit some features which are reminiscent of resonance production, the properties of the data do not add up to convincing evidence for such a process. ZEUS and H1 therefore set upper limits on the cross section for resonant production of LQs decaying subsequently into eq or $\nu q'$ (the limits are thus on $\sigma_{\mathrm{LQ}}\beta$, where β is the appropriate LQ branching ratio). In these analyses it is presupposed that the LQs behave according to the models discussed in Sect. 2.2.2 and that not more than one species at a time contributes to the ep cross section. For a given value of M_{LQ}, the limits are derived from the numbers of observed and expected events in the associated search window, assuming Poissonian statistics. The 95% C.L. ZEUS limits [76] are reproduced in Fig. 4.41 as a typical example; similar results are obtained by H1 [84] (not shown). Note that these limits are independent of the quark flavors to which the LQ is assumed to couple, except at the highest M_{LQ}, where production from sea quarks would necessitate a coupling too large for the NWA to be valid. No prominent structure is present in the ZEUS limit curves for $M_{\mathrm{LQ}} \gtrsim 200\,\mathrm{GeV}$. In this region, production cross sections larger than 0.3 pb (0.6 pb) are excluded for scalar (vector) LQs. The "excess cross section" of 0.4 pb observed in the 1994–96 data (see Sects. 2.2.2 and 4.1.1) is thus meanwhile beyond the exclusion limit, at least for the scalar scenario.

Upper limits for the Yukawa couplings, λ, are derived from the LQ cross section limits for the 1994–97 e^+p data sets, taking into account an increase of the LO cross sections by $\mathcal{O}(20-50\%)$ due to QCD corrections of $\mathcal{O}(\alpha_s)$ [251, 266]. The 95% C.L. exclusion curves obtained by ZEUS and H1 for the

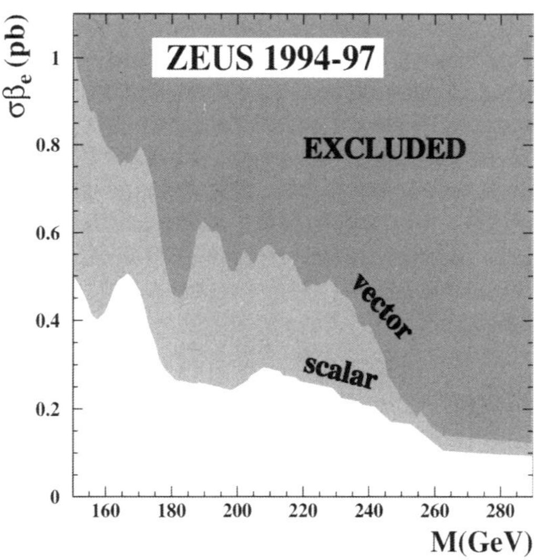

Fig. 4.41. Upper limits of the cross section for resonant LQ production with subsequent LQ decay to eq (ZEUS 1994–97 e^+p data). The *light* (*dark*) *shaded* areas are excluded at 95% C.L. for scalar (vector) LQs. The plot has been reproduced from [76]

LQs with $F=0$ listed in Table 2.3 are shown as functions of M_{LQ} in Figs. 4.42 and 4.43, respectively. All limits are for LQs which couple to u or d quarks, i.e. can be produced from valence quarks in e^+p scattering. The results of ZEUS and H1 are very similar for $M_{\mathrm{LQ}} \lesssim 280\,\mathrm{GeV}$, whereas H1 gains additional sensitivity at higher masses from including the SM $\times$ LQ interference and also the u-channel terms in the analysis (see above). If e^+u or e^+d LQs exist in the mass range $M_{\mathrm{LQ}} \lesssim 230\,\mathrm{GeV}$, their λ values cannot exceed a few times 10^{-2}; couplings of electromagnetic strength ($\lambda = \sqrt{4\pi\alpha} \approx 0.3$) are excluded for masses up to about 270 – 280 GeV. In Fig. 4.42, the ZEUS limits are compared with constraints from the search for LQ pair production in $p\bar{p}$ reactions at the Tevatron, yielding upper limits[18] on M_{LQ} which depend on β_e but not on λ (see Sect. 2.2.2), and from measurements of fermion pair production at LEP, where LQs would induce a CI-type modification of the SM cross sections. In contrast to the HERA limits, the Tevatron and LEP results are independent of assumptions on the quark flavor to which the LQs couple, except that top quarks in the final states are excluded (LEP) or strongly phase-space suppressed (Tevatron). It becomes obvious that the HERA experiments have a unique discovery window for eu or ed resonance production with $M_{\mathrm{LQ}} \gtrsim 250\,\mathrm{GeV}$, i.e. close to the HERA kinematic limit. This window grows for $\beta_e < 1$ or in the case of a possible future increase of the HERA center-of-mass energy achieved by higher beam energies [447].

The coupling limits presented in Figs. 4.42 and 4.43 are for LQs with $F=0$ which could be produced from the fusion of positrons and valence quarks in e^+p scattering. As discussed in Sect. 2.2.2, an alternative explanation for the HERA high-Q^2 events might be an eq resonance with $|F|=2$ and $\lambda = \mathcal{O}(0.25)$, which would have to be produced from antiquarks. The 95% C.L. limits derived by H1 from the e^+p data for such a process [84] are shown in Fig. 4.44. These constraints are much weaker than those for $F=0$ since the corresponding cross sections involve antiquark PDFs, which are strongly suppressed at high x. Assuming that the LQ signatures at LEP and the Tevatron are equal for $F=0$ and $|F|=2$, we conclude from Figs. 4.44 and 4.42 that the current HERA discovery potential for $e^+\bar{q}$ resonances (i.e. $|F|=2$) is competitive for masses around 220 GeV and inferior elsewhere. The HERA sensitivity in this channel is expected to be approximately the same as for e^+q resonances with $F=0$ which do not couple to valence quarks, for example stop production according to (2.57); neither ZEUS nor H1 have published results for such scenarios. For $|F|=2$ LQs coupling to first-generation quarks, however, much more stringent constraints than in Fig. 4.44 can be obtained from e^-p scattering, where such LQs can be produced from valence quarks. The preliminary results of a resonance search performed by ZEUS in the 1998–99 e^-p data are shown in Fig. 4.45. These exclusion limits are of similar magnitude to those for $F=0$ and exclude values of $\lambda = \mathcal{O}(0.25)$ for $M_{\mathrm{LQ}} \lesssim 270\,\mathrm{GeV}$.

[18] Note that the combined CDF and DØ limit [282] ($M_{\mathrm{LQ}} > 242\,\mathrm{GeV}$ for $\beta_e = 1$ at 95% C.L.) is stronger than that indicated in Fig. 4.42.

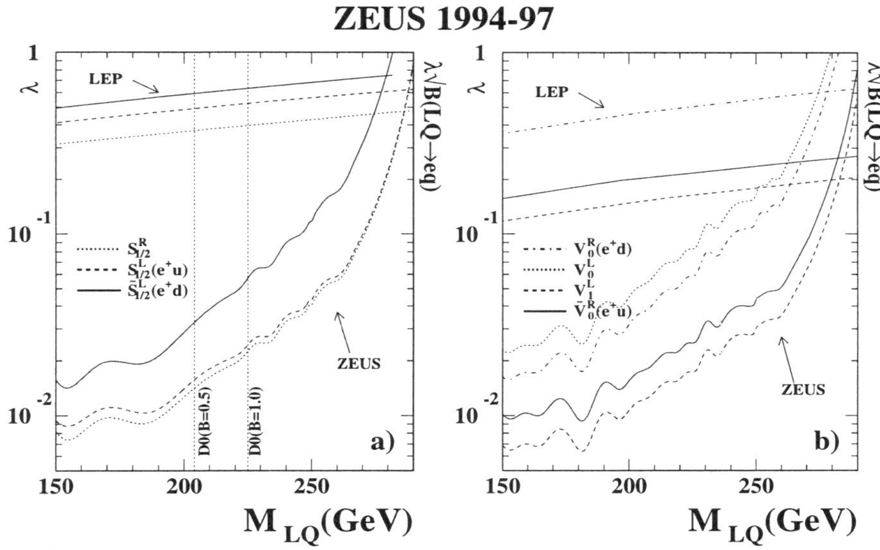

Fig. 4.42. ZEUS results for the upper limits on the Yukawa couplings, λ, of different types of (**a**) scalar and (**b**) vector LQs with $F=0$, as functions of their masses. The LQ symbols are those introduced in Table 2.3, with the chirality of the coupling added as a superscript. For LQ-type objects with a branching ratio β_e different from the "standard value" β_e^0, deduced from the couplings in Table 2.3, the limits can be interpreted in terms of $\lambda\sqrt{\beta_e/\beta_e^0}$. The ZEUS results are compared with representative limits from LEP [349] and with the λ-independent mass limits from DØ [281] for different values of β_e (see Sect. 2.2.2). The regions above the ZEUS and LEP curves, and left of the DØ bounds, are excluded at 95% C.L. The plots have been reproduced from [76]

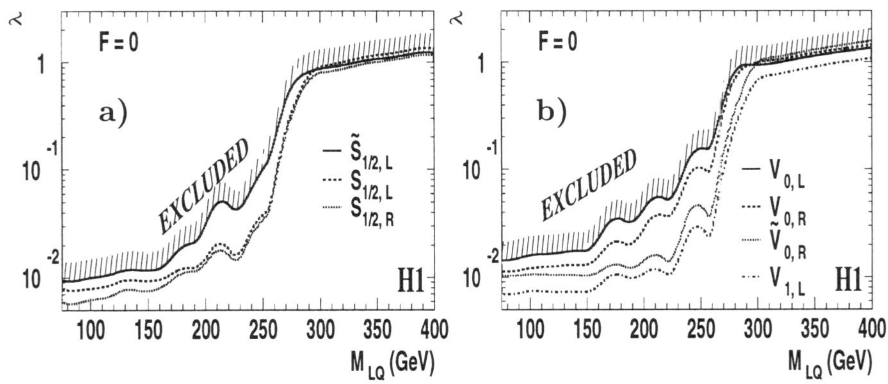

Fig. 4.43. H1 results for the upper limits on the Yukawa couplings, λ, for different types of (**a**) scalar and (**b**) vector LQs with $F=0$, as functions of their masses. The LQ symbols are those introduced in Table 2.3, with the chirality of the coupling added as an additional subscript. The region above the curves is excluded at 95% C.L. The interference of the LQ s-channel and the SM amplitudes yields sensitivity to high masses, $M_{\mathrm{LQ}} \gtrsim \sqrt{s}$. The plots have been reproduced from [84]

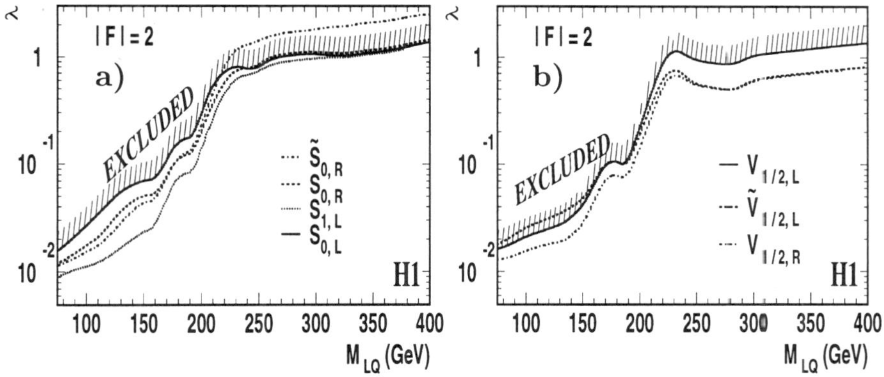

Fig. 4.44. H1 results for the upper limits on the Yukawa couplings, λ, for different types of (**a**) scalar and (**b**) vector LQs with $|F| = 2$, as functions of their masses. The LQ symbols are those introduced in Table 2.3, with the chirality of the coupling added as an additional subscript. The region above the curves is excluded at 95% C.L. The interference of the LQ s-channel and the SM amplitudes yields sensitivity to high masses, $M_{\mathrm{LQ}} \gtrsim \sqrt{s}$. The plots have been reproduced from [84]

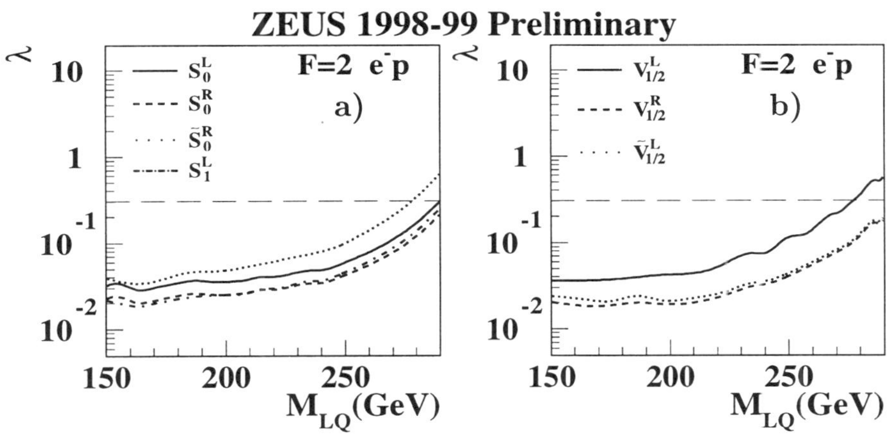

Fig. 4.45. Preliminary ZEUS results for the upper limits on the Yukawa couplings, λ, for different types of (**a**) scalar and (**b**) vector LQs with $|F| = 2$, as functions of their masses (1998–99 e^-p data). The LQ symbols are those introduced in Table 2.3, with the chirality of the coupling added as a superscript. The *dashed horizontal lines* indicate the electromagnetic coupling strength, $\lambda = \sqrt{4\pi\alpha}$. The region above the curves is excluded at 95% C.L. The plots have been reproduced from [268]

From the above, it seems fair to conclude that the LQ interpretation of the HERA data can only be reconciled with the experimental constraints from the Tevatron and LEP for scalar objects with $F = 0$ and $\beta_e < 1$.[19] None of the LQ

[19] LQs which do not couple to first-generation quarks also appear to be marginally allowed. We will, however, not discuss this possibility here.

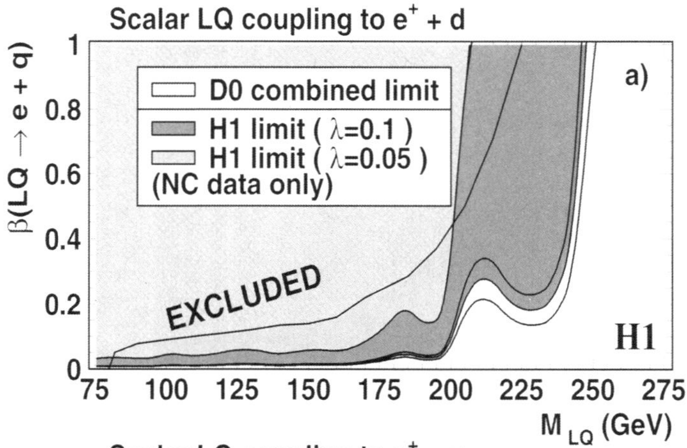

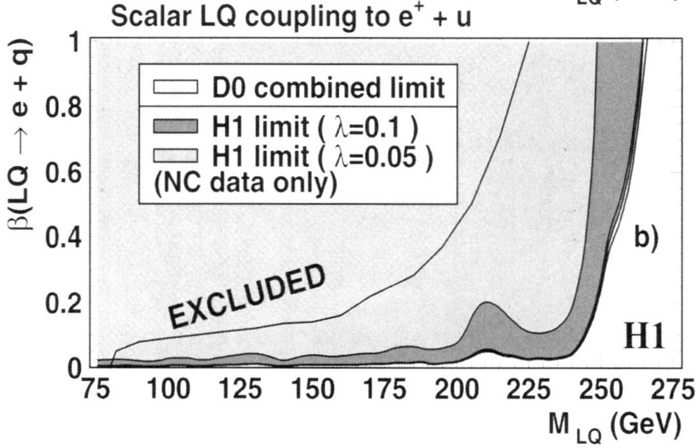

Fig. 4.46. H1 exclusion limits for $\beta_e = \text{BR}(\text{LQ} \to eq)$ as functions of M_{LQ} for scalar LQs coupling (**a**) to e^+d and (**b**) to e^+u. The *light* (*dark*) *shaded regions* are excluded at 95% C.L. for Yukawa couplings of $\lambda = 0.05$ (0.1). The bands around the limit curves for $\lambda = 0.1$ indicate the variation due to PDF uncertainties. The *hatched area* represents the region excluded by DØ (cf. Fig. 2.20). The plots have been reproduced from [84]

species in Table 2.3 fulfills these requirements. However, possible candidates are R_P-violating squarks, which have $\beta_e < 1$ owing to their R_P-conserving cascade decays (see next subsection). The exact value of β_e can vary in a broad range, depending on the exact values of the SUSY parameters. H1 has therefore presented the resonance constraints as exclusion limits at 95% C.L. on β_e as a function of M_{LQ} for fixed values of the Yukawa coupling, $\lambda = 0.05$ and $\lambda = 0.1$ (Fig. 4.46). This representation can be directly compared with the result of the DØ search for LQ pair production shown in Fig. 2.20. The following observations can be made in Fig. 4.46:

- The bump in the limit curves around $M_{\rm LQ} = 210\,{\rm GeV}$ indicates the position of the H1 event excess. LQs with couplings as large as 0.1 would be compatible both with the H1 and with the DØ data if they have $\beta_e \lesssim 0.4$.
- For low values of β_e, the HERA limits are more restrictive than those from DØ. The HERA experiments thus have a unique sensitivity in this region, providing in particular a discovery potential for R_P-violating squarks with masses below the HERA center-of-mass energy.

Searches for Other Decay Modes of R_P-Violating Squarks

In view of the fact that the HERA data allow an interpretation in terms of R_P-violating squarks without coming into conflict with existing experimental bounds, the search for non-LQ-like decays of such objects gains particular priority. Squarks can undergo different cascade decays, leading to a variety of final-state topologies. Which of the decay modes are allowed, and occur with which branching ratios, depends on the SUSY parameters and on the mass hierarchy of supersymmetric particles. Feynman diagrams of the two decay modes for which preliminary results have recently been reported by ZEUS [78] and H1 [85] are shown in Fig. 4.47. In these analyses it was assumed that the lightest neutralino, χ^0, (i.e. the lightest mass eigenstate of the mixed neutral gaugino and higgsino states) is the lightest supersymmetric particle,

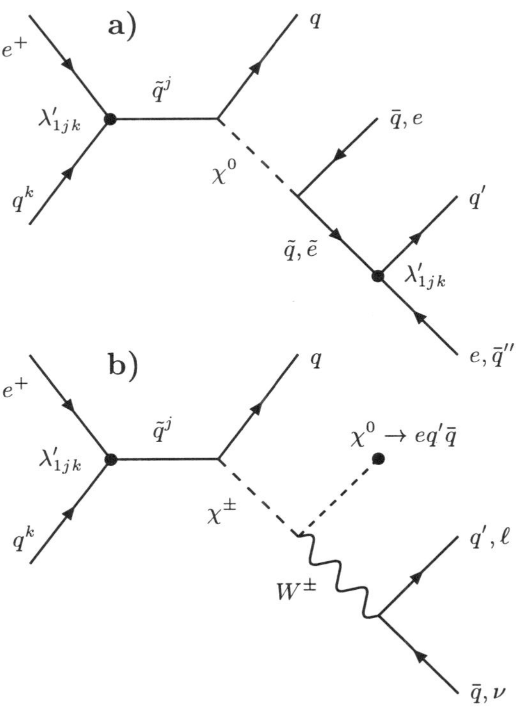

Fig. 4.47. Two examples of formation and subsequent cascade decay of an R_P-violating squark in e^+q scattering: (**a**) $\tilde{q} \to q\chi^0$ with a subsequent neutralino decay $\chi^0 \to eq'\bar{q}$ and (**b**) $\tilde{q} \to q\chi^\pm \to q'W^\pm\chi^0$, where the neutralino decays as in (**a**). The R_P-violating vertices are marked with a *filled circle*

that the gluino is heavier than the squark (so that the decay $\tilde{q} \rightarrow q\tilde{g}$ is prohibited) and that only one of the λ'_{1jk} (i.e. one flavor combination $1jk$, where the first index is fixed by the flavor of the initial-state electron) contributes significantly. H1 adjusted the SUSY parameters M_2, $\tan\beta$ and μ such that χ^0 is either photino- or zino-dominated and has a mass between 40 and 150 GeV (see [65, 85] and references therein), whereas ZEUS presupposed that the χ^0 is a pure photino with a mass of 50 – 150 GeV and that the squark decays predominantly according to the mechanism of Fig. 4.48a.

The decays in Fig. 4.47 produce, amongst others, final states containing three or more jets plus, with equal probability, an electron or positron. The SM background mainly comes from NC DIS and from photoproduction events in which a part of the hadronic final state is falsely identified as an electron. The "wrong-sign" signature, e^- + jets, is almost background-free. In Fig. 4.48, the spectrum of the invariant e^+ + jets mass spectrum observed in the H1 search [85] is compared with the SM expectation. Neither H1 nor ZEUS find a significant excess of events in this or in the e^- + jets channel. This result, together with the e + jet spectra investigated in the LQ searches, yield mass-dependent 95% C.L. upper limits of 0.2 – 0.5 pb [78] on the production cross section of squarks in SUSY models adhering to the assumptions specified above. The corresponding upper limits derived by both collaborations for the coupling strengths λ'_{1j1} of squarks resonantly produced in e^+u or e^+d fusion are shown in Figs. 4.49 and 4.50. Taking into account that the branching ratio

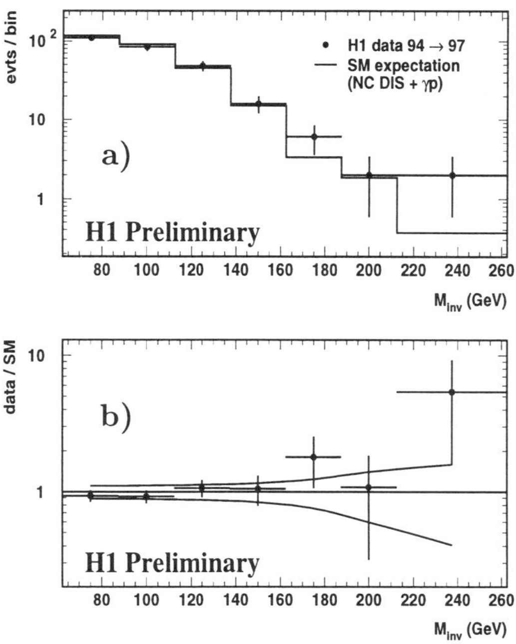

Fig. 4.48. Search for resonant production of R_P-violating squarks which subsequently decay to a positron and three or more jets (preliminary results from H1, 1994–97 e^+p data). (**a**) Distribution of the invariant e + jets mass (*filled circles*), compared with the SM background expectation (*histogram*). (**b**) Ratio of measured and expected distributions; the *upper* and *lower curves* indicate the total 1σ uncertainty of the SM prediction. The plots have been reproduced from [85]

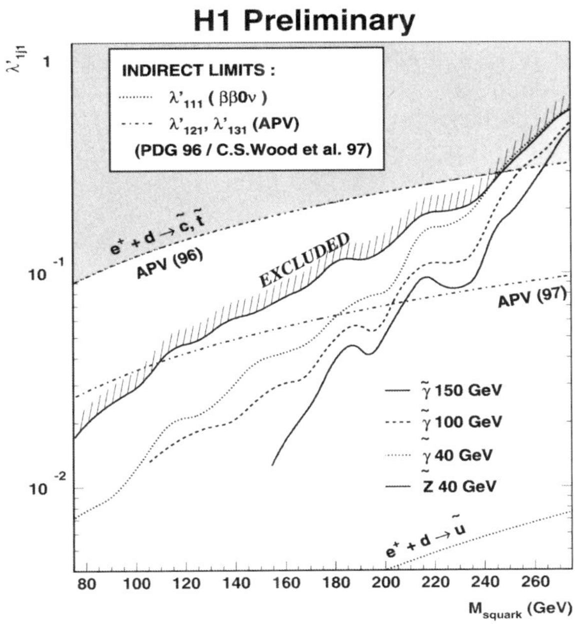

Fig. 4.49. Exclusion limits for the Yukawa couplings λ'_{1j1} of R_P-violating squarks as functions of the squark mass (preliminary H1 results, 1994–97 e^-p data). The *wiggly lines* represent the H1 limits for different masses and mixtures of the lightest neutralino (the *top line* is for a zino-like neutralino with a mass of 40 GeV). Also indicated are the most stringent limits from low-energy phenomena from neutrinoless double-beta decay ($\beta\beta0\nu$) for λ'_{111} and from atomic parity violation (APV) for λ'_{121} and λ'_{131}. The difference between both APV curves reflects the recent improvement of the corresponding measurements [289, 448]. In all cases the region above the curves are excluded at 95% C.L. The plot has been reproduced from [85]

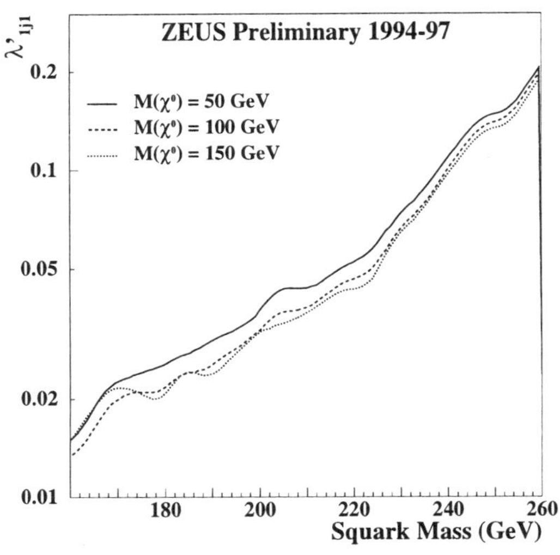

Fig. 4.50. Exclusion limits for the Yukawa couplings λ'_{1j1} of R_P-violating squarks as functions of the squark mass, for different masses of the lightest neutralino, which is assumed to be a pure photino (preliminary ZEUS results, 1994–97 e^+p data). In all cases the regions above the curves are excluded at 95% C.L. The plot has been reproduced from [78]

β_e is below about 0.2 for $M_{\tilde{q}} \lesssim 240\,\text{GeV}$ [78], these constraints are found to be more stringent than those in Figs. 4.42 and 4.43, interpreted in terms of $\lambda\sqrt{\beta_e}$. No evidence is hence found for squark production in the investigated scenario, although, owing to the larger branching ratio, one would expect a higher statistical significance of a possible signal in the e + multiple jets mode than in the LQ-type decay channel.

The exclusion limits derived from low-energy measurements indicated in Fig. 4.43 demonstrate that λ'_{111} is constrained by the nonobservation of neutrinoless double-beta decay [321] far beyond the reach of the HERA experiments, whereas the HERA sensitivity to λ'_{121} and λ'_{131} is similar to that of atomic parity violation measurements, which in turn yield the strongest low-energy limits. However, future HERA results can be expected to gain additional significance by exploiting all possible search channels. For example, the studies can be extended to signatures caused by the alternative neutralino decay $\chi^0 \to \nu q'\bar{q}$, which were already investigated in an H1 analysis of the 1994 data [65], or to leptonic decays of the W in Fig. 4.47b. In addition, a complete scan of the parameter space of R_P-violating SUSY has to replace the specific assumptions under which the current analyses have been performed in order to cover all possible squark scenarios with their respective decay channels and branching ratios.

For the particular case of a stop squark produced according to (2.56), the decay $\tilde{t} \to t\chi^0$ is kinematically prohibited for most of the HERA mass range owing to the large top mass. In this case, the H1 analysis would still be sensitive to the gauge decay $\tilde{t} \to b\chi^+$. However, if this decay was suppressed because the χ^+ was too heavy and in addition the sbottom was lighter than the stop squark, the e + multiple jets signature would not occur and instead the decay $\tilde{t} \to \tilde{b}W^+$ might have a substantial branching ratio [310, 311]. One of the experimental signatures for this channel would be the $\not{P}_\text{t}\ell j$ topology caused by the secondary decays $\tilde{b} \to d\nu$ (R_P-violating, via λ'_{131}) and $W \to \ell\nu$.

Actually, H1 has observed six $\not{P}_\text{t}\ell j$ events (five with $\ell = \mu^\pm$, one with an electron) in the 1994–97 e^+p data [88], where 0.8 ± 0.2 (2.4 ± 0.5) events from SM processes are expected for the muon (electron) channel. An event display of one of the muon events is shown in Fig. 4.51. In Fig. 4.52, the kinematic properties of these events are compared with simulations of the production and subsequent leptonic decay of W bosons and of reactions of the type $ep \to \mu^+\mu^-X$ (Bethe–Heitler muon pair production), which are the dominant sources of $\not{P}_\text{t}\ell j$ topologies in the SM. At least two of the H1 events (labeled $\mu 1$ and $\mu 2$) have larger hadronic transverse momentum and a lower value of the transverse lepton–neutrino mass,

$$M_\text{T}^{\ell\nu} = \sqrt{\left(\left|\boldsymbol{P}_\text{t}^\text{miss}\right| + \left|\boldsymbol{P}_\text{t}^\ell\right|\right)^2 - \left(\boldsymbol{P}_\text{t}^\text{miss} + \boldsymbol{P}_\text{t}^\ell\right)^2}\,, \tag{4.11}$$

than expected for SM processes ($\boldsymbol{P}_\text{t}^\text{miss}$ and $\boldsymbol{P}_\text{t}^\ell$ are the vectors of the transverse missing and lepton momenta, respectively). The exceptionality of these events may be illustrated by the fact that already the observation of the

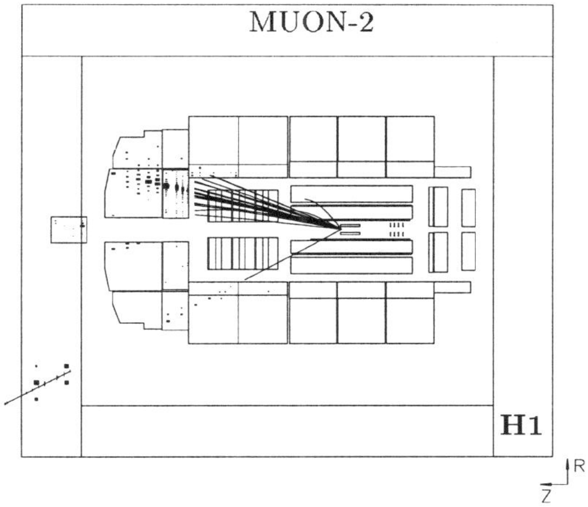

Fig. 4.51. One of the events with $\not{P}_{\mathrm{t}}\ell j$ topology observed by H1. The isolated track pointing downwards is identified as muon with positive charge and a transverse momentum of $p_{\mathrm{t}} = 28\,\mathrm{GeV}$; the jet in the upper hemisphere has a transverse momentum of $P_{\mathrm{T}}^{X} = 68\,\mathrm{GeV}$. The missing P_{T} amounts to 43 GeV. The figure has been reproduced from [88]

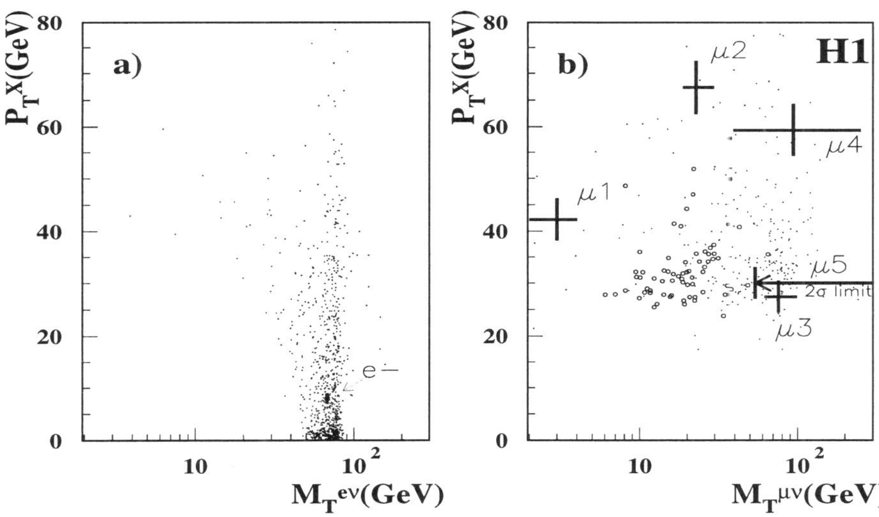

Fig. 4.52. Distribution of the H1 events with $\not{P}_{\mathrm{t}}\ell j$ topology in the plane of hadronic transverse momentum, P_{T}^{X}, vs. neutrino–lepton transverse mass, $M_{\mathrm{T}}^{\ell\nu}$; (**a**) for $\ell = e$ and (**b**) for $\ell = \mu$. The *crosses* indicate the position and 1σ uncertainties of the observed events; the *dots* (*open circles*) represent the expected distributions for W production (Bethe–Heitler muon pair production). The latter two reactions are taken from simulations corresponding to the 500-fold integrated luminosity of the data. The plots have been reproduced from [88]

first muon event ($\mu1$) in 1994 [449] had triggered a lot of discussion. However, although it may seem that the $\not{P}_{\mathrm{t}}\ell j$ events support the hypothesis of R_P-violating squark production at HERA, there are two facts that contradict this interpretation: (i) the rate of such events observed by ZEUS does not exceed the SM expectation [83]; (ii) a search for the $\tilde{t} \to \tilde{b}W^{+}$ decay in the channel $\tilde{b}W^+ \to d\nu q'\bar{q}$ in the H1 data did not yield evidence for an anomalous rate of such reactions, restricting the expected number of events with a $(W \to \nu\mu) + \text{jet}$ final state to below 0.36 at 95% C.L. for the specific case of $M_{\tilde{t}} = 200\,\text{GeV}$ and $M_{\tilde{b}} = 100\,\text{GeV}$ [450, 451]. The latter observation generally disfavors explanations of the $\not{P}_\text{t}\ell j$ events involving a W decay as the source of the lepton. This applies in particular to the decay $t \to bW$ of top quarks, which might be singly produced in CC reactions, $e^+b \to \bar{\nu}t$, or in quark-flavor-changing neutral currents, $ec \to et$, as speculated in [452].[20]

In summary we conclude that no convincing evidence has been found at HERA for the production of R_P-violating squarks. Although there are observations which intriguingly point into this direction, like the resonance-like pattern of the H1 (and, to some degree, also ZEUS) high-Q^2 data or the H1 events with $\not{P}_\text{t}\ell j$ topology, these pieces of evidence are not corroborated by further investigations and do not form a consistent picture.

4.4.2 Contact Interactions

The search for contact interactions (CIs) at HERA is based on the comparison of measured cross sections or event distributions with the predictions derived from the (DIS + CI) cross section obtained by using (2.60). The main information is contained in the Q^2 dependence of $\text{d}\sigma/\text{d}Q^2$, which is directly affected by the terms of $\mathcal{O}(Q^2/\Lambda^2)$. The cross section modification induced by a CI becomes most prominent at high Q^2 (see Fig. 2.23), where, however, the statistical precision of the data is lowest. As a consequence, CI analyses are most sensitive if they use the full Q^2 spectrum above a few hundred GeV^2. CI searches are usually performed by fitting the simulated expectation for a given CI scenario to the observed data, and by determining the most likely value of $1/\Lambda^2$ and its confidence interval from the χ^2 or likelihood returned by the fit.

In 1997, when the HERA experiments reported the high-Q^2 excess and CIs were conjectured as a possible explanation [227, 265, 329–335], initial CI exclusion limits had already been published by H1 ([61], using the $e^\pm p$ data collected until 1994) and by ZEUS ([57], 1993–95 $e^\pm p$ data). Both collaborations had investigated purely chiral CI scenarios coupling to u or d quarks, i.e. those for which the η^q_{mn} in Table 2.4 are nonzero for only one combination of m and n. No indication for the existence of such CIs had been found, and corresponding 95% C.L. lower limits on Λ of typically $1-2\,\text{TeV}$ had been placed by both collaborations. Despite the fact that these CIs are parity-violating and that the resulting limits are much weaker than those from atomic parity violation, they indicate the sensitivity of the analyses and in particular demonstrate that the early data would not have permitted the detection of

[20] These reactions would require the presence of an intrinsic bottom or charm component, respectively, of the proton to occur with rates detectable at HERA (see Sect. 2.2.1).

CIs with effective mass scales of $\mathcal{O}(3-4\,\mathrm{TeV})$, as suggested as explanations for the HERA high-Q^2 excess (see Sect. 2.2.2).

Meanwhile, CI results based on the full set of 1994–97 e^+p data are available from ZEUS [79] and H1 [86]. In the following, we will discuss the ZEUS analysis in some detail to illustrate the methods and statistical issues relevant for CI studies at HERA.

The analysis method pursued by ZEUS is based on a comparison of the distribution of "measured" MC-simulated quantities with the corresponding experimental data, i.e. it involves neither unfolding nor acceptance corrections. This approach automatically takes correctly into account the correlations between the hypothesized cross section modifications and the pattern of event migrations. The CIs are included in the simulation by applying to each MC event a weight

$$w = \frac{\mathrm{d}^2\sigma_{\mathrm{SM+CI}}}{\mathrm{d}x\,\mathrm{d}Q^2} \Bigg/ \frac{\mathrm{d}^2\sigma_{\mathrm{SM}}}{\mathrm{d}x\,\mathrm{d}Q^2} , \tag{4.12}$$

evaluated at the true kinematic variables characterizing the hard subprocess.[21] The cross sections in (4.12) are calculated in LLA; higher-order loop corrections to the numerator are undefined since CIs are a nonrenormalizable effective theory (see Sect. 2.2.2). It is expected that the dominant radiative corrections, i.e. those involving radiation of one or more real photons, mostly cancel in the cross section ratio.

The most likely value of Λ is determined for each CI scenario using an unbinned log-likelihood technique. The log-likelihood function, L, is calculated from the event distribution in x and y (which contains the full experimental information on the doubly differential cross section):

$$L(\epsilon/\Lambda^2) = -\sum_i \ln p_i(x_i, y_i; \epsilon/\Lambda^2) , \tag{4.13}$$

where the sum runs over all selected events inside the kinematic region inside the bounds $Q^2 > 500\,\mathrm{GeV}^2$ and $0.04 < x, y < 0.95$. The value of the probability density $p(x, y; \epsilon/\Lambda^2)$ is determined for each observed event from the local density of MC events around its reconstructed (x, y) position. The (x, y) integral over $p(x, y; \epsilon/\Lambda^2)$ is normalized to unity for each value of ϵ/Λ^2, whereby the experimental information contained in the total number of observed events is discarded and L is determined from the shape of the experimental (x, y) distribution alone. This procedure is justified a posteriori by the fact that the modification of the total number of expected events in the kinematic region under study which is induced by a CI with a strength corresponding to the 95% exclusion limit is typically below the systematic

[21] Q^2 and x are calculated from the propagator four-momentum q according to (1.2) and (1.3), and the electron beam energy entering the calculation of y is reduced by E_γ if a photon with this energy is radiated in a narrow cone around the e beam direction.

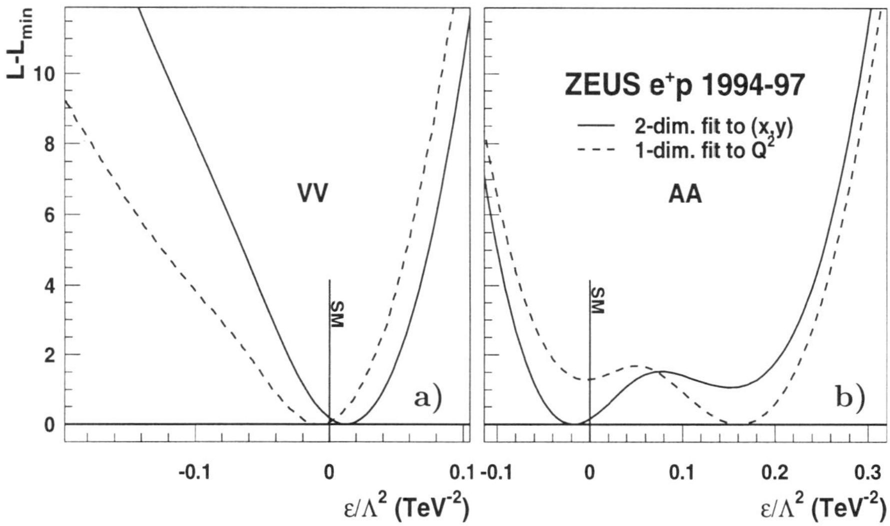

Fig. 4.53. Log-likelihood functions for (**a**) the VV and (**b**) the AA CI scenarios (see Table 2.4), obtained by ZEUS in the unbinned analysis of the (x, y) distribution (*solid lines*) and by the binned analysis of the Q^2_{DA} distribution (*dashed lines*). Both methods use different statistical approaches and differ in the treatment of systematic effects. In each figure, the regions given by $\epsilon = +1$ and $\epsilon = -1$ represent two different CI scenarios. The SM expectation, $\epsilon/\Lambda^2 = 0$, is indicated by the *vertical lines*. The plots have been reproduced from [79]

uncertainty of the luminosity measurement. As a cross-check of the unbinned analysis, a second independent log-likelihood function is calculated from the observed and expected event numbers in intervals of Q^2_{DA}, assuming that the probability in each interval is given by Poissonian statistics. This method makes use of the normalization information.[22] The results of both methods are generally in good agreement, as demonstrated by the example shown in Fig. 4.53. However, for a few scenarios, such as $X2^-$ and also VV^-, the log-likelihood function resulting from the (x, y)-based analysis rises faster with $1/\Lambda^2$ (and therefore yields stronger exclusion limits on Λ) than that resulting from the analysis of the Q^2_{DA} distribution. This observation is understood to reflect the fact that more input information is available to the (x, y) fit.

Note that the likelihood curves in Fig. 4.53 extend from negative to positive values of ϵ/Λ^2, although different ϵ correspond to different physics models. The most likely values of ϵ/Λ^2, ϵ/Λ_0^2, and the corresponding 95% C.L. confidence intervals defining the lower limits on Λ are therefore determined

[22] In this second method, systematic uncertainties such as that of the luminosity measurement are taken into account by including Gaussian errors of their amplitudes into the likelihood function, which is then minimized with respect to these amplitudes for each value of ϵ/Λ^2. No systematic effects are taken into account in the unbinned analysis at this stage.

separately for the positive and negative regions of ϵ/Λ^2 (referred to as "hemispheres" in the following). Whereas ϵ/Λ_0^2 is simply taken as the position of the absolute minimum of L in the respective hemisphere, the Λ limits cannot directly be inferred from the likelihood functions since the two-minimum structure occurring for some CI scenarios (see Fig. 4.53b) prohibits the interpretation of L in terms of a probability density. ZEUS instead uses Monte Carlo experiments (MCEs), i.e. samples of MC events corresponding to the data luminosity in which CIs are simulated by reweighting the events according to (4.12). For each MCE, the log-likelihood analysis is performed as a function of the assumed "true" value of Λ, Λ_{true}, for each of the CI scenarios under study. The lower limit of Λ at 95% C.L. for a given CI scenario is determined as the value of Λ_{true} at which 95% of the MCEs produce most likely values of $|\epsilon/\Lambda^2|$ larger than that found in the data. This technique is also used to take into account systematic effects by including appropriately modified MCEs in the limit-setting procedure.

The CI analysis by H1 [86] uses the measured differential cross section $\mathrm{d}\sigma/\mathrm{d}Q^2$ as input. A χ^2 function, $\chi^2(\epsilon/\Lambda^2)$, is determined from the comparison of measured and expected cross section values for each CI scenario. Uncorrelated and correlated systematic uncertainties (see Sect. 4.1.2) are taken into account appropriately in the calculation of χ^2. In deriving exclusion limits from $\chi^2(\epsilon/\Lambda^2)$, H1 pursues a different approach from ZEUS: both ϵ hemispheres of a pair of CI scenarios are interpreted simultaneously by determining one global minimum, ϵ/Λ_0^2, and assigning to it a two-sided confidence interval corresponding to 95% C.L. (one-sigma, two-sigma) significances by solving the equation $\chi^2(\epsilon/\Lambda^2) - \chi^2(\epsilon/\Lambda_0^2) = 3.84\,(1,\,4)$ for the respective values of ϵ/Λ^2. Note that the interpretation of these confidence intervals in terms of Λ limits must be performed with care if ϵ/Λ_0^2 deviates significantly from zero, i.e. if $\chi^2(0) - \chi^2(\epsilon/\Lambda_0^2) \gtrsim 1$. The H1 analysis was performed with the CTEQ5D PDFs [139] and cross-checked with the MRST [141] and GRV94 [117] sets. The values of ϵ/Λ_0^2 obtained with the different PDFs differ by less than 20% of the corresponding one-sigma confidence intervals. The quoted CI exclusion limits are the weakest resulting from any of these PDF sets.

The most likely values of ϵ/Λ^2 and the CI limits obtained in the ZEUS and H1 analyses are indicated in Figs. 4.54 and 4.55, respectively. In order to illustrate the experimental sensitivity, the modifications of the NC DIS data which would be expected in presence of VV CIs with a strength corresponding to the 95% C.L. exclusion limits are shown in Fig. 4.56 (ZEUS) and Fig. 4.57 (H1). In the cases where $\epsilon/\Lambda_0^2 \neq 0$, ZEUS has applied the MCE technique to estimate the probability, p_{SM}, that a statistical fluctuation in an experiment with the SM cross section would produce a value of $|\epsilon/\Lambda_0^2|$ larger than that found in the data. For all CI scenarios in Fig. 4.54, p_{SM} is found to exceed 15%, indicating that no deviation from the SM expectation is present in the ZEUS data which carries significant evidence for CIs. The same conclusion can be reached for the H1 data from the fact that the SM

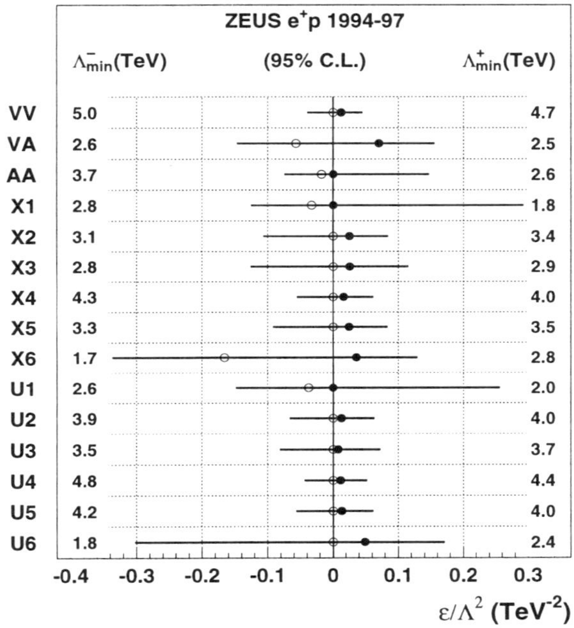

Fig. 4.54. CI exclusion limits from the ZEUS analysis (1994–97 e^+p data). The CI scenarios are defined in Table 2.4. The *horizontal bars* indicate the 95% C.L. confidence intervals on ϵ/Λ^2; the *filled* (*open*) *circles* show the values of ϵ/Λ_0^2 for $\epsilon = +1$ ($\epsilon = -1$). The lower limits on Λ are indicated at the margins. The figure has been reproduced from [79]

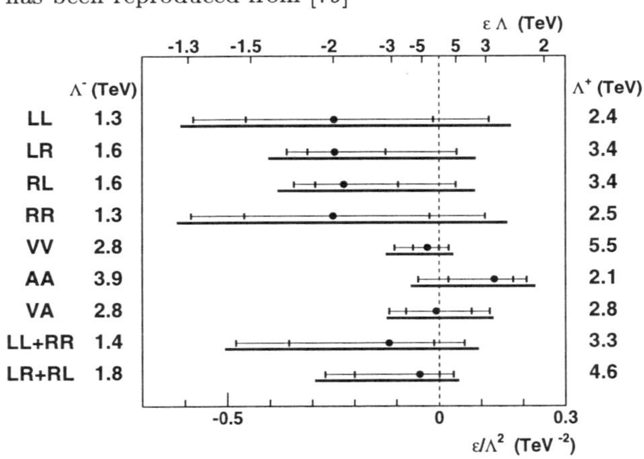

Fig. 4.55. CI limits from the H1 analysis (1994–97 e^+p data). The *thick horizontal bars* represent the 95% C.L. confidence intervals of ϵ/Λ^2 including PDF uncertainties; the *horizontal error bars* indicate the one- and two-sigma confidence intervals around ϵ/Λ_0^2 (*filled circles*) obtained using the CTEQ5D PDF set. The CI scenarios LL + RR and LR + RL correspond to X3 and X4, respectively, in the notation of this book. The lower limits on Λ are indicated at the margins. The figure has been reproduced from [86]

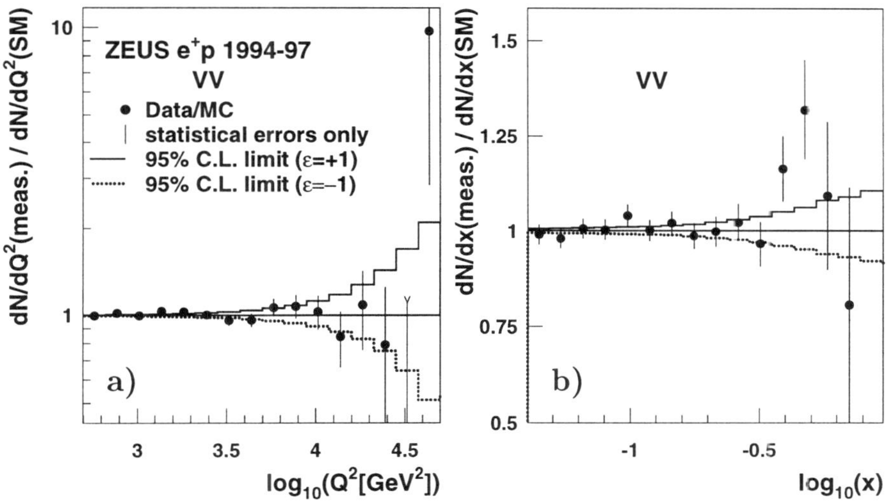

Fig. 4.56. Sensitivity of (**a**) the Q^2 and (**b**) the x distributions to CIs of the VV type (ZEUS results, 1994–97 e^+p data). The *filled circles* show the ratio of the observed distributions to the SM predictions, with the *error bars* indicating statistical uncertainties only. The *full* (*dotted*) *lines* represent the expected modifications due to CIs with a strength corresponding to the 95% C.L. exclusion limits for $\epsilon = +1$ ($\epsilon = -1$) (see Table 4.6). Note that the underlying binned distribution is used for presentational purposes only. The plots have been reproduced from [79]

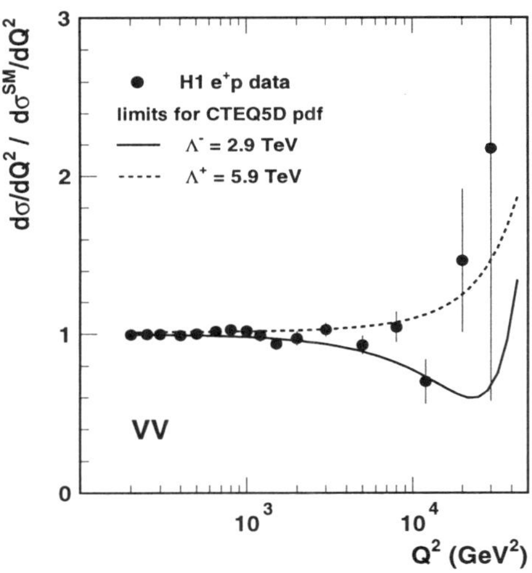

Fig. 4.57. Sensitivity of $\mathrm{d}\sigma/\mathrm{d}Q^2$ to CIs of the VV type (H1 results, 1994–97 e^+p data). The *filled circles* show the ratio of the measured cross sections to the SM predictions, with the *error bars* indicating statistical and uncorrelated systematic uncertainties combined in quadrature. The *full line* (*dashed line*) represents the expected modifications due to CIs with a strength corresponding to the 95% C.L. exclusion limits for $\epsilon = -1$ ($\epsilon = +1$) (see Table 4.6). The plot has been reproduced from [86]

expectation, $\epsilon/\Lambda^2 = 0$, is inside the two-sigma confidence intervals for all CIs under study. In particular, CI scenarios like VV with $\Lambda = 3-4\,\mathrm{TeV}$, which have been suggested as explanations of the HERA high-Q^2 excess events (see Sect. 2.2.2), are not compatible with the results of both experiments.

Table 4.5. Lower limits at 95% C.L. on the ratio $M_{\rm LQ}/\lambda$ derived for high-mass LQs in a CI-type analysis (H1 results, 1994–97 e^+p data). The LQ names and quantum numbers correspond to those listed in Table 2.3. The chiralities of the LQ couplings to e and q and their relative strengths are specified by $m, n = \rm L, R$ and $\tilde{\eta}^{u,d}_{mn}$, respectively (see text). The table has been reproduced from [86]

LQ	mn	$\epsilon\,\tilde{\eta}^u_{mn}$	$\epsilon\,\tilde{\eta}^d_{mn}$	$\frac{M_{\rm LQ}}{\lambda}\Big\vert_{\rm min}$	LQ	mn	$\epsilon\,\tilde{\eta}^u_{mn}$	$\epsilon\,\tilde{\eta}^d_{mn}$	$\frac{M_{\rm LQ}}{\lambda}\Big\vert_{\rm min}$
$S_0^{\rm L}$	LL	+1/2	—	0.62 TeV	$V_0^{\rm L}$	LL	—	−1	0.67 TeV
$S_0^{\rm R}$	RR	+1/2	—	0.57 TeV	$V_0^{\rm R}$	RR	—	−1	0.55 TeV
$\tilde{S}_0^{\rm R}$	RR	—	+1/2	0.22 TeV	$\tilde{V}_0^{\rm R}$	RR	−1	—	0.41 TeV
$S_{1/2}^{\rm L}$	LR	−1/2	—	0.34 TeV	$V_{1/2}^{\rm L}$	LR	—	+1	0.38 TeV
$S_{1/2}^{\rm R}$	RL	−1/2	−1/2	0.32 TeV	$V_{1/2}^{\rm R}$	RL	+1	+1	0.96 TeV
$\tilde{S}_{1/2}^{\rm L}$	LR	—	−1/2	0.45 TeV	$\tilde{V}_{1/2}^{\rm L}$	LR	+1	—	1.06 TeV
$S_1^{\rm L}$	LL	+1/2	+1	0.42 TeV	$V_1^{\rm L}$	LL	−2	−1	0.45 TeV

A direct interpretation of CI limits in terms of high-mass LQ phenomenology, i.e. for LQs with masses exceeding the HERA center-of-mass energy, has been reported by H1 [86]. For e^+p scattering, each of the LQs listed in Table 2.3 would contribute to the cross section, predominantly via coupling to the valence quarks in s-channel processes for $F=0$ and in the u-channel for $|F|=2$. The effective CI mass scale and interference sign are related to the LQ parameters via

$$\epsilon\,\eta^q_{mn}\,\frac{4\pi}{\Lambda^2} = \epsilon\,\tilde{\eta}^q_{mn}\left(\frac{\lambda}{M_{\rm LQ}}\right)^2 . \tag{4.14}$$

The chirality indices $m, n = \rm L, R$ as well as the sign ϵ and the coefficients $\tilde{\eta}^q_{mn}$ are fixed by the LQ quantum numbers. They are listed in Table 4.5, together with the 95% C.L. lower limits on $M_{\rm LQ}/\lambda$ inferred by H1 from the CI analysis under the assumption that the cross section only receives contributions from one LQ type, and that different charge states of a given LQ type are mass-degenerate. Note that all CIs in Table 4.5 are parity-violating, i.e. would contribute to the weak nucleon charge in (2.63) and are therefore severely constrained by atomic parity violation experiments; an $S_0^{\rm L}$ LQ with $M_{\rm LQ}/\lambda = 1\,\rm TeV$ would e.g. induce a modification of the weak charge of $\Delta^{\rm CI}Q_{\rm w} = 5.7$, almost 10 standard deviations above the experimental measurement (2.64). Nevertheless, the H1 results in Table 4.5 indicate the HERA sensitivity to LQs in the region of parameter space where $M^2_{\rm LQ} \gg s$ and λ is sufficiently small to render higher-order corrections small.

Finally, Table 4.6 presents a comparison of the ZEUS and H1 Λ limits shown in Figs. 4.54 and 4.55 with recent results obtained by the LEP experiments ALEPH [347], DELPHI [348], L3 [349] and OPAL [351] (from cross section measurements of fermion pair production in e^+e^- collisions) and by the Tevatron experiments CDF [343] and DØ [344] (from measurements of

Table 4.6. Lower 95% C.L. exclusion limits on the effective CI mass scales (in TeV) reported by the experiments at HERA (ep DIS), at the Tevatron (Drell–Yan e^+e^- production in $p\bar{p}$ collisions) and at LEP (fermion pair production in e^+e collisions) for the scenarios defined in Table 2.4. The HERA and Tevatron experiments are mainly sensitive to CIs coupling to u and d quarks, whereas the LEP results (except those from DELPHI and for U3 and U4) assume flavor-symmetric CIs

		HERA		Tevatron		LEP			
CI	ϵ	ZEUS [79]	H1 [86]	CDF [343]	DØ [344]	ALEPH [347][a]	DELPHI [348][b]	L3 [349][a]	OPAL [351]
VV	+	4.7	5.5	3.5	4.9	9.6	—	3.8	4.7
	−	5.0	2.8	5.2	6.1	9.2	—	5.0	7.2
AA	+	2.6	2.1	3.8	4.7	10.5	—	5.6	8.1
	−	3.7	3.9	4.8	5.5	10.4	—	3.5	4.2
VA	+	2.5	2.8	—	—	—	—	—	—
	−	2.6	2.8	—	—	—	—	—	—
X1	+	1.8	—	—	3.9	—	—	—	—
	−	2.8	—	—	4.5	—	—	—	—
X2	+	3.4	—	—	—	—	—	—	—
	−	3.1	—	—	—	—	—	—	—
X3	+	2.9	3.3	—	4.2	10.0	—	3.2	5.4
	−	2.8	1.4	—	5.1	9.7	—	2.8	4.4
X4	+	4.0	4.6	—	3.9	3.3	—	2.4	3.4
	−	4.3	1.8	—	4.4	5.6	—	3.7	7.1
X5	+	3.5	—	—	—	—	—	—	—
	−	3.3	—	—	—	—	—	—	—
X6	+	2.8	—	—	4.0	—	—	—	—
	−	1.7	—	—	4.3	—	—	—	—
U1	+	2.0	—	—	—	—	—	—	—
	−	2.6	—	—	—	—	—	—	—
U2	+	4.0	—	—	—	—	—	—	—
	−	3.9	—	—	—	—	—	—	—
U3	+	3.7	—	—	—	—	—	4.3	6.4
	−	3.5	—	—	—	—	—	3.5	7.8
U4	+	4.4	—	—	—	—	—	1.8	2.5
	−	4.8	—	—	—	—	—	2.2	4.2
U5	+	4.0	—	—	—	—	—	—	—
	−	4.2	—	—	—	—	—	—	—
U6	+	2.4	—	—	—	—	—	—	—
	−	1.8	—	—	—	—	—	—	—
LL	+	—	2.4	2.5	3.3	8.0	1.4/3.4	3.8	5.5
	−	—	1.3	3.7	4.2	8.0	3.4/4.0	2.5	3.1
LR	+	—	3.4	2.8	3.4	3.8	2.1/2.4	3.0	3.8
	−	—	1.6	3.3	3.6	4.1	2.5/2.1	3.2	4.4
RL	+	—	3.4	2.9	3.3	2.3	2.3/2.0	2.3	2.7
	−	—	1.6	3.2	3.7	5.0	2.3/2.7	4.3	6.4
RR	+	—	2.5	2.6	3.3	6.0	1.7/2.7	2.8	3.5
	−	—	1.3	3.6	4.0	5.5	3.0/2.0	3.5	4.9

[a] Results are (partly) preliminary.

[b] Results are for CIs coupling to $u\,/\,(d \text{ or } s)$ quarks and have been scaled by $\sqrt{2}$ to compensate for the value of $\eta_{mn} = 1/2$ used by DELPHI

the Drell–Yan electron pair production cross section in $p\bar{p}$ reactions). With the few exceptions that are explicitly stated, all limits in this table are for CIs which couple equally strongly to all quark flavors. The relative sensitivity to different CI scenarios depends on the SM $\times$ CI interference sign, which is opposite in e^+p scattering on the one hand from that in e^+e^- and $p\bar{p}$ scattering on the other. The consistency of the LEP, Tevatron and HERA limits with respect to the sign convention used has not been verified in producing Table 4.6; the assignment of the limits to ϵ has been adopted from the original papers.

For most of the CI types, the HERA limits are of similar magnitude to the corresponding results from LEP and the Tevatron, although a few of the LEP limits (in particular those from ALEPH [347]) are significantly stronger than the results from HERA. It should be noted, however, that this is only true for flavor-symmetric CIs. Limits for CIs which couple only to first-generation quarks would differ only by a small amount from those reported here for ep or $p\bar{p}$ scattering, but would be significantly weaker in the case of LEP (e.g. the results reported by OPAL [351] for CIs coupling only to u and d quarks).

In the future, enhanced sensitivity to CIs can be expected at HERA owing to increased data samples, to the complementary information contained in high-statistics e^-p data sets and to the availability of longitudinal e polarization. A detailed study of the corresponding future discovery potential can be found in [353].

5. Outlook

We have now investigated the phenomenology of high-Q^2 NC and CC ep reactions at HERA in quite some detail. We have seen how both the strong and the electroweak sectors of the standard model (SM) come into play when deriving precise predictions for the cross sections in this kinematic regime, and how hypothesized phenomena beyond the SM could modify these predictions. In the last sections we have confronted the SM predictions and various theoretical scenarios beyond the SM with the experimental data. The precision with which these theoretical conjectures can be tested has increased in an impressive way since HERA came into operation, and a lot has been learned since that time. In general, we can draw one central conclusion: the SM predictions and the basic concepts of describing deep inelastic scattering (DIS) hold in the HERA kinematic region, at least to the level of accuracy achieved with the existing data.

Yet, many questions still await conclusive answers. Are the few peculiar deviations of the HERA data from the SM predictions just statistical fluctuations, or are they the first signs of new physics phenomena, and – if so – of what nature are they? Are the standard assumptions used for the QCD-based parameterizations of parton distribution functions (PDFs) correct, or are modifications required at high x? To what extent can the HERA experiments contribute to the spectrum of high-precision measurements of the electroweak SM ingredients? Will the measurements in the perturbative regime of QCD help us to understand better the transition region to low Q^2, where pQCD methods are inapplicable? Trying to predict the developments of the forthcoming years in the exciting field of ep physics is of course difficult and, concerning the results, almost impossible. Nevertheless, some contours of the future of particle physics in general and HERA operation in particular are already discernible, and we want to summarize them briefly in the following. A detailed account of the physics potential of HERA is given in [43].

The data discussed in this report have mainly been recorded in the years 1994–97. Since then, the proton beam energy has been raised to 920 GeV (corresponding to a 6% increase of the ep center-of-mass energy), and the HERA vacuum system has been improved significantly, allowing the collider to run efficiently with electrons as well as positrons. Roughly 15 pb^{-1} of e^-p data have been delivered to both ZEUS and H1 in 1998–99, of which a few initial, preliminary results have been shown. Since May 1999, another period of e^+p data taking has been in progress, scheduled to last until September 2000. All in all, substantially increased e^+p data sets and about $15\,\text{pb}^{-1}$ of e^-p data per experiment will be available by the end of 2000 for precise combined e^-p and e^+p analyses. These studies will increase the accuracy of the current measurements and in addition allow us to disentangle the structure functions $\mathcal{F}_2$ and $\mathcal{F}_3$ for NC and CC scattering, which in turn will constitute an important input for future PDF fits and QCD analyses. We can thus hope that the new data will contribute significantly to our understanding of the PDFs and in particular may shed light on the high-x behavior of the u- and d-quark distributions. In addition, the searches for anomalies in these data will help to investigate the origin of some interesting (some may say, intriguing or exciting) findings described in this book. They will explore new regions of the parameter space for various hypothesized phenomena beyond the SM and will in particular profit from the complementary sensitivity of the e^+p and e^-p data to many of these scenarios. The running period in 2000 will be the last one in the first phase of HERA operation. It is hoped that a refined understanding of high-Q^2 DIS at HERA will emerge from these data and serve as the solid fundament needed for the interpretation of the high-statistics and high-precision measurements of ep scattering with longitudinally polarized e beams expected from the second HERA phase.

For this second phase of HERA operation (HERA2000), a major upgrade of the accelerator complex is foreseen. In a nine-month shutdown after the 2000 data-taking period, the ZEUS and H1 interaction zones will be rebuilt, and the spin rotators needed to provide longitudinal electron polarization to both experiments will be installed. The goal is to increase the instantaneous luminosity in the HERA2000 phase by a factor of about five and to allow each experiment to collect a total of $1000\,\text{pb}^{-1}$ of ep data with longitudinally polarized e^+ and e^- beams until about 2005. This would correspond to an increase of the event samples by an order of magnitude and an associated gain of precision, which reflects not only the increase of statistical accuracy but also the detailed control of systematic effects which will become possible with these large data sets.

Both collider experiments will undergo major modifications before resuming data taking in 2001. In order to achieve the luminosity increase, magnets will have to be introduced into the central detector volumes close to the beam pipe, causing acceptance losses in the forward and backward directions. On the other hand, ZEUS plans to improve the tracking-measurement precision and acceptance by the installation of a silicon microvertex detector (MVD) and of a new forward straw tube tracking detector (STT) which will replace the TRDs. Both detector components are scheduled to be operational from 2001 onwards. The data-taking conditions will be substantially different in the HERA2000 phase owing to the high reaction rates; it will in particular not be possible to trigger on processes with high cross sections. Consequently, it is planned to concentrate on investigations of rare reaction classes, for which the results after the first HERA phase are still statistics-limited. Typically, such processes have high transverse momenta and little relevant particle flow in the forward and backward directions, where the acceptance will be restricted by the magnets. High-Q^2 DIS will be one of the central issues, and the longitudinal e polarization will open additional degrees of freedom for extracting SM parameters and searching for new phenomena. In the case of finding evidence for physics beyond the SM, the e polarization will be extremely helpful for distinguishing different phenomenological scenarios, e.g. by giving direct access to the chirality of the couplings involved.

We have seen that many hypothesized scenarios of new physics in ep scattering imply observable signatures also in other reactions, and that a coherent picture only arises from the complete set of all experimental data. We therefore want to complement the outlook for the forthcoming years of ep physics at HERA by a glance at the projected future of e^+e^- and $p\bar{p}$ experiments. At the Tevatron, a new main injector is currently being installed, and both experiments (CDF and DØ) will undergo major upgrades. From 2001 onwards, they will take high-quality $p\bar{p}$ data at large rates. Within a few years from 2001, data samples corresponding to integrated luminosities of several inverse femtobarns are expected to be available, extending significantly the sensitivity and kinematic range of the Tevatron physics program. After 2005, the pp collider LHC at CERN will deliver the first data at a center-of-mass energy of 14 TeV, which will access an unexplored, new kinematic region of parton–parton reactions. In the e^+e^- sector, LEP will terminate its successful operation in September 2000, delivering data at the record center-of-mass energy of presumably 206 GeV until then. Thereafter, e^+e^- scattering at high energies will have a break. Its future depends on the decision as to which of the options for linear e^+e^- colliders at center-of-mass energies of $\mathcal{O}(0.5-1\,\mathrm{TeV})$ will be realized. Intense technological research and development programs are under way for several such projects, e.g. at DESY for the TESLA collider. The TESLA technical design report is expected to be completed by the beginning of 2001. In addition to the e^+e^- mode, TESLA might also be operated in combination with HERA to provide ep collisions at a center-of-mass energy of roughly 1 TeV. A similar option for the CERN LEP/LHC ring, which has been discussed previously, appears unrealistic today since the LEP accelerator will be dismantled before the installation of the LHC.

Different ideas exist on possible particle accelerators in the even more distant future, amongst them multi-TeV e^+e^- colliders, pp or $p\bar{p}$ synchrotrons of even higher energies and luminosities than LHC, and also muon accelerators, including the use of the high-intensity neutrino beams they produce. Also, it might become feasible to an increasing extent to exploit measurements of cosmic particle fluxes and reactions for extracting particle physics results. We will conclude the outlook with this wide spectrum of possible future activities before completely getting lost in speculations ...

6. Summary

The electron–proton collider HERA at DESY in Hamburg has opened a new chapter in the exploration of deep inelastic lepton–nucleon scattering (DIS) and continues the success story of this field of particle physics, which had before paved the way for our current understanding of the interplay of the fundamental forces in the standard model (SM) and of the structure of the nucleon.

By the end of 1996, the two multipurpose *ep* experiments ZEUS and H1 at HERA had collected a total of more than $35\,\mathrm{pb}^{-1}$ of e^+p data and small samples of e^-p data. These data allowed for the first time statistically significant investigations of both neutral-current (NC, $ep \to eX$) and charged-current (CC, $ep \to \nu X$) DIS at very high momentum transfers, $Q^2 \gtrsim 1000\,\mathrm{GeV}^2$. New kinematic regions, where the DIS cross sections are in the subpicobarn regime, became accessible to exploration, thus opening the door to a wide spectrum of possible new findings and discoveries. A lot of excitement was triggered when both experiments reported an excess of NC events over the SM prediction at $Q^2 \gtrsim 15\,000\,\mathrm{GeV}^2$ and at high values of the Bjorken scaling variable, $x \gtrsim 0.4$. The evidence for these claims was based on a few instances of *ep* reactions and the comparison of their rate with the corresponding expectations, which are determined by extrapolating lower-energy measurements into the HERA regime of kinematics by using the SM concepts.

Meanwhile both experiments have analyzed e^+p data corresponding to a total luminosity of more than $80\,\mathrm{pb}^{-1}$. The data taken in 1997 neither confirm nor exclude the existence of a high-Q^2 NC anomaly. Slight excesses in high-Q^2 charged-current scattering are intriguing but not conclusive and can be explained within the SM framework, e.g. by a deviation of parton distributions at high x from the standard parameterizations. Nevertheless, the original observation of a discrepancy between data and theoretical expectation has initiated a lot of ongoing activity. The experiments substantially improved their understanding of systematic uncertainties and investigated biases introduced by the specific analysis procedures applied. On the theoretical side, the sources of uncertainty for the high-Q^2 cross section prediction were studied in detail, many suggestions were made as to which as yet neglected effects might modify the SM prediction and various scenarios of phenomena beyond the SM were related to the HERA excess events.

What we have summarily called the SM prediction above is in fact a complex interplay of experimental measurements, the electroweak sector of the SM and perturbative applications of QCD (pQCD). Our contemporary understanding of deep inelastic lepton–nucleon scattering is based on the concept that it proceeds via a hard subprocess ($eq \to eq$ scattering in the simplest case) described by matrix elements involving the electroweak part of the SM Lagrangian, and that the flux of quarks is given by process-independent parton distribution functions (PDFs). The x dependence of the PDFs includes all nonperturbative aspects of QCD dynamics inside the nucleon, while their Q^2 dependence follows from the DGLAP evolution equations, which result from pQCD.

The usual procedure to evaluate DIS cross sections is to suitably parameterize the PDFs as functions of x, perform the QCD evolution, calculate the predictions for the experimentally observed quantities and determine iteratively the PDF parameters by χ^2-minimizing the difference between predictions and measurements. Uncertainties of the resulting cross section expectations for the HERA high-Q^2 regime come from various sources, such as the statistical and systematic errors of the experimental input data, effects of neglected higher orders in the perturbation series, the treatment of electroweak radiative corrections, the experimental uncertainties of the SM parameters and of the strong coupling constant, the theoretical treatment of the heavy charm and bottom quarks in the initial and final states, and the procedure applied to derive nucleon PDFs from data involving heavier nuclei. For NC ep reactions, the dominant contribution to cross section uncertainties comes from the errors of the experimental measurements at lower Q^2 and amounts to typically 3 – 8% in the region of high x and Q^2. Higher orders of pQCD, electroweak radiative corrections, heavy-quark effects and errors of SM parameters are demonstrated to affect the NC cross sections at the percent level or below, except in some restricted kinematic regions which are not of particular relevance for the physics discussed in this book. For CC ep reactions, the SM prediction requires us to disentangle the u and d distributions (which contribute to the low-energy data mainly in the combination $4u + d$). As a consequence, the resulting uncertainties are larger than in the NC case and exceed 20% at $Q^2 \gtrsim 2 \times 10^4\,\mathrm{GeV}^2$ or $x \gtrsim 0.5$ for e^+p scattering. In addition, the determination of the d and u PDFs to date depends sensitively on fixed-target data taken with deuterium targets, and, as pointed out recently, a consistent relativistic treatment of the deuteron may imply that the d/u PDF ratio does not tend to zero as x approaches unity (which is assumed in most contemporary PDF analyses). This effect may cause a significant increase of the predicted e^+p CC cross section at high x (more than 20% for $x \gtrsim 0.5$).

Variations of the standard PDF scheme which have been suggested and could modify the high-x, high-Q^2 cross sections include the addition of a small "hard" component $\propto (1 - x)^{0.1}$ to the PDFs, and the concept of intrinsic charm, which conjectures that the proton wave function contains a small fraction of valence-like (anti)charm quarks. These scenarios do not change the NC cross sections by more than $\mathcal{O}(10\%)$, but they may yield significant enhancements of the CC cross sections at high x, which are at the verge of the sensitivity reached by the HERA experiments.

An exciting option is the interpretation of discrepancies between experimental data and the SM prediction in terms of new physics phenomena which are not included in the SM. Various such scenarios have been discussed in detail in their impact on the HERA cross sections and on a large spectrum of other processes, from parity violation in atoms at energy scales of eV to $p\bar{p}$ scattering at a center-of-mass energy of 1.8 TeV at the Tevatron. The most prominent of these scenarios are the resonant production of a new state (generically called a leptoquark, LQ) coupling to both electrons and quarks, the special case in which this LQ is the supersymmetric partner of a quark in an extension of the minimal supersymmetric model allowing for violation of the R-parity, and a generalized class of new processes at mass scales beyond the HERA domain which effectively can be described in terms of contact interaction contributions to the HERA cross sections.

Resonant LQ production at HERA causes a narrow peak in $\mathrm{d}\sigma/\mathrm{d}x$ and is furthermore characterized by a harder distribution of the Bjorken variable y than DIS reactions with the same x, yielding DIS-like events with particularly high Q^2. An LQ interpretation of the HERA excess events would imply an LQ with a mass of 200 − 225 GeV coupling to valence quarks and positrons, i.e. having a fermion number $F = 0$. The LQ–quark–positron Yukawa coupling of such an LQ would be a few times 0.01. Severe constraints on LQ parameters from low-energy experiments and in particular from direct searches for LQ pair production at the Tevatron suggest that such an LQ should be scalar and have a low branching ratio for a decay into electron–quark final states. These requirements almost rule out any of the "classical" LQ types which respect $\mathrm{SU}(2)_\mathrm{L} \times \mathrm{U}(1)_\mathrm{Y}$ symmetry and only couple to leptons, quarks and gauge bosons.

Squarks in supersymmetric models with R-parity violation can have LQ-type couplings but are different from the LQs described above because they also decay in R-parity-conserving modes and can hence have low eq branching ratios. The preferred scenarios would be the production of $\tilde{c}$ or $\tilde{t}$ squarks from d quarks or, less likely, the production of $\tilde{t}$ squarks from s quarks. No CC-type events with the typical (1+1)-jet plus missing-P_t signature would be expected in these reactions.

Contact interaction (CI) contributions to the ep cross sections are calculated from effective four-fermion interactions characterized by the ratio of a coupling constant, g, and a mass scale, Λ. They represent a valid asymptotic low-energy approximation to the exact description of a wide class of hypothesized phenomena, including the exchange of new heavy gauge bosons or of LQ-type states, or compositeness at large mass scales. Any CI modifying

$eq \to eq$ matrix elements also affects $q\bar{q} \to e^+e^-$ and $e^+e^- \to q\bar{q}$ scattering and can hence be investigated at the Tevatron and LEP experiments. CIs are classified according to the chiralities and flavors of the fermions involved, and according to whether their interference with the SM amplitude is constructive or destructive. They mainly modify $\mathrm{d}\sigma/\mathrm{d}Q^2$, typically giving rise to a marked enhancement at the highest Q^2 and to a moderate depletion or excess at intermediate Q^2, depending on the interference sign. CIs are quantified by fixing $g^2 = 4\pi$ and determining the effective mass scale Λ. Scenarios which have been suggested as explanations for the HERA excess events are parity-conserving (to avoid stringent constraints from measurements of atomic parity violation) and have $\Lambda = \mathcal{O}(3-4\,\mathrm{TeV})$. However, the LEP and Tevatron experiments, which have recently increased their sensitivity to effective mass scale values of typically $5-10\,\mathrm{TeV}$, do not corroborate the existence of such CIs.

The ZEUS and H1 detectors have been designed with the goal to investigate ep reactions in all their manifestations, with a high priority assigned to DIS at high Q^2. Both detectors consequently provide comprehensive measurements of the final states of these reactions, as far as they are accessible to detection techniques. The detectors have almost hermetic calorimetric coverage, and yield precise tracking information in the central region. Nevertheless, it required long and tedious work to understand the detector responses in detail, to calibrate the individual components and to optimize the analysis methods for accuracy and efficiency. Many detailed investigations only became possible with the large data volumes recorded in 1996 and 1997. Both ZEUS and H1 are now able to reconstruct the kinematic variables of DIS events with a precision at the few-percent level for NC reactions, and of typically 25% for CC reactions. Event selection efficiencies, potential backgrounds, detector-related reconstruction biases and the effect of initial-state photon radiation have been investigated in detail and in general are found to be well reproduced by the Monte Carlo simulation.

Experimental results on high-Q^2 DIS reported by ZEUS and H1 comprise the distributions of kinematic variables and their comparison with SM expectations, the measurements and QCD analysis of singly and doubly differential NC and CC cross sections, tests of the electroweak aspects of the SM predictions, and searches for eq resonance production and for CI signatures. In addition, searches for R_P-violating squarks in decay modes other than to eq have been performed.

Event distributions have played a central rôle in the analyses investigating the high-Q^2 NC event excess in the 1994–96 data. Good agreement between measurements and predictions was found for $Q^2 \lesssim 15\,000\,\mathrm{GeV}^2$, but both ZEUS and H1 observed more events than expected at higher Q^2. The H1 events clustered around $x \approx 0.45$, corresponding to an eq invariant mass of about 200 GeV. The probability to observe an excess of the observed or larger significance anywhere in the kinematic region under study was estimated to be 1% for H1 and 7% for ZEUS. The data taken in 1997 confirmed neither the patterns observed before nor the existence of an excess at all. The excess cross section is about 0.4 pb in the 1994–96 data and less than half of this value in the combined 1994–97 data sample. Significances for the 1994–97 data sets have not been evaluated.

Differential NC cross sections $\mathrm{d}\sigma/\mathrm{d}Q^2$, $\mathrm{d}\sigma/\mathrm{d}x$ and $\mathrm{d}\sigma/\mathrm{d}y$ have been reported by ZEUS and H1 and agree within errors with each other and with the SM predictions. A slight and not particularly significant excess is seen in $\mathrm{d}\sigma/\mathrm{d}Q^2$ for $Q^2 > 20\,000\,\mathrm{GeV}^2$. The cross section $\mathrm{d}\sigma/\mathrm{d}x$ shows perfect agreement with the SM prediction for $Q^2 > 400\,\mathrm{GeV}^2$, up to Q^2 above $10\,000\,\mathrm{GeV}^2$. However, the x intervals in these measurements are large, and would tend to wash out resonance-like structures. The doubly differential cross section $\mathrm{d}^2\sigma/(\mathrm{d}x\,\mathrm{d}Q^2)$ has been measured by H1 in the kinematic region $150\,\mathrm{GeV}^2 < Q^2 < 30\,000\,\mathrm{GeV}^2$ and is found to be in good agreement with QCD fits based mainly on lower-Q^2 fixed-target data.

In the CC analyses, no significant discrepancies between prediction and observation have been found in the event distributions. Moderate data excesses are present in the tails of the x and Q^2 distributions. The singly differential cross sections with respect to Q^2, x and y, as well as $\mathrm{d}^2\sigma/(\mathrm{d}x\,\mathrm{d}Q^2)$, have been measured by ZEUS and H1. The results of both experiments agree with each other and with the SM prediction, except at high x and Q^2, where the same slight mismatch between data and expectation as in the event distributions is found. This discrepancy appears to be consistently explainable by an increased d quark distribution at high x, in accordance with a nonzero d/u ratio at $x \to 1$ as suggested by reanalyses of deuterium structure function data and also by QCD arguments.

Both collaborations have performed QCD fits with the twofold goal to test the agreement of their new high-Q^2 data with those at lower Q^2 and from fixed-target experiments, and to determine the uncertainty in the SM cross section prediction induced by the parton distributions. Very good agreement of the fit results with the measured NC and CC cross sections is found, both in the case where the input data to the fits are restricted to $Q^2 \lesssim 200\,\mathrm{GeV}^2$ and if the complete region covered by measurements is used. Over the full kinematic range, the HERA DIS data thus follow the pQCD evolution of structure functions measured at much lower energy scales. Even without using the CC data as input, the ZEUS fit results in an enlarged d quark density at high x which brings the CC data and QCD fit into good agreement also at the highest x and Q^2. In a complementary approach, H1 has extracted the u and d PDFs at $x = 0.25$ and 0.4 from the measured doubly differential NC and CC cross sections, using the result of the QCD fit to correct for small contributions of other quark flavors.

In addition to the QCD fits, ZEUS and H1 have also extracted information about the lepton–quark couplings and their helicity structure and about the gauge boson masses from the DIS cross sections. The agreement of $\mathrm{d}\sigma/\mathrm{d}y$

and of the $(1-y)^2$ dependence of the doubly differential cross sections with the corresponding SM expectations confirms the angular dependence of the electron–quark scattering amplitude as predicted in the SM from the spins and chiral couplings of the reacting particles. Clear evidence for the presence of the Z propagator is found in the highest-Q^2 NC data. The statistical precision of these data is, however, not yet sufficient to determine the Z propagator mass from the subtle cross section modifications it induces. In CC DIS, the situation is different since the cross section is directly proportional to the square of the W propagator. ZEUS has determined a W mass of $81.4^{+2.7}_{-2.6}$ (stat.) $\pm$ 2.0 (syst.) $^{+3.3}_{-3.0}$ (PDF) GeV from $\mathrm{d}\sigma/\mathrm{d}Q^2$; H1 finds a value of 80.9 GeV with similar errors. The good agreement of these results with the world average ($M_W = 80.371 \pm 0.03\,\mathrm{GeV}$) adds to the consistent picture and constitutes an important new test of the SM in the kinematic region of large space-like virtualities. Whereas M_W is inferred from the shape of the differential cross section $\mathrm{d}\sigma/\mathrm{d}Q^2$, its normalization (i.e. the total CC cross section) allows us to determine the Fermi constant, G_F. ZEUS has presented both measurements as an error ellipse in the (M_W, G_F) plane, which is found to be fully consistent with the corresponding world averages and the SM constraint relating G_F and M_W. In summary, all experimental evidence supports the validity of the SM in the kinematic realm explored by HERA. The HERA data are, however, not yet accurate enough to contribute significantly to the precision measurements of electroweak parameters performed at LEP, SLC and the Tevatron.

The resonance-like pattern of the high-Q^2 excess reported in 1997, which was particularly prominent in the H1 data, has triggered a lot of detailed studies of the experimental and phenomenological aspects of eq resonance production at HERA, resulting in improved search analyses. QCD and QED radiation are found to introduce sizable biases in the reconstructed mass values which strongly depend on y and on the reconstruction method. These biases tend to reduce the difference of the average masses assigned to the ZEUS and H1 excess events. Nevertheless, neither ZEUS nor H1 find convincing evidence for eq resonances in the NC- or CC-type decay channels, and report upper limits on the Yukawa coupling vs. mass of such objects. The scenario which has been suggested to explain the high-Q^2 excess events (leptoquarks with a mass around 200 GeV and a coupling of a few times 0.01) is almost excluded by these new studies, even though these events still induce bumps in the H1 limit curves. The e^+p data are particularly sensitive to resonances with fermion number $F=0$, which can be excluded up to masses of $270-280\,\mathrm{GeV}$ if they have an electromagnetic coupling strength and exclusively decay to eq. Similar results are obtained by ZEUS from the 1998–99 e^-p data for resonances with $|F|=2$. The HERA sensitivity to LQs is found to exceed that of the Tevatron and LEP experiments for LQs with masses beyond about 240 GeV or with low branching ratios to eq.

LQ resonances with other than eq decay modes could occur in supersymmetric models with violation of R-parity, in which squarks could couple to quarks and leptons. No evidence for squark production was found in searches for final states consisting of a lepton and multiple jets, which would be expected for several squark decay modes, with branching ratios depending on the parameters of supersymmetry. Some excitement was caused by the H1 observation of an unexpectedly high rate of events with an isolated muon, missing transverse momentum and at least one jet, which could be assigned to stop production with subsequent decay into a sbottom (going to νq) and a W disintegrating into $\mu\nu$. However, this interpretation is contradicted by the absence of signal events in the corresponding hadronic W decay channel and in the ZEUS data.

Searches for indications of CIs in the NC data have been performed by ZEUS and H1, based on the event distribution in (x, y) and on the cross section $\mathrm{d}\sigma/\mathrm{d}Q^2$, respectively. No significant evidence for the existence of eq CIs was found, and 95% C.L. exclusion limits in the range 1.3 – 5.5 TeV have been placed on the effective mass scales. None of the investigated CI scenarios is able to explain the high-Q^2 data while simultaneously remaining compatible with the lower-Q^2 data. The HERA CI limits are similar in magnitude to those recently published by the Tevatron and the LEP experiments. However, these results are complementary with respect to the quark flavors to which the CIs couple: the HERA and Tevatron analyses are predominantly sensitive to CIs of first-generation quarks, whereas the LEP studies assume quark flavor symmetry.

To conclude, the SM once more proves to be successful in predicting cross sections in unexplored regions. All HERA high-Q^2 data are in good agreement with the expectations, except at the highest x and Q^2, where moderately significant indications of a cross section excess are observed. Currently, it is impossible to decide whether ZEUS and H1 have had a first glimpse at new physics phenomena, or whether they observed a conspiracy of statistical fluctuations. Further clarification is expected from enlarged data samples to be available in the forthcoming years, in particular from the year 2001 onwards, when HERA will resume operation after a major upgrade and increased event rates as well as longitudinal electron beam polarization will become available to the experiments. We look forward with enthusiasm and excitement to the future of particle physics research in the realm of high-Q^2 deep inelastic scattering, to possible discoveries and surprises, and to high-precision measurements assessing those effects which are currently still hidden behind statistical uncertainties. We hope that this book may serve as a sound basis and guide in pursuing these investigations, and are curious to find out whether the results available in a few years' time will fit into the framework it describes.

List of Abbreviations, Acronyms and Symbols

Frequently Used Abbreviations

APV	atomic parity violation
CC	charged current
CI(s)	contact interaction(s)
CKM	Cabibbo–Kobayashi–Maskawa
C.L.	confidence level
DIS	deep inelastic scattering
FFNS	fixed-flavor-number scheme
GSW	Glashow–Salam–Weinberg
IC	intrinsic charm
IP	interaction point
ISR	initial-state radiation
LLA	leading-log approximation
LO	leading order
LQ(s)	leptoquark(s)
MC	Monte Carlo simulation
MCE(s)	Monte Carlo experiment(s)
NC	neutral current
NLLA	next-to-leading-log approximation
NLO	next-to-leading order
NWA	narrow-width approximation
PDF(s)	parton distribution function(s)
PDG	Particle Data Group
PM(s)	photomultiplier(s)
pQCD	perturbative quantum chromodynamics
QCD	quantum chromodynamics
QED	quantum electrodynamics
QPM	quark–parton model
RGE	renormalization group equation
R.M.S.	root mean square
SM	standard model
SUSY	supersymmetry
VFNS	variable-flavor-number scheme

Acronyms of Laboratories, Accelerators and Experiments

ALEPH	e^+e^- experiment, LEP
BCDMS	fixed-target μ–nucleus experiment, SPS
BEBC	Big European Bubble Chamber, fixed-target '$\overline{\nu}$'–nucleus experiments, SPS
CCFR	fixed-target '$\overline{\nu}$'–nucleus experiment, Tevatron
CDF	$p\bar{p}$ experiment, Tevatron
CDHS(W)	fixed-target '$\overline{\nu}$'–nucleus experiment, SPS
CERN	European Laboratory for Particle Physics, Geneva, Switzerland
CHARM	fixed-target '$\overline{\nu}$'–nucleus experiment, SPS
DØ	$p\bar{p}$ experiment, Tevatron
DELPHI	e^+e^- experiment, LEP
DESY	Deutsches Elektronen-Synchrotron, Hamburg, Germany
E605	fixed-target p–nucleus experiment, Tevatron
E665	fixed-target μ–nucleus experiment, Tevatron
E706	fixed-target π–, K– and p–nucleus experiment, Tevatron
E772	fixed-target p–nucleus experiment, Tevatron
E866	fixed-target pp and pd experiment, Tevatron
EMC	European Muon Collaboration, fixed-target μ–nucleus experiment, SPS
FNAL	Fermi National Accelerator Laboratory (Fermilab), Chicago, IL, USA
H1	ep experiment, HERA
HERA	ep collider ring, DESY
HERA-B	internal-target p–nucleus experiment, HERA
HERMES	internal-target e–nucleus experiment, HERA
L3	e^+e^- experiment, LEP
LEP	Large Electron Positron collider ring, CERN
LHC	Large Hadron Collider (pp), under construction, CERN
NA51	fixed-target pp and pd experiment, SPS
NMC	New Muon Collaboration, fixed-target μ–nucleus experiment, SPS
PETRA	e^+e^- collider ring, DESY
OPAL	e^+e^- experiment, LEP
SLAC	Stanford Linear Accelerator Center, CA, USA
SLC	Stanford Linear Collider (e^+e^-), SLAC
SLD	e^+e^- experiment, SLC
S$p\bar{p}$S	Super $p\bar{p}$ Synchrotron, $p\bar{p}$ collider, CERN
SPS	Super Proton Synchrotron, p accelerator, CERN
TESLA	projected linear e^+e^- collider, DESY
Tevatron	$p\bar{p}$ collider ring, FNAL
UA1, UA2	$p\bar{p}$ experiments, S$p\bar{p}$S
WA70	fixed-target πp and pp experiment, SPS
ZEUS	ep experiment, HERA

Frequently Used Symbols

$a \cdot b$	Lorentz-invariant scalar product of four-vectors a and b
α	fine-structure constant
α_{s}	QCD coupling constant
$\beta_{\ell,\nu}$	branching ratio $\mathrm{BR}(\mathrm{LQ} \to \ell q, \text{`}\bar{\nu}\text{'} q)$
BR(...)	branching ratio of the decay quoted in parentheses
$\chi^0, \chi^{\pm}$	neutralino, chargino (SUSY partners of gauge and Higgs bosons)
d	deuteron
e	electron (used generically for e^- and e^+)
$\tilde{e}$	selectron (SUSY partner of e)
ϵ	sign of CI amplitude
η^q_{mn}	coefficient of CI vector coupling of electron with chirality m to quark q with chirality n ($m, n = \mathrm{L}, \mathrm{R}$)
$\mathcal{F}^r_i$	proton structure function ($i = 1, 2, 3, \mathrm{L}$; $r = \mathrm{NC}, \mathrm{CC}$)
f	elementary fermion (lepton or quark)
G_{F}	Fermi constant
g	gluon
h	hadron
L	chirality index for left-handed fermion fields or couplings
ℓ	charged lepton, $\ell = e, \mu, \tau$
Λ	effective mass scale of contact ineractions
λ	lepton–quark–LQ Yukawa coupling (also denoted by $g_{\mathrm{R,L}}$)
λ'_{ijk}	lepton–quark–squark Yukawa coupling in R_P-violating SUSY (i, j, k are flavor indices)
M_h	mass of particle h (for hadrons or bosons)
m_f	mass of fermion f (for leptons or quarks)
$\mu_{\mathrm{f,r}}$	QCD factorization and renormalization scales
ν	neutrino (used generically for ν and $\bar{\nu}$ and for all flavors)
$\tilde{\nu}$	sneutrino (SUSY partner of ν)
$\mathcal{O}(\ldots)$	the order of ...
$\not{P}_{\mathrm{t}}\ell j$	event topology with missing transverse momentum, an isolated high-p_{t} lepton and a jet
p	proton
p_{t}	transverse momentum
Q^2	negative square of four-momentum tranfer between e and p
Q_f	charge of fermion f in units of the positron charge
q	quark ($q = d, u, s, c, b, t$ indicating the flavor)
$\tilde{q}$	squark (SUSY partner of q)
$q(x, Q^2)$	PDF of quark q ($q = d, u, s, c, b, t$)
R	chirality index for right-handed fermion fields or couplings
R_P	R-parity
θ_{W}	weak mixing angle (Weinberg angle)
x	Bjorken scaling variable x
y	Bjorken scaling variable y

References

1. ZEUS Collaboration, M. Derrick et al., Phys. Lett. **B 316** (1993) 412;
   ZEUS Collaboration, M. Derrick et al., Phys. Lett. **B 345** (1995) 576;
   ZEUS Collaboration, M. Derrick et al., Z. Phys. **C 69** (1996) 607.
2. ZEUS Collaboration, M. Derrick et al., Z. Phys. **C 65** (1994) 379.
3. ZEUS Collaboration, M. Derrick et al., Z. Phys. **C 72** (1996) 399.
4. ZEUS Collaboration, J. Breitweg et al., Phys. Lett. **B 407** (1997) 402;
   ZEUS Collaboration, J. Breitweg et al., Eur. Phys. J. **C 12** (2000) 35.
5. ZEUS Collaboration, J. Breitweg et al., Phys. Lett. **B 407** (1997) 432.
6. ZEUS Collaboration, J. Breitweg et al., Eur. Phys. J. **C 7** (1999) 609.
7. H1 Collaboration, I. Abt et al., Nucl. Phys. **B 407** (1993) 515;
   H1 Collaboration, I. Abt et al., Phys. Lett. **B 321** (1994) 161;
   H1 Collaboration, S. Aid et al., Phys. Lett. **B 354** (1995) 494;
   H1 Collaboration, S. Aid et al., Nucl. Phys. **B 449** (1995) 3;
   H1 Collaboration, C. Adloff et al., Phys. Lett. **B 393** (1997) 452;
   H1 Collaboration, C. Adloff et al., Nucl. Phys. **B 545** (1999) 21.
8. H1 Collaboration, T. Ahmed et al., Nucl. Phys. **B 439** (1995) 471.
9. H1 Collaboration, S. Aid et al., Nucl. Phys. **B 470** (1996) 3.
10. H1 Collaboration, C. Adloff et al., Z. Phys. **C 72** (1996) 593.
11. H1 Collaboration, C. Adloff et al., Nucl. Phys. **B 497** (1997) 3.
12. ZEUS Collaboration, M. Derrick et al., Phys. Lett. **B 297** (1992) 404;
    ZEUS Collaboration, M. Derrick et al., Phys. Lett. **B 322** (1994) 287;
    ZEUS Collaboration, M. Derrick et al., Phys. Lett. **B 342** (1995) 417;
    ZEUS Collaboration, M. Derrick et al., Z. Phys. **C 67** (1995) 81;
    ZEUS Collaboration, M. Derrick et al., Phys. Lett. **B 348** (1995) 665;
    ZEUS Collaboration, M. Derrick et al., Phys. Lett. **B 363** (1995) 201;
    ZEUS Collaboration, M. Derrick et al., Phys. Lett. **B 369** (1996) 55;
    ZEUS Collaboration, M. Derrick et al., Phys. Lett. **B 384** (1995) 401;
    ZEUS Collaboration, J. Breitweg et al., Eur. Phys. J. **C 4** (1998) 591;
    ZEUS Collaboration, J. Breitweg et al., Eur. Phys. J. **C 6** (1999) 67
    ZEUS Collaboration, J. Breitweg et al., Eur. Phys. J. **C 11** (1999) 35.
13. ZEUS Collaboration, M. Derrick et al., Phys. Lett. **B 306** (1993) 158.
14. ZEUS Collaboration, M. Derrick et al., Phys. Lett. **B 354** (1995) 163.
15. ZEUS Collaboration, J. Breitweg et al., Eur. Phys. J. **C 2** (1998) 61.
16. ZEUS Collaboration, J. Breitweg et al., Eur. Phys. J. **C 1** (1998) 109.
17. ZEUS Collaboration, J. Breitweg et al., Eur. Phys. J. **C 2** (1998) 77.
18. ZEUS Collaboration, J. Breitweg et al., Eur. Phys. J. **C 6** (1999) 239.
19. ZEUS Collaboration, J. Breitweg et al., Phys. Lett. **B 443** (1998) 394.
20. H1 Collaboration, T. Ahmed et al., Phys. Lett. **B 297** (1992) 205;
    H1 Collaboration, I. Abt et al., Phys. Lett. **B 314** (1993) 436
    H1 Collaboration, T. Ahmed et al., Phys. Lett. **B 346** (1995) 415;
    H1 Collaboration, S. Aid et al., Phys. Lett. **B 356** (1995) 118;

H1 Collaboration, C. Adloff et al., Phys. Lett. **B 415** (1997) 418;
H1 Collaboration, C. Adloff et al., Phys. Lett. **B 415** (1997) 418;
H1 Collaboration, C. Adloff et al., Eur. Phys. J. **C 5** (1998) 625;
H1 Collaboration, C. Adloff et al., Eur. Phys. J. **C 6** (1999) 575;
H1 Collaboration, C. Adloff et al., Eur. Phys. J. **C 13** (2000) 397.

21. H1 Collaboration, I. Abt et al., Z. Phys. **C 61** (1994) 59.
22. H1 Collaboration, S. Aid et al., Z. Phys. **C 70** (1996) 17.
23. H1 Collaboration, C. Adloff et al., Eur. Phys. J. **C 1** (1998) 97.
24. H1 Collaboration, C. Adloff et al., Eur. Phys. J. **C 13** (2000) 415.
25. H1 Collaboration, C. Adloff et al., Nucl. Phys. **B 538** (1999) 3.
26. H1 Collaboration, C. Adloff et al., Nucl. Phys. **B 545** (1999) 3.
27. ZEUS Collaboration, M. Derrick et al., Phys. Lett. **B 315** (1993) 481;
ZEUS Collaboration, M. Derrick et al., Phys. Lett. **B 332** (1994) 228;
ZEUS Collaboration, M. Derrick et al., Phys. Lett. **B 346** (1995) 399;
ZEUS Collaboration, M. Derrick et al., Z. Phys. **C 68** (1995) 569;
ZEUS Collaboration, M. Derrick et al., Phys. Lett. **B 356** (1995) 129;
ZEUS Collaboration, J. Breitweg et al., Z. Phys. **C 75** (1997) 421;
ZEUS Collaboration, J. Breitweg et al., Eur. Phys. J. **C 2** (1998) 237;
ZEUS Collaboration, J. Breitweg et al., Eur. Phys. J. **C 5** (1998) 41;
ZEUS Collaboration, J. Breitweg et al., Eur. Phys. J. **C 6** (1999) 43.
28. ZEUS Collaboration, M. Derrick et al., Phys. Lett. **B 338** (1994) 483.
29. ZEUS Collaboration, J. Breitweg et al., Eur. Phys. J. **C 1** (1998) 81.
30. ZEUS Collaboration, J. Breitweg et al., Phys. Lett. **B 421** (1998) 368.
31. H1 Collaboration, T. Ahmed et al., Nucl. Phys. **B 429** (1994) 477;
H1 Collaboration, T. Ahmed et al., Phys. Lett. **B 348** (1995) 681;
H1 Collaboration, C. Adloff et al., Z. Phys. **C 74** (1997) 221;
H1 Collaboration, C. Adloff et al., Z. Phys. **C 76** (1997) 613;
H1 Collaboration, C. Adloff et al., Eur. Phys. J. **C 6** (1999) 421;
H1 Collaboration, C. Adloff et al., Eur. Phys. J. **C 6** (1999) 587.
32. H1 Collaboration, S. Aid et al., Z. Phys. **C 70** (1996) 609.
33. H1 Collaboration, C. Adloff et al., Eur. Phys. J. **C 1** (1998) 495.
34. H1 Collaboration, C. Adloff et al., Phys. Lett. **B 428** (1998) 206.
35. H1 Collaboration, C. Adloff et al., Eur. Phys. J. **C 5** (1998) 439.
36. ZEUS Collaboration, M. Derrick et al., Phys. Lett. **B 350** (1995) 120;
ZEUS Collaboration, M. Derrick et al., Phys. Lett. **B 356** (1995) 601;
ZEUS Collaboration, M. Derrick et al., Z. Phys. **C 69** (1995) 39;
ZEUS Collaboration, M. Derrick et al., Phys. Lett. **B 377** (1996) 259;
ZEUS Collaboration, M. Derrick et al., Phys. Lett. **B 380** (1996) 220;
ZEUS Collaboration, M. Derrick et al., Z. Phys. **C 73** (1997) 253;
ZEUS Collaboration, J. Breitweg et al., Z. Phys. **C 75** (1997) 215;
ZEUS Collaboration, J. Breitweg et al., Z. Phys. **C 76** (1997) 599;
ZEUS Collaboration, J. Breitweg et al., Eur. Phys. J. **C 2** (1998) 247;
ZEUS Collaboration, J. Breitweg et al., Phys. Lett. **B 437** (1998) 432;
ZEUS Collaboration, J. Breitweg et al., Eur. Phys. J. **C 6** (1999) 603;
ZEUS Collaboration, J. Breitweg et al., Eur. Phys. J. **C 12** (2000) 393.
37. ZEUS Collaboration, M. Derrick et al., Z. Phys. **C 73** (1996) 73.
38. H1 Collaboration, T. Ahmed et al., Phys. Lett. **B 338** (1994) 507;
H1 Collaboration, S. Aid et al., Nucl. Phys. **B 463** (1996) 3;
H1 Collaboration, S. Aid et al., Nucl. Phys. **B 468** (1996) 3;
H1 Collaboration, S. Aid et al., Nucl. Phys. **B 472** (1996) 3;
H1 Collaboration, C. Adloff et al., Z. Phys. **C 75** (1997) 607;
H1 Collaboration, C. Adloff et al., Phys. Lett. **B 421** (1998) 385;
H1 Collaboration, C. Adloff et al., Eur. Phys. J. **C 13** (2000) 371.

39. H1 Collaboration, C. Adloff et al., Eur. Phys. J. **C 10** (1999) 373.
40. ZEUS Collaboration, M. Derrick et al., Phys. Lett. **B 293** (1992) 465; ZEUS Collaboration, M. Derrick et al., Z. Phys. **C 63** (1994) 391.
41. H1 Collaboration, T. Ahmed et al., Phys. Lett. **B 299** (1993) 374; H1 Collaboration, S. Aid et al., Z. Phys. **C 69** (1995) 27; H1 Collaboration, S. Aid et al., Phys. Lett. **B 392** (1997) 234.
42. *Proceedings of the 1991 Workshop on Physics at HERA*, ed. W. Buchmüller and G. Ingelman, Hamburg, Germany, 1991.
43. *Proceedings of the 1995/96 Workshop on Future Physics at HERA*, ed. G. Ingelman, A. De Roeck and R. Klanner, Hamburg, Germany, 1996.
44. H. Abramowicz and A.C. Caldwell, Rev. Mod. Phys. **71** (1999) 1275.
45. B. Foster, Int. J. Mod. Phys. **A 13** (1998) 1543.
46. A.M. Cooper-Sarkar, R.C.E. Devenish and A. De Roeck, Int. J. Mod. Phys. **A 13** (1998) 3385.
47. M. Erdmann, *The Partonic Structure of the Photon*, Springer Tracts in Modern Physics, Volume 138 (Springer, Berlin, Heidelberg, 1997).
48. M. Kuhlen, *QCD at HERA*, Springer Tracts in Modern Physics, Volume 150 (Springer, Berlin, Heidelberg, 1999).
49. J.A. Crittenden, *Exclusive Production of Neutral Vector Mesons at the Electron–Proton Collider HERA*, Springer Tracts in Modern Physics, Volume 140 (Springer, Berlin, Heidelberg, 1997).
50. ZEUS Collaboration, M. Derrick et al., Phys. Rev. Lett. **75** (1995) 1006.
51. ZEUS Collaboration, M. Derrick et al., Z. Phys. **C 72** (1996) 47.
52. H1 Collaboration, T. Ahmed et al., Phys. Lett. **B 324** (1994) 241; H1 Collaboration, Charged current differential cross sections for e^-p and e^+p interactions at very high Q^2, Abstract pa02-078, *XXVIII International Conference on High Energy Physics, Warsaw*, July 25–31, 1996.
53. H1 Collaboration, S. Aid et al., Phys. Lett. **B 379** (1996) 319.
54. H1 Collaboration, S. Aid et al., Z. Phys. **C 67** (1995) 565.
55. ZEUS Collaboration, M. Derrick et al., Phys. Lett. **B 306** (1993) 173.
56. ZEUS Collaboration, M. Derrick et al., Search for leptoquarks in *ep* collisions at $\sqrt{s} = 296\,\text{GeV}$, in *Proc. 27th Int. Conf. on High-Energy Physics (ICHEP94), Glasgow*, ed. P.J. Bussey and I.G. Knowles, Institute of Physics, Philadelphia, PA, 1995, p. 797.
57. ZEUS Collaboration, Limits on new physics from neutral current *ep* deep inelastic scattering at high Q^2, Abstract pa04-036, *XXVIII International Conference on High-Energy Physics, Warsaw*, July 25–31, 1996.
58. ZEUS Collaboration, Search for lepton flavor violation in *ep* collisions at 300 GeV center of mass energy, Abstract pa11-019, *XXVIII International Conference on High-Energy Physics, Warsaw*, July 25–31, 1996.
59. ZEUS Collaboration, M. Derrick et al., Z. Phys. **C 73** (1997) 613.
60. H1 Collaboration, I. Abt et al., Nucl. Phys. **B 396** (1993) 3.
61. H1 Collaboration, S. Aid et al., Phys. Lett. **B 353** (1995) 578.
62. H1 Collaboration, S. Aid et al., Phys. Lett. **B 380** (1996) 461.
63. H1 Collaboration, T. Ahmed et al., Z. Phys. **C 64** (1994) 545.
64. H1 Collaboration, S. Aid et al., Phys. Lett. **B 369** (1996) 173.
65. H1 Collaboration, S. Aid et al., Z. Phys. **C 71** (1996) 211.
66. ZEUS Collaboration, M. Derrick et al., Phys. Lett. **B 316** (1993) 207.
67. ZEUS Collaboration, M. Derrick et al., Z. Phys. **C 65** (1995) 627.
68. ZEUS Collaboration, J. Breitweg et al., Z. Phys. **C 76** (1997) 631.
69. H1 Collaboration, T. Ahmed et al., Phys. Lett. **B 340** (1994) 205.
70. H1 Collaboration, S. Aid et al., Nucl. Phys. **B 483** (1997) 44.
71. ZEUS Collaboration, J. Breitweg et al., Z. Phys. **C 74** (1997) 207.

72. H1 Collaboration, C. Adloff et al., Z. Phys. **C 74** (1997) 191.
73. ZEUS Collaboration, J. Breitweg et al., Eur. Phys. J. **C 11** (1999) 427.
74. ZEUS Collaboration, J. Breitweg et al., Eur. Phys. J. **C 12** (2000) 411.
75. H1 Collaboration, C. Adloff et al., Eur. Phys. J. **C 13** (2000) 609.
76. ZEUS Collaboration, J. Breitweg et al., Preprint DESY-00-023, DESY (2000), hep-ex/0002038, accepted by Eur. Phys. J.
77. ZEUS Collaboration, Study of high mass $\bar{\nu}$–jet systems in positron–proton scattering at HERA, Abstract 546, *International Europhysics Conference on High-Energy Physics, Tampere*, July 15–21, 1999.
78. ZEUS Collaboration, Search for squarks from R-parity violating supersymmetry in e^+p collisions at HERA, Abstract 548, *International Europhysics Conference on High-Energy Physics, Tampere*, July 15–21, 1999.
79. ZEUS Collaboration, J. Breitweg et al., Eur. Phys. J. **C 14** (2000) 239.
80. ZEUS Collaboration, Search for excited fermions in *ep* collisions at HERA, Abstract 555, *International Europhysics Conference on High-Energy Physics, Tampere*, July 15–21, 1999.
81. ZEUS Collaboration, Search for lepton flavor violation in *ep* collisions at HERA, Abstract 551, *International Europhysics Conference on High-Energy Physics, Tampere*, July 15–21, 1999.
82. ZEUS Collaboration, J. Breitweg et al., Phys. Lett. **B 434** (1998) 214.
83. ZEUS Collaboration, J. Breitweg et al., Phys. Lett. **B 471** (2000) 411.
84. H1 Collaboration, C. Adloff et al., Eur. Phys. J. **C 11** (1999) 447.
85. H1 Collaboration, A search for squarks of R-parity violating SUSY at HERA, Abstract 580, *XXIX International Conference on High Energy Physics, Vancouver*, July 23–29, 1998.
86. H1 Collaboration, C. Adloff et al., Phys. Lett. **B 479** (2000) 358.
87. H1 Collaboration, Search for excited fermions at HERA, Abstract 581, *XXIX International Conference on High Energy Physics, Vancouver*, July 23–29, 1998.
88. H1 Collaboration, C. Adloff et al., Eur. Phys. J. **C 5** (1998) 575.
89. E. Rutherford, Phil. Mag. **21** (1911) 669;
H. Geiger and E. Marsden, Proc. Roy. Soc. (London) **82** (1909) 495;
H. Geiger, Proc. Roy. Soc. (London) **83** (1910) 492.
90. C.S. Wu et al., Phys. Rev. **105** (1957) 1413.
91. E. Fermi, Z. Phys. **88** (1934) 161;
E. Fermi, Nuovo Cimento **11** (1934) 1.
92. M. Gell-Mann and Y. Ne'eman, *The Eightfold Way* (Benjamin, New York, 1964).
93. *Developments in the Quark Theory of Hadrons,* Vol. I: 1964–1978, ed. D.B. Lichtenberg and S.P. Rosen (Hadronic Press, Palm Harbor, FL, USA, 1980).
94. E.D. Bloom et al., Phys. Rev. Lett. **23** (1969) 930;
M. Breidenbach et al., Phys. Rev. Lett. **23** (1969) 935.
95. J.D. Bjorken, Phys. Rev. **179** (1969) 1547.
96. R.P. Feynman, Phys. Rev. Lett. **23** (1969) 1415.
97. Gargamelle Neutrino Collaboration, F.J. Hasert et al., Phys. Lett. **B 46** (1973) 138;
W.F. Fry and D. Haidt, Report CERN-75-1, CERN (1975);
G. Myatt, Neutral currents, in *Proc. 6th Int. Symp. Electron and Photon Interactions at High Energies (1973), Bonn, Germany*, ed. H. Rollnik and W. Pfeil, North-Holland, Amsterdam, 1973, p. 389;
A. Benvenuti et al., Phys. Rev. Lett. **32** (1974) 800.

98. S.L. Glashow, Nucl. Phys. **22** (1961) 579;
S. Weinberg, Phys. Rev. Lett. **19** (1967) 1264;
A. Salam, Weak and electromagnetic interactions, in *Elementary Particle Theory, Relativistic Groups and Analyticity*, ed. N. Svartholm, Almquist and Wissell, Stockholm, 1968, p. 367.
99. UA1 Collaboration, G. Arnison et al., Phys. Lett. **122 B** (1983) 103;
UA1 Collaboration, G. Arnison et al., Phys. Lett. **126 B** (1983) 398.
100. C.H. Llewellyn Smith, Phys. Rep. **3 C** (1972) 261.
101. TASSO Collaboration, R. Brandelik et al., Phys. Lett. **B 86** (1979) 243;
MARKJ Collaboration, D.P. Barber et al., Phys. Rev. Lett. **43** (1979) 830;
PLUTO Collaboration, Ch. Berger et al., Phys. Lett. **86 B** (1979) 418;
JADE Collaboration, W. Bartel et al., Phys. Lett. **91 B** (1980) 142.
102. V.N. Gribov and L.N. Lipatov, Sov. J. Nucl. Phys. **15** (1972) 438;
L.N. Lipatov, Yad. Fiz. **20** (1974) 181 [also in Sov. J. Nucl. Phys **20** (1975) 94];
G. Altarelli and G. Parisi, Nucl. Phys. **B 126** (1977) 298;
Yu.L. Dokshitser, Sov. Phys. JETP **46** (1977) 641.
103. *Proceedings of the Study of an ep facility for Europe*, ed. U. Amaldi, DESY Report 79/48, Hamburg, Germany, 1979.
104. HERA – a proposal for a large electron proton colliding beam facility at DESY, DESY Report HERA 81-10, Hamburg, Germany, 1981;
B.H. Wiik, HERA, in *Proc. 11th Int. Conf. on High Energy Accelerators (1980), Geneva, Switzerland*, ed. W.S. Newman, Birkhäuser, CERN, 1980, p. 120.
105. J.D. Bjorken, Phys. Rev. **185** (1969) 1975.
106. R.P. Feynman, *Photon–Hadron Interactions* (Benjamin, New York, 1972).
107. W. Hollik et al., Electroweak parameters at HERA: theoretical aspects, in *Proc. Workshop on Physics at HERA*, ed. W. Buchmüller and G. Ingelman, DESY, Hamburg, Germany, 1991, p. 923.
108. Particle Data Group, C. Caso et al., Eur. Phys. J. **C 3** (1998) 1.
109. Particle Data Group, C. Caso et al., Eur. Phys. J. **C 3** (1998) 1, and 1999 off-year partial update for the 2000 edition available on the PDG WWW pages (`http://pdg.lbl.gov/`).
110. G. Altarelli, Phys. Rep. **81** (1982) 1.
111. A.J. Buras, Rev. Mod. Phys. **52** (1980) 199.
112. E. Reya, Phys. Rep. **69** (1981) 195.
113. *Perturbative Quantum Chromodynamics*, Physics Reports Reprint Book Series, Volume 5, ed. M. Jacob (North-Holland, Amsterdam, 1982).
114. *Perturbative Quantum Chromodynamics*, Advanced Series on Directions in High Energy Physics, Volume 5, ed. A.H. Mueller (World Scientific, Singapore, 1989).
115. R.G. Roberts, *The Structure of the Proton* (Cambridge University Press, Cambridge, 1990).
116. ZEUS Collaboration, Measurement of the proton structure function F_2 at very low Q^2 and very low x, Abstract 493, *International Europhysics Conference on High-Energy Physics, Tampere*, July 15–21, 1999.
117. M. Glück, E. Reya and A. Vogt, Z. Phys. **C 67** (1995) 433.
118. W.A. Bardeen et al., Phys. Rev. **D 18** (1978) 3998.
119. B. Badełek et al., Rev. Mod. Phys. **64** (1992) 927.
120. É.A. Kuraev, L.N. Lipatov and V.S. Fadin, Sov. Phys. JETP **45** (1977) 199;
Ya.Ya. Balitskiĭ and L.N. Lipatov, Sov. J. Nucl. Phys. **28** (1978) 822;
L.N. Lipatov, Sov. Phys. JETP **63** (1986) 904;
L.N. Lipatov, Pomeron in quantum chromodynamics, in *Perturbative*

Quantum Chromodynamics, ed. A.H. Mueller, World Scientific, Singapore, 1989.
121. W.L. van Neerven, Review of higher order QCD corrections to structure functions, in *Proc. Workshop on Future Physics at HERA*, ed. G. Ingelman, A. De Roeck and R. Klanner, DESY, Hamburg, Germany, 1996, p. 56.
122. J. Blümlein et al., A detailed comparison of NLO QCD evolution codes, in *Proc. Workshop on Future Physics at HERA*, ed. G. Ingelman, A. De Roeck and R. Klanner, DESY, Hamburg, Germany, 1996, p. 23.
123. S.A. Larin and J.A.M. Vermaseren, Phys. Lett. **B 303** (1993) 334;
S.A. Larin et al., The next-to-next-to-leading QCD calculation of the moments of deep inelastic structure functions, in *Proc. Workshop on Future Physics at HERA*, ed. G. Ingelman, A. De Roeck and R. Klanner, DESY, Hamburg, Germany, 1996, p. 67;
S.A. Larin et al., Nucl. Phys. **B 492** (1997) 338;
A.L. Kataev et al., Nucl. Phys. Proc. Suppl. **B 64** (1998) 138;
S. Catani, M. Grazzini, Phys. Lett. **B 446** (1999) 143.
124. H. Georgi and H.D. Politzer, Phys. Rev. **D 14** (1976) 1829;
O. Nachtmann, Nucl. Phys. **B 63** (1973) 237.
125. A.D. Martin and M.G. Ryskin, Phys. Lett. **B 431** (1998) 395.
126. U.K. Yang and A. Bodek, Phys. Rev. Lett. **82** (1999) 2467.
127. A.D. Martin et al., Phys. Lett. **B 443** (1998) 301.
128. W.J. Marciano, Phys. Rev. **D 29** (1984) 580.
129. A.D. Martin, W.J. Stirling and R.G. Roberts, Phys. Rev. **D 47** (1993) 867.
130. A.D. Martin, W.J. Stirling and R.G. Roberts, Phys. Rev. **D 50** (1994) 6734.
131. M. Glück, E. Reya and A. Vogt, Z. Phys. **C 48** (1990) 471;
M. Glück, E. Reya and A. Vogt, Z. Phys. **C 53** (1992) 127;
M. Glück, E. Reya and M. Stratmann, Nucl. Phys. **B 422** (1994) 37.
132. M. Glück, E. Reya and A. Vogt, Eur. Phys. J. **C 5** (1998) 461.
133. J.P. Leveille and T. Weiler, Nucl. Phys. **B 147** (1979) 147.
134. E. Laenen et al., Phys. Lett. **B 291** (1992) 325;
E. Laenen et al., Nucl. Phys. **B 392** (1993) 162;
E. Laenen et al., Nucl. Phys. **B 392** (1993) 229;
S. Riemersma, J. Smith and W.L. van Neerven, Phys. Lett. **B 347** (1995) 143.
135. A.D. Martin et al., Eur. Phys. J. **C 2** (1998) 287;
C.R. Schmidt, QCD phenomenology of charm production at HERA, in *Proc. 5th Int. Workshop on Deep Inelastic Scattering and QCD (DIS97), Chicago, IL, USA*, ed. J. Repond and D. Krakauer, American Institute of Physics, Woodbury, NY, 1997, p. 381, hep-ph/9706496;
M. Buza et al., Eur. Phys. J. **C 1** (1998) 301;
W.-K. Tung, The heavy quark parton oxymoron – a mini-review of heavy quark production theory in PQCD, in *Proc. 5th Int. Workshop on Deep Inelastic Scattering and QCD (DIS97), Chicago, IL, USA*, ed. J. Repond and D. Krakauer, American Institute of Physics, Woodbury, NY, 1997, p. 1014, hep-ph/9706480;
F.M. Steffens, W. Melnitchouk, A.W. Thomas, Eur. Phys. J. **C 11** (1999) 673.
136. H.L. Lai and W.K. Tung, Z. Phys. **C 74** (1997) 463.
137. M.A.G. Aivazis et al., Phys. Rev. **D 50** (1994) 3102.
138. J.C. Collins, Phys. Rev. **D 58** (1998) 094002.
139. CTEQ Collaboration, H.L. Lai et al., Eur. Phys. J. **C 12** (2000) 375.
140. R.S. Thorne and R.G. Roberts, Phys. Rev. **D 57** (1998) 6871;
R.S. Thorne and R.G. Roberts, Phys. Lett. **B 421** (1998) 303.

141. A.D. Martin et al., Eur. Phys. J. **C 4** (1998) 463;
A.D. Martin et al., Nucl. Phys. Proc. Suppl. **79** (1999) 105.
142. M. Böhm and H. Spiesberger, Nucl. Phys. **B 294** (1987) 1081;
D.Yu. Bardin et al., Z. Phys. **C 42** (1989) 679;
H. Spiesberger, Radiative corrections to deep inelastic *ep* scattering at HERA, in *Proc. 13th Int. School of Theoretical Physics (1989), Szczyrk, Poland*, ed. M. Zralek, Silesian Univ., Katowice, 1989, p. 145, also in Report DESY-89-175.
143. M. Böhm and H. Spiesberger, Nucl. Phys. **B 304** (1988) 749;
D.Yu. Bardin et al., Z. Phys. **C 44** (1989) 149;
H. Spiesberger, Nucl. Phys. **B 349** (1991) 109.
144. G. Wolf, Preprint DESY-97-047, DESY (1997), hep-ex/9704006.
145. H. Spiesberger et al., Radiative corrections at HERA, in *Proc. Workshop on Physics at HERA*, ed. W. Buchmüller and G. Ingelman, DESY, Hamburg, Germany, 1991, p. 798.
146. CDF Collaboration, F. Abe et al., Phys. Rev. Lett. **82** (1999) 271; erratum, Phys. Rev. Lett. **82** (1999) 2808;
DØ Collaboration, B. Abbott et al., Phys. Rev. **D 60** (1999) 052001.
147. Particle Data Group, L. Montanet et al., Phys. Rev. **D 50** (1994) 1173, and 1995 off-year partial update for the 1996 edition available on the PDG WWW pages (`http://pdg.lbl.gov/`).
148. A. Kwiatkowski, H. Spiesberger and H.-J. Möhring, Z. Phys. **C 50** (1991) 165;
A. Kwiatkowski, H. Spiesberger and H.-J. Möhring, Comp. Phys. Commun. **69** (1992) 155, also in *Proc. Workshop on Physics at HERA*, ed. W. Buchmüller and G. Ingelman, DESY, Hamburg, Germany, 1991, p. 1294.
149. LEP Collaborations, LEP Electroweak Working Group and SLD Heavy Flavour and Electroweak Groups, D. Abbaneo et al., Preprint CERN-EP/99-015, CERN (1999).
150. S. Eidelmann and F. Jegerlehner, Z. Phys. **C 67** (1995) 585;
M. Steinhauser, Phys. Lett. **B 429** (1998) 158.
151. BCDMS Collaboration, A.C. Benvenuti et al., Phys. Lett. **195 B** (1987) 91.
152. BCDMS Collaboration, A.C. Benvenuti et al., Phys. Lett. **B 223** (1989) 485.
153. BCDMS Collaboration, A.C. Benvenuti et al., Phys. Lett. **B 237** (1990) 592.
154. NMC Collaboration, M. Arneodo et al., Phys. Lett. **B 364** (1995) 107;
NMC Collaboration, M. Arneodo et al., Nucl. Phys. **B 483** (1997) 3.
155. E665 Collaboration, M.R. Adams et al., Phys. Rev. **D 54** (1996) 3006.
156. L.W. Whitlow, Phys. Lett. **B 282** (1992) 475;
L.W. Whitlow, *Deep Inelastic Structure Functions from Electron Scattering on Hydrogen, Deuterium, and Iron at 0.6 $GeV^2 \leq Q^2 \leq 30\,GeV^2$*, Ph.D. thesis, Stanford Univ., 1990, Report SLAC-357.
157. T. Sloan, G. Smadja and R. Voss, Phys. Rep. **162** (1988) 45;
S.R. Mishra and F. Sciulli, Ann. Rev. Nucl. Part. Sci. **39** (1989) 259.
158. EMC Collaboration, J.J. Aubert et al., Nucl. Phys. **B 213** (1983) 31.
159. NMC Collaboration, M. Arneodo et al., Phys. Rev. **D 50** (1994) 1.
160. CHARM Collaboration, F. Bergsma et al., Phys. Lett. **123 B** (1983) 269.
161. CHARM Collaboration, J.V. Allaby et al., Phys. Lett. **197 B** (1987) 281.
162. CDHSW Collaboration, H. Abramowicz et al., Z. Phys. **C 25** (1984) 29;
CDHSW Collaboration, P. Berge et al., Z. Phys. **C 49** (1991) 187.
163. BEBC WA59 Collaboration, K. Varvell et al., Z. Phys. **C 36** (1987) 1;
BEBC WA21 Collaboration, G.T. Jones et al., Z. Phys. **C 44** (1989) 379;
BEBC WA21/WA25 Collaboration, G.T. Jones et al., Z. Phys. **C 62** (1994) 575.

164. CCFR Collaboration, P.Z. Quintas et al., Phys. Rev. Lett. **71** (1993) 1307.
165. CCFR Collaboration, W.C. Leung et al., Phys. Lett. **B 317** (1993) 655.
166. CCFR Collaboration, W.G. Seligman et al., Phys. Rev. Lett. **79** (1997) 1213.
167. CCFR Collaboration, A.O. Bazarko et al., Z. Phys. **C 65** (1995) 189.
168. UA2 Collaboration, J. Alitti et al., Phys. Lett. **B 276** (1992) 354;
UA2 Collaboration, J. Alitti et al., Phys. Lett. **B 276** (1992) 365.
169. CDF Collaboration, F. Abe et al., Phys. Rev. Lett. **76** (1996) 3070.
170. CDF Collaboration, F. Abe et al., Phys. Rev. Lett. **74** (1995) 850;
CDF Collaboration, F. Abe et al., Phys. Rev. Lett. **81** (1998) 5754.
171. E605 Collaboration, G. Moreno et al., Phys. Rev. **D 43** (1991) 2815.
172. E772 Collaboration, P.L. McGaughey et al., Phys. Rev. **D 50** (1994) 3038.
173. NA51 Collaboration, A. Baldit et al., Phys. Lett. **B 332** (1994) 244.
174. FNAL E866/NuSea Collaboration, E.A. Hawker et al., Phys. Rev. Lett. **80** (1998) 3715;
FNAL E866/NuSea Collaboration, J.C. Peng et al., Phys. Rev. **D 58** (1998) 092004.
175. WA70 Collaboration, M. Bonesini et al., Z. Phys. **C 38** (1988) 371.
176. E706 Collaboration, L. Apanasevich et al., Phys. Rev. Lett. **81** (1998) 2642.
177. L. Apanasevich et al., Phys. Rev. **D 59** (1999) 074007;
P. Aurenche et al., Eur. Phys. J. **C 9** (1999) 107.
178. CDF Collaboration, F. Abe et al., Phys. Rev. Lett. **77** (1996) 438.
179. C. Mesropian, for the CDF and DØ Collaborations, Preprint FERMILAB-CONF-99-177-E, Fermilab (1999), to appear in *Proc. 34th Rencontres de Moriond: QCD and Hadronic Interactions (1999), Les Arcs, France.*
180. DØ Collaboration, B. Abbott et al., Phys. Rev. Lett. **82** (1999) 2451.
181. S. Bethge, Jet physics at LEP and world summary of α_s, in *Proc. 4th Int. Symp. on Radiative Corrections (RADCOR98), Barcelona, Spain*, ed. J. Sola, World Scientific, Singapore, 1999, p. 243, hep-ex/9812026.
182. A.D. Martin, W.J. Stirling and R.G. Roberts, MRS parton distributions, in *Proc. Workshop on Quantum Field Theoretical Aspects of High-Energy Physics (1993), Bad Frankenhausen, Germany*, ed. B. Geyer and E.-M. Ilgenfritz, Univ. Leipzig, 1993, p. 11, also in Report RAL-93-077, Rutherford Lab.
183. A.D. Martin, R.G. Roberts and W.J. Stirling, Phys. Lett. **B 387** (1996) 419.
184. CTEQ Collaboration, H.L. Lai et al., Phys. Rev. **D 51** (1995) 4763.
185. CTEQ Collaboration, H.L. Lai et al., Phys. Rev. **D 55** (1997) 1280.
186. A.M. Cooper-Sarkar et al., J. Phys. **G 25** (1999) 1387.
187. M. Botje, Eur. Phys. J. **C 14** (2000) 285.
188. W. Melnitchouk and A. Thomas, Acta Phys. Pol. **B 27** (1996) 1407.
189. E665 Collaboration, M.R. Adams et al., Phys. Rev. Lett. **75** (1995) 1466.
190. NMC Collaboration, M. Arneodo et al., Nucl. Phys. **B 487** (1997) 3;
W. Melnitchouk and A.W. Thomas, Phys. Lett. **B 377** (1996) 11;
W. Melnitchouk and J.C. Peng, Phys. Lett. **B 400** (1997) 220.
191. G.R. Farrar and D.R. Jackson, Phys. Rev. Lett. **35** (1975) 1416.
192. S. Kuhlmann et al., Phys. Lett. **B 476** (2000) 291.
193. BEBC WA21/WA25 Collaboration, G.T. Jones et al., Z. Phys. **C 62** (1994) 601.
194. E. Mirkes, Report TTP-97-39, Univ. Karlsruhe (1997), hep-ph/9711224.
195. M. Bengtsson and T. Sjöstrand, Z. Phys. **C 37** (1988) 465.

196. G. Ingelman, LEPTO VERSION 6.1 – The Lund Monte Carlo for deep inelastic lepton–nucleon scattering, in *Proc. Workshop on Physics at HERA*, ed. W. Buchmüller and G. Ingelman, DESY, Hamburg, Germany, 1991, p. 1366; G. Ingelman, A. Edin and J. Rathsman, Comp. Phys. Commun. **101** (1997) 108.
197. G. Gustafson, Phys. Lett. **175 B** (1986) 453;
G. Gustafson and U. Pettersson, Nucl. Phys. **B 306** (1988) 746;
B. Andersson et al., Z. Phys. **C 43** (1989) 625;
B. Andersson, G. Gustafson and L. Lönnblad, Nucl. Phys. **B 339** (1990) 393.
198. L. Lönnblad, ARIADNE VERSION 4, in *Proc. Workshop on Physics at HERA*, ed. W. Buchmüller and G. Ingelman, DESY, Hamburg, Germany, 1991, p. 1440;
L. Lönnblad, Comp. Phys. Commun. **71** (1992) 15.
199. B. Andersson et al., Phys. Rep. **97** (1983) 31.
200. T. Sjöstrand, Comp. Phys. Commun. **39** (1986) 347.
201. T. Sjöstrand, Comp. Phys. Commun. **82** (1994) 74;
T. Sjöstrand, Report LU-TP-95-20, Lund Univ. (1995), hep-ph/9508391.
202. B.R. Webber, Nucl. Phys. **B 238** (1984) 492;
G. Marchesini and B.R. Webber, Nucl. Phys. **B 310** (1988) 461.
203. G. Marchesini et al., Comp. Phys. Commun. **67** (1992) 465.
204. Ya.I. Azimov et al., Z. Phys. **C 27** (1985) 65;
B. Andersson, P. Dahlqvist and G. Gustafson, Z. Phys. **C 44** (1989) 455;
B. Andersson, P. Dahlqvist and G. Gustafson, Z. Phys. **C 44** (1989) 461.
205. ZEUS Collaboration, J. Breitweg et al., Eur. Phys. J. **C 8** (1999) 367.
206. H1 Collaboration, T. Ahmed et al., Phys. Lett. **B 298** (1993) 469
207. H1 Collaboration, I. Abt et al., Z. Phys. **C 63** (1994) 377.
208. H1 Collaboration, C. Adloff et al., Phys. Lett. **B 406** (1997) 256.
209. H1 Collaboration, C. Adloff et al., Eur. Phys. J. **C 12** (2000) 595.
210. ZEUS Collaboration, M. Derrick et al., Z. Phys. **C 59** (1993) 231.
211. H1 Collaboration, I. Abt et al., Phys. Lett. **B 328** (1994) 176.
212. H1 Collaboration, T. Ahmed et al., Nucl. Phys. **B 445** (1995) 195.
213. H1 Collaboration, C. Adloff et al., Z. Phys. **C 76** (1997) 213.
214. H1 Collaboration, C. Adloff et al., Eur. Phys. J. **C 10** (1999) 363.
215. ZEUS Collaboration, M. Derrick et al., Z. Phys. **C 67** (1995) 227.
216. H1 Collaboration, S. Aid et al., Nucl. Phys. **B 445** (1995) 3.
217. H1 Collaboration, S. Aid et al., Nucl. Phys. **B 480** (1996) 3.
218. H1 Collaboration, S. Aid et al., Z. Phys. **C 72** (1996) 573.
219. H1 Collaboration, C. Adloff et al., Nucl. Phys. **B 485** (1997) 3
220. H1 Collaboration, C. Adloff et al., Phys. Lett. **B 462** (1999) 440.
221. ZEUS Collaboration, M. Derrick et al., Z. Phys. **C 67** (1995) 93.
222. ZEUS Collaboration, M. Derrick et al., Z. Phys. **C 68** (1995) 29.
223. ZEUS Collaboration, M. Derrick et al., Z. Phys. **C 70** (1996) 1
224. ZEUS Collaboration, J. Breitweg et al., Phys. Lett. **B 414** (1997) 428.
225. ZEUS Collaboration, J. Breitweg et al., Eur. Phys. J. **C 11** (1999) 251.
226. ZEUS Collaboration, J. Breitweg et al., Eur. Phys. J. **C 12** (2000) 53.
227. G. Altarelli et al., Nucl. Phys. **B 506** (1997) 3.
228. K.S. Babu et al., Phys. Lett. **B 402** (1997) 367.
229. J. Kalinowski et al., Z. Phys. **C 74** (1997) 595.
230. R. Rückl and H. Spiesberger, Theoretical interpretations of the HERA high-Q^2 events, in *Proc. Workshop on New Trends in HERA Physics (1997), Ringberg Castle, Germany*, ed. B.A. Kniehl, G. Kramer and A. Wagner, World Scientific, Singapore, 1998, p. 113, hep-ph/9710327;
R. Rückl and H. Spiesberger, Leptoquarks and the HERA high Q^2 events,

in *Proc. Workshop on Physics Beyond the Desert (1997), Ringberg Castle, Germany*, ed. H.V. Klapdor-Kleingrothaus and H. Päs, Institute of Physics, Bristol, 1998, p. 304, hep-ph/9711352.
231. G. Altarelli, Nucl. Phys. Proc. Suppl. **B 62** (1998) 3.
232. J. Ellis, Hunting down interpretations of the HERA large-Q^2 data, in *Proc. Int. Europhysics Conf. on High Energy Physics (HEP97), Jerusalem, Israel*, ed. D. Lellouch, G. Mikenberg and E. Rabinovici, Springer, Berlin, Heidelberg, 1999, p. 1263, hep-ph/9712340.
233. S. Kuhlmann, H.L. Lai and W.K. Tung, Phys. Lett. **B 409** (1997) 271.
234. K.S. Babu, C. Kolda and J. March-Russel, Phys. Lett. **B 408** (1997) 268.
235. ZEUS Collaboration, Measurement of high-Q^2 charged-current e^-p deep inelastic scattering cross sections, Abstract 558, *International Europhysics Conference on High-Energy Physics, Tampere*, July 15–21, 1999.
236. H1 Collaboration, Measurement of the neutral and charged current cross-sections in electron–proton collisions at high Q^2 at HERA, Abstract 157b, *International Europhysics Conference on High Energy Physics, Tampere*, July 15–21, 1999.
237. S. Rock and P. Bosted, Preprint hep-ph/9706436, American Univ., Washington D.C. (1997).
238. E.D. Bloom and F.J. Gilman, Phys. Rev. Lett. **25** (1970) 1140.
239. W. Melnitchouk and A.W. Thomas, Phys. Lett. **B 414** (1997) 134.
240. S.J. Brodsky et al., Phys. Lett. **93 B** (1980) 451;
S.J. Brodsky, C. Peterson and N. Sakai, Phys. Rev. **D 23** (1981) 2745.
241. EMC Collaboration, J.J. Aubert et al., Phys. Lett. **110 B** (1982) 73.
242. R608 Collaboration, P. Chauvat et al., Phys. Lett. **199 B** (1987) 304.
243. J.F. Gunion and R. Vogt, Preprint UCD-97-14, UC, Davis (1997), hep-ph/9706252.
244. L.F. Abbott and E. Farhi, Phys. Lett. **101 B** (1981) 69;
L.F. Abbott and E. Farhi, Nucl. Phys. **B 189** (1981) 547;
W.A. Bardeen and V. Višnjić, Nucl. Phys. **B 194** (1982) 422;
W. Buchmüller, R.D. Peccei and T. Yanagida, Phys. Lett. **124 B** (1983) 67;
R. Barbieri, A. Masiero and G. Veneziano, Phys. Lett. **128 B** (1983) 179;
B. Schrempp and F. Schrempp, Phys. Lett. **153 B** (1985) 101.
245. K. Lane, An introduction to technicolor, in *Proc. Building Blocks of Creation – from Microfermis to Megaparsecs (TASI93), Boulder, CO, USA*, ed. S. Raby and T. Walker, World Scientific, River Edge, NJ, USA, 1994, p. 381, hep-ph/9401324.
246. H. Georgi and S.L. Glashow, Phys. Rev. Lett. **32** (1974) 438.
247. J.C. Pati and A. Salam, Phys. Rev. **D 10** (1974) 275;
G. Senjanović and A.Šokorac, Z. Phys. **C 20** (1983) 255.
248. P.H. Frampton and B.-H. Lee, Phys. Rev. Lett. **64** (1990) 619;
P.H. Frampton and T.W. Kephart, Phys. Rev. **D 42** (1990) 3892.
249. E. Witten, Nucl. Phys. **B 258** (1985) 75;
M. Dine et al., Nucl. Phys. **B 259** (1985) 549;
J.D. Breit, B.A. Ovrut and G.C. Segrè, Phys. Lett. **158 B** (1985) 33;
J.L. Hewett and T.G. Rizzo, Phys. Rep. **183** (1989) 193.
250. J.L. Hewett and T.G. Rizzo, Phys. Rev. **D 56** (1997) 5709.
251. Z. Kunszt and W.J. Stirling, Z. Phys. **C 75** (1997) 453.
252. N.G. Deshpande and B. Dutta, Phys. Lett. **B 424** (1998) 313.
253. J.K. Elwood and A.E. Faraggi, Nucl. Phys. **B 512** (1998) 42.
254. O.E. Evnin, Preprint hep-ph/9711433, Rostov State Univ. (1997).
255. J.L. Hewett and T.G. Rizzo, Phys. Rev. **D 58** (1998) 055005.
256. S.F. King and G.K. Leontaris, Phys. Lett. **B 406** (1997) 309.

257. A. Kusenko, V. Kuzmin and I.I. Tkachev, Phys. Lett. **B 432** (1998) 361.
258. G.K. Leontaris and J.D. Vergados, Phys. Lett. **B 409** (1997) 283.
259. E. Ma, M. Raidal and U. Sarkar, Eur. Phys. J. **C 8** (1999) 301.
260. I. Montvay, Phys. Lett. **B 407** (1997) 22.
261. M. Sekiguchi, H. Wada and S. Ishida, Prog. Theor. Phys. **99** (1998) 707.
262. H. Terazawa, Mod. Phys. Lett. **A 13** (1998) 2427.
263. Z. Xiao, Preprint RAL-TR-97-043, Rutherford Lab. (1997), hep-ph/9709324.
264. W. Buchmüller, R. Rückl and D. Wyler, Phys. Lett. **191 B** (1987) 442; erratum, Phys. Lett. **B 448** (1999) 320.
265. G. Altarelli, G.F. Giudice and M.L. Mangano, Nucl. Phys. **B 506** (1997) 29.
266. T. Plehn et al., Z. Phys. **C 74** (1997) 611.
267. J. Blümlein, Z. Phys. **C 74** (1997) 605.
268. ZEUS Collaboration, Study of high mass e–jet systems in electron–proton scattering at HERA, Abstract 552, *International Europhysics Conference on High-Energy Physics, Tampere*, July 15–21, 1999.
269. C. Friberg, E. Norrbin and T. Sjöstrand, Phys. Lett. **B 403** (1997) 329.
270. S. Jadach, W.Płaczek and B.F.L. Ward, Preprint UTHEP-97-0103, Univ. Tennessee (1997), hep-ph/9705395.
271. M. Heyssler and W.J. Stirling, Phys. Lett. **B 407** (1997) 259.
272. V.V. Kiselev, Phys. Rev. **D 58** (1998) 054008.
273. T.K. Kuo and T. Lee, Mod. Phys. Lett. **A 12** (1997) 2367.
274. D. Bowser-Chao et al., Phys. Lett. **B 432** (1998) 167.
275. S. Davidson, D. Bailey and B.A. Campbell, Z. Phys. **C 61** (1994) 613; M. Leurer, Phys. Rev. **D 49** (1994) 333; W. Buchmüller and D. Wyler, Phys. Lett. **177 B** (1986) 377.
276. D.E. Acosta and S.K. Blessing, Ann. Rev. Nucl. Part. Sci. **49** (1999) 389.
277. B. Dion, L. Marleau and G. Simon, Phys. Rev. **D 59** (1999) 015001.
278. M. Krämer et al., Phys. Rev. Lett. **79** (1997) 341.
279. J. Blümlein, E. Boos and A. Kryukov, Z. Phys. **C 76** (1997) 137.
280. X. Wu, for the CDF Collaboration, Leptoquarks and R-parity violating supersymmetry search at CDF, in *Proc. Int. Europhysics Conf. on High-Energy Physics (HEP97), Jerusalem, Israel*, ed. D. Lellouch, G. Mikenberg and E. Rabinovici, Springer, Berlin, Heidelberg, 1999, p. 1229, FERMILAB-CONF-97-400-E.
281. DØ Collaboration, B. Abbott et al., Phys. Rev. Lett. **80** (1998) 2051.
282. C. Grosso-Pilcher, G. Landsberg and M. Paterno, for the CDF and DØ Collaborations, Preprint FERMILAB-PUB-98-312-E, Fermilab (1998), hep-ex/9810015.
283. J.A. Valls, for the CDF and DØ Collaborations, W', Z', leptoquarks and bump hunting at the Tevatron, in *Proc. 32. Recontres de Moriond: QCD and High-Energy Hadronic Interactions (1997), Les Arcs, France*, ed. J. Tran Thanh Van, Ed. Frontieres, Paris, 1997, p. 59, FERMILAB-CONF-97-135-E.
284. M. Suzuki, Phys. Rev. **D 57** (1998) 541.
285. CDF Collaboration, F. Abe et al., Phys. Rev. Lett. **81** (1998) 4806.
286. CDF Collaboration, F. Abe et al., Phys. Rev. Lett. **78** (1997) 2906; CDF Collaboration, F. Abe et al., Phys. Rev. Lett. **82** (1999) 3206.
287. DØ Collaboration, B. Abbott et al., Phys. Rev. Lett. **83** (1999) 2896; DØ Collaboration, B. Abbott et al., Phys. Rev. Lett. **84** (2000) 2088.
288. DØ Collaboration, B. Abbott et al., Phys. Rev. Lett. **81** (1998) 38.
289. C.S. Wood et al., Science **275** (1997) 1759.
290. A. Deandrea, Phys. Lett. **B 409** (1997) 277; L. Giusti and A. Strumia, Phys. Lett. **B 410** (1997) 229; M.J. Ramsey-Musolf, Phys. Rev. **C 60** (1999) 015501.

291. R. Casalbuoni et al., Phys. Lett. **B 460** (1999) 135.
292. K.S. Babu, C. Kolda and J. March-Russel, Phys. Lett. **B 408** (1997) 261.
293. P.H. Frampton and M. Harada, Phys. Rev. **D 58** (1998) 095013.
294. E. Keith and E. Ma, Phys. Rev. Lett. **79** (1997) 4318.
295. H.P. Nilles, Phys. Rep. **110** (1984) 1.
296. B. Dutta, R.N. Mohapatra and S. Nandi, Phys. Lett. **B 412** (1997) 337.
297. J. Butterworth and H. Dreiner, Nucl. Phys. **B 397** (1993) 3.
298. A.Yu. Smirnov and F. Vissani, Phys. Lett. **B 380** (1996) 317.
299. V. Barger, G.F. Giudice and T. Han, Phys. Rev. **D 40** (1989) 2987.
300. R.M. Godbole, P. Roy and X. Tata, Nucl. Phys. **B 401** (1993) 67;
J. Erler, J.L. Feng and N. Polonsky, Phys. Rev. Lett. **78** (1997) 3063;
J. Kalinowski, Acta Phys. Pol. **B 28** (1997) 2423;
J. Kalinowski et al., Phys. Lett. **B 414** (1997) 297.
301. M. Carena et al., Phys. Rev. **D 58** (1998) 095003.
302. E. Asakawa, J. Kamoshita and A. Sugamoto, Phys. Lett. **B 451** (1999) 105.
303. A.S. Belyaev and A.V. Gladyshev, Preprint hep-ph/9704343, Moscow State Univ. and JINR, Dubna (1997);
A.S. Belyaev and A.V. Gladyshev, Preprint hep-ph/9807547, IFT, Sao Paulo and JINR, Dubna (1998).
304. M. Carena et al., Phys. Lett. **B 414** (1997) 92.
305. D. Choudhury and S. Raychaudhuri, Phys. Lett. **B 401** (1997) 54.
306. H. Dreiner and P. Morawitz, Nucl. Phys. **B 503** (1997) 55.
307. J. Ellis, S. Lola and K. Sridhar, Phys. Lett. **B 408** (1997) 252.
308. J. Ellis, S. Lola and G.G. Ross, Nucl. Phys. **B 526** (1998) 115.
309. J.E. Kim and P. Ko, Phys. Rev. **D 57** (1998) 489.
310. T. Kon and T. Kobayashi, Phys. Lett. **B 409** (1997) 265.
311. T. Kon, T. Matsushita and T. Kobayashi, Mod. Phys. Lett. **A 12** (1997) 3143.
312. B.C. Allanach, A. Dedes and H.K. Dreiner, Phys. Rev. **D 60** (1999) 075014.
313. S. Barshay and G. Kreyerhoff, Mod. Phys. Lett. **A 12** (1997) 2765.
314. G. Bhattacharyya, Phys. Rev. **D 57** (1998) 3944.
315. K. Cheung, D.A. Dicus and B. Dutta, Phys. Rev. **D 58** (1998) 057705.
316. G.F. Giudice and R. Rattazzi, Phys. Lett. **B 406** (1997) 321.
317. A.S. Joshipura, V. Ravindran and S.K. Vempati, Phys. Rev. **D 57** (1998) 5327.
318. R.N. Mohapatra, Phys. Atom. Nucl. **61** (1998) 963 [also in Yad. Fiz. **61** (1998) 1056].
319. S. Dimopoulos and L.J. Hall, Phys. Lett. **207 B** (1988) 210.
320. G. Bhattacharyya and D. Choudhury, Mod. Phys. Lett. **A 10** (1995) 1699;
G. Bhattacharyya, J. Ellis and K. Sridhar, Mod. Phys. Lett. **A 10** (1995) 1583.
321. M. Hirsch, H.V. Klapdor-Kleingrothaus and S.G. Kovalenko, Phys. Rev. **D 53** (1996) 1329.
322. A.S. Joshipura, V. Ravindran and S.K. Vempati, Phys. Lett. **B 451** (1999) 98.
323. E. Perez, Y. Sirois and H. Dreiner, R-parity violating supersymmetry at HERA, in *Proc. Workshop on Future Physics at HERA*, ed. G. Ingelman, A. De Roeck and R. Klanner, DESY, Hamburg, Germany, 1996, p. 295.
324. ALEPH Collaboration, R. Barate et al., Eur. Phys. J. **C 7** (1999) 383;
ALEPH Collaboration, R. Barate et al., Eur. Phys. J. **C 13** (2000) 29.
325. OPAL Collaboration, G. Abbiendi et al., Eur. Phys. J. **C 12** (2000) 1.
326. CDF Collaboration, F. Abe et al., Phys. Rev. Lett. **83** (1999) 2133.
327. DØ Collaboration, B. Abbott et al., Phys. Rev. Lett. **83** (1999) 4476.

328. K. Akama, K. Katsuura and H. Terazawa, Phys. Rev. **D 56** (1997) 2490.
329. V. Barger et al., Phys. Lett. **B 404** (1997) 147.
330. V. Barger et al., Phys. Rev. **D 57** (1998) 391.
331. P. Chiappetta and J.M. Virey, Phys. Lett. **B 389** (1996) 89.
332. F. Cornet and J. Rico, Phys. Lett. **B 412** (1997) 343.
333. N.G. Deshpande, B. Dutta and X.-G. He, Phys. Lett. **B 408** (1997) 288.
334. N. Di Bartolomeo and M. Fabbrichesi, Phys. Lett. **B 406** (1997) 237.
335. K. Hagiwara and S. Matsumoto, Phys. Lett. **B 424** (1998) 362.
336. W. Buchmüller and D. Wyler, Phys. Lett. **B 407** (1997) 147.
337. F. Caravaglios, Phys. Lett. **B 423** (1998) 395.
338. A.E. Nelson, Phys. Rev. Lett. **78** (1997) 4159.
339. S. Godfrey, Mod. Phys. Lett. **A 12** (1997) 1859.
340. R. Rückl, Phys. Lett. **129 B** (1983) 363.
341. J.A. Grifols, E. Massó and R. Toldrà, Phys. Rev. **D 57** (1998) 2005.
342. F. Cornet and J. Rico, Acta Phys. Pol. **B 28** (1997) 2439.
343. CDF Collaboration, F. Abe et al., Phys. Rev. Lett. **79** (1997) 2198.
344. DØ Collaboration, B. Abbott et al., Phys. Rev. Lett. **82** (1999) 4769.
345. CCFR/NuTeV Collaboration, K.S. McFarland et al., Eur. Phys. J. **C 1** (1998) 509.
346. ALEPH Collaboration, R. Barate et al., Eur. Phys. J. **C 12** (2000) 183.
347. ALEPH Collaboration, M.N. Minard et al., Preprint CERN-OPEN-99-270, CERN (1999), Abstract 694, *International Europhysics Conference on High Energy Physics, Tampere*, July 15–21, 1999.
348. DELPHI Collaboration, P. Abreu et al., Eur. Phys. J. **C 11** (1999) 383.
349. L3 Collaboration, M. Acciarri et al., Phys. Lett. **B 433** (1998) 163; D. Bourilkov, Preprint ETHZ-IPP PR-98-02, ETH Zürich (1998) hep-ex/9806027.
350. OPAL Collaboration, K. Ackerstaff et al., Eur. Phys. J. **C 2** (1998) 441; OPAL Collaboration, G. Abbiendi et al., Eur. Phys. J. **C 6** (1999) 1
351. OPAL Collaboration, G. Abbiendi et al., Eur. Phys. J. **C 13** (2000) 553.
352. M.C. Gonzalez-Garcia and S.F. Novaes, Phys. Lett. **B 407** (1997) 255.
353. A.F. Żarnecki, Eur. Phys. J. **C 11** (1999) 539.
354. S.A. Blundell, J. Sapirstein and W.R. Johnson, Phys. Rev. **D 45** (1992) 1602.
355. S.C. Bennett and C.E. Wieman, Phys. Rev. Lett. **82** (1999) 2484; errata, Phys. Rev. Lett. **82** (1999) 4153 and Phys. Rev. Lett. **83** (1999) 889.
356. V. Barger et al., Phys. Rev. **D 57** (1998) 3833.
357. J.L. Rosner, Phys. Rev. **D 61** (2000) 016006.
358. D. Zeppenfeld, K. Cheung, Preprint MADPH-98-1081, Wisconsin Univ. (1998), to appear in *Proc. 5th Int. WEIN Symposium on Physics Beyond the Standard Model (WEIN98), Santa Fe, NM, USA*, hep-ph/9810277.
359. S.L. Adler, Preprint IASSNS-HEP-97/12, IAS, Princeton (1997), hep-ph/9702378.
360. B.A. Arbuzov, HERA high Q^2 events as indications of excited leptons with weak isotopic spin 3/2, in *Proc. 5th Int. Conf. on Physics Beyond the Standard Model (1997), Balholm, Norway*, ed. G. Eigen, P. Osland and B. Stugu, American Institute of Physics, Woodbury, NY, 1997, p. 411, hep-ph/9703460.
361. G. Köpp et al., Z. Phys. **C 65** (1995) 545.
362. N. Brümmer, Search for exotics at HERA, in *Proc. 10th Rencontres de Physique de la Vallée d'Aoste: Results and Perspectives in Particle Physics, La Thuile, Italy, 1996*, ed. M. Greco, Instituto Naz. Fis. Nucl., Frascati Physics Series, Frascati, Italy, 1996, p. 553.

363. N. Arkani-Hamed, S. Dimopoulos and G. Dvali, Phys. Lett. **B 429** (1998) 263;
N. Arkani-Hamed, S. Dimopoulos and G. Dvali, Phys. Rev. **D 59** (1999) 086004.
364. J.C. Long, H.W. Chan and J.C. Price, Nucl. Phys. **B 539** (1999) 23.
365. G.F. Giudice, R. Rattazzi and J.D. Wells, Nucl. Phys. **B 544** (1999) 3.
366. L3 Collaboration, M. Acciarri et al., Phys. Lett. **B 464** (1999) 135;
L3 Collaboration, M. Acciarri et al., Phys. Lett. **B 470** (1999) 281.
367. ALEPH Collaboration, A. Litke et al., Preprint CERN-OPEN-99-269, CERN (1999), Abstract 417, *International Europhysics Conference on High Energy Physics, Tampere*, July 15–21, 1999.
368. A.R. White, Preprint ANL-HEP-PR-97-09, ANL, Argonne (1997), hep-ph/9704248.
369. M.V. Chizhov, Preprint MIH-97-04, Nordita (1997), hep-ph/9704409.
370. ZEUS Collaboration, ed. U. Holm, *The ZEUS Detector*, Status report (unpublished), DESY, 1993.
371. H1 Collaboration, I. Abt et al., Report DESY-93-103, DESY (1993).
372. H1 Collaboration, I. Abt et al., Nucl. Instrum. Methods **A 386** (1997) 310.
373. N. Harnew et al., Nucl. Instrum. Methods **A 279** (1989) 290;
ZEUS Collaboration, UK Group, C.B. Brooks et al., Nucl. Instrum. Methods **A 283** (1989) 477;
B. Foster et al., Nucl. Phys. Proc. Suppl. **B 32** (1993) 181;
ZEUS Collaboration, B. Foster et al., Nucl. Instrum. Methods **A 338** (1994) 254;
D.S. Bailey et al., Nucl. Instrum. Methods **A 396** (1997) 320.
374. C. Alvisi et al., Nucl. Instrum. Methods **A 305** (1991) 30.
375. P. Dobberstein, F. Krawczyk and M. Schäfer-Jotter, Nucl. Instrum. Methods **A 306** (1991) 474;
B. Bock et al., Nucl. Instrum. Methods **A 344** (1994) 335.
376. V. Beitzel et al., Nucl. Instrum. Methods **A 323** (1992) 135.
377. M. Derrick et al., Nucl. Instrum. Methods **A 309** (1991) 77;
ZEUS Collaboration, Calorimeter Group, A. Andresen et al., Nucl. Instrum. Methods **A 309** (1991) 101;
ZEUS Collaboration, Barrel Calorimeter Group, A. Bernstein et al., Nucl. Instrum. Methods **A 336** (1993) 23.
378. ZEUS Collaboration, FPC Group, A. Bamberger et al., Preprint DESY-99-194, DESY (1999).
379. ZEUS Collaboration, SRTD Group, A. Bamberger et al., Nucl. Instrum. Methods **A 401** (1997) 63.
380. ZEUS Collaboration, Presampler Group, A. Bamberger et al., Nucl. Instrum. Methods **A 382** (1996) 419.
381. A. Dwurazny et al., Nucl. Instrum. Methods **A 277** (1989) 176.
382. I. Kudla et al., Nucl. Instrum. Methods **A 300** (1991) 480;
ZEUS Collaboration, BAC Group, H. Abramowicz et al., Nucl. Instrum. Methods **A 313** (1992) 126;
M. Bobrowski et al., Nucl. Instrum. Methods **A 323** (1992) 309.
383. G. Abbiendi et al., Nucl. Instrum. Methods **A 333** (1993) 342;
M. De Giorgi et al., Nucl. Instrum. Methods **A 378** (1996) 472.
384. D. Kisielewska et al., Nukleonika **31** (1986) 205, Report DESY-HERA-85-25;
ZEUS Luminosity Monitor Group, J. Andruszków et al., Report DESY-92-066, DESY (1992).
385. J. Bürger et al., Nucl. Instrum. Methods **A 279** (1989) 217;
M.W. Schulz, Nucl. Instrum. Methods **A 315** (1992) 449.

386. K. Müller et al., Nucl. Instrum. Methods **A 312** (1992) 457.
387. H. Bärwolff et al., Nucl. Instrum. Methods **A 283** (1989) 467.
388. S. Egli et al., Nucl. Instrum. Methods **A 283** (1989) 487.
389. D. Pitzl et al., Nucl. Instrum. Methods **A 348** (1994) 454;
J. Bürger et al., Nucl. Instrum. Methods **A 367** (1995) 422;
W. Erdmann et al., Nucl. Instrum. Methods **A 372** (1996) 188;
W. Eick et al., Nucl. Instrum. Methods **A 386** (1997) 81.
390. G.A. Beck et al., Nucl. Instrum. Methods **A 283** (1989) 471;
J.M. Bailey et al., Nucl. Instrum. Methods **A 323** (1992) 184.
391. H. Grässler et al., Nucl. Instrum. Methods **A 323** (1992) 401.
392. S. Burke et al., Nucl. Instrum. Methods **A 373** (1996) 227.
393. S. Eichenberger et al., Nucl. Instrum. Methods **A 323** (1992) 532.
394. H1 Collaboration, Calorimeter Group, W. Braunschweig et al., Nucl. Instrum. Methods **A 275** (1989) 246;
W. Flauger, Nucl. Instrum. Methods **A 289** (1990) 446;
H1 Collaboration, Calorimeter Group, B. Andrieu et al., Nucl. Instrum. Methods **A 336** (1993) 460.
395. H1 Collaboration, Calorimeter Group, B. Andrieu et al., Nucl. Instrum. Methods **A 344** (1994) 492.
396. P. Schacht, Nucl. Instrum. Methods **A 327** (1993) 212.
397. E. Fretwurst et al., Nucl. Instrum. Methods **A 288** (1990) 1.
398. H1 Collaboration, BEMC Group, J. Bán et al., Nucl. Instrum. Methods **A 372** (1996) 399.
399. H1 Collaboration, SPACAL Group, T. Nicholls et al., Nucl. Instrum. Methods **A 374** (1996) 149;
H1 Collaboration, SPACAL Group, R.-D. Appuhn et al., Nucl. Instrum. Methods **A 382** (1996) 395;
H1 Collaboration, SPACAL Group, R.-D. Appuhn et al., Nucl. Instrum. Methods **A 386** (1997) 397;
H1 Collaboration, SPACAL Group, R.-D. Appuhn et al., Nucl. Instrum. Methods **A 404** (1998) 265.
400. H. Cronström et al., Nucl. Instrum. Methods **A 340** (1994) 304.
401. T. Ahmed et al., Nucl. Instrum. Methods **A 364** (1995) 456.
402. R. Schattevoy, *Optimierung, Validation und Analyse der ZEUS-Ereignisselektion unter besonderer Berücksichtigung der Kalorimetrie*, Ph.D. thesis, Univ. Bonn, 1994, Report BONN-IR-94-08.
403. ZEUS Collaboration, A. Bamberger et al., Nucl. Instrum. Methods **A 277** (1989) 46.
404. J. Krüger, Nucl. Instrum. Methods **A 315** (1992) 311.
405. U. Behrens et al., Nucl. Instrum. Methods **A 323** (1992) 611.
406. M. Fleischer, The performance of the H1 liquid argon calorimeter, in *Proc. 7th Int. Conf. on Calorimetry in High-Energy Physics (ICCHEP97), Tucson, AZ, USA*, ed. E. Cheu, T. Embry, J. Rutherfoord and R. Wigmans, World Scientific, Singapore, 1998, p. 327, also in report DESY-98-005A.
407. ZEUS Collaboration, Calorimeter Group, H. Tiecke et al., Nucl. Instrum. Methods **A 263** (1988) 94;
R. Klanner, Nucl. Instrum. Methods **A 265** (1988) 200;
ZEUS Collaboration, Calorimeter Group, G. D'Agostini et al., Nucl. Instrum. Methods **A 274** (1989) 134;
ZEUS Collaboration, Calorimeter Group, H. Tiecke et al., Nucl. Instrum. Methods **A 277** (1989) 42;
ZEUS Collaboration, Calorimeter Group, U. Behrens et al., Nucl. Instrum. Methods **A 289** (1990) 115;

ZEUS Collaboration, Calorimeter Group, A. Andresen et al., Nucl. Instrum. Methods **A 290** (1990) 95;
I. Ambats et al., Nucl. Instrum. Methods **A 320** (1992) 161.

408. ZEUS Collaboration, Calorimeter Group, F. Barreiro et al., Nucl. Instrum. Methods **A 292** (1990) 259.
409. H1 Collaboration, Calorimeter Group, B. Andrieu et al., Nucl. Instrum. Methods **A 336** (1993) 499;
H1 Collaboration, Calorimeter Group, B. Andrieu et al., Nucl. Instrum. Methods **A 350** (1994) 57.
410. H. Bethe and W. Heitler, Proc. Roy. Soc. Lond. **A 146** (1934) 83.
411. K.J.F. Gaemers and M. van der Horst, Nucl. Phys. **B 316** (1989) 269; erratum, Nucl. Phys. **B 336** (1990) 184;
M. van der Horst, Nucl. Phys. **B 347** (1990) 149;
M. van der Horst, Phys. Lett. **B 244** (1990) 107;
A.A. Akhundov, D.Yu. Bardin and L.V. Kalinovskaya, Z. Phys. **C 51** (1991) 557.
412. W.H. Smith et al., Nucl. Instrum. Methods **A 355** (1995) 278;
G.P. Heath et al., Nucl. Instrum. Methods **A 315** (1992) 431.
413. U. Behrens, L. Hagge and W.O. Vogel, Nucl. Instrum. Methods **A 332** (1993) 253.
414. T. Wolff et al., Nucl. Instrum. Methods **A 323** (1992) 537.
415. H. Spiesberger, HERACLES. *An Event Generator for ep Interactions at HERA Including Radiative Processes (*VERSION 4.6*)*, 1996, available on WWW: `http://www.desy.de/~hspiesb/heracles.html`.
416. G.A. Schuler and H. Spiesberger, DJANGO: The interface for the event generators HERACLES and LEPTO, in *Proc. Workshop on Physics at HERA*, ed. W. Buchmüller and G. Ingelman, DESY, Hamburg, Germany, 1991, p. 1419;
H. Spiesberger, DJANGO6 VERSION 2.4 – *A Monte Carlo Generator for Deep Inelastic Lepton Proton Scattering Including QED and QCD Radiative Effects*, 1996, available on WWW:
`http://www.desy.de/~hspiesb/django6.html`.
417. U. Baur, J.A.M. Vermaseren and D. Zeppenfeld, Nucl. Phys. **B 375** (1992) 3.
418. J.A.M. Vermaseren, Nucl. Phys. **B 229** (1983) 347;
S.P. Baranov et al., LPAIR – a generator for lepton pair production, in *Proc. Workshop on Physics at HERA*, ed. W. Buchmüller and G. Ingelman, DESY, Hamburg, Germany, 1991, p. 1478.
419. K. Rosenbauer, *Search for Leptoquarks and Leptogluons in the H1 Experiment at HERA* [in German], Ph.D. thesis, RWTH Aachen, 1995, Report PITHA 95/16.
420. S. Mrenna, Comp. Phys. Commun. **101** (1997) 232;
R. Galea and T.W. Jones, An update of the SUSANA generator and the use of SPYTHIA for R_P violation generation, in *Proc. Workshop on Monte-Carlo Generators for HERA Physics*, ed. A.T. Doyle, G. Grindhammer, G. Ingelman and H. Jung, DESY, Hamburg, Germany, 1999, p. 652, available on WWW: `http://www.desy.de/~heramc/proceedings/wg80`.
421. R. Brun et al., Report CERN-DD/EE/84-1, CERN (1987).
422. A. Kappes, *Verwendung von Neuronalen Netzen zur Identifikation des gestreuten Elektrons in ep-Ereignissen mit hohem Q^2 bei ZEUS*, Diploma thesis, Univ. Bonn, 1997, Report BONN-IB-97-28.

423. H. Abramowicz, A. Caldwell and R. Sinkus, Nucl. Instrum. Methods **A 365** (1995) 508;
R. Sinkus and T. Voss, Nucl. Instrum. Methods **A 391** (1997) 360.
424. ZEUS Collaboration, J. Breitweg et al., Phys. Lett. **B 413** (1997) 201.
425. S. Bentvelsen, J. Engelen and P. Kooijman, Reconstruction of (x, Q^2) and extraction of structure functions in neutral current scattering at HERA, in *Proc. Workshop on Physics at HERA*, ed. W. Buchmüller and G. Ingelman, DESY, Hamburg, Germany, 1991, p. 23.
426. U. Bassler and G. Bernardi, Nucl. Instrum. Methods **A 361** (1995) 197.
427. U. Bassler and G. Bernardi, Nucl. Instrum. Methods **A 426** (1999) 583.
428. A. Quadt, for the ZEUS Collaboration, High-Q^2 cross sections from ZEUS at HERA, talk at DESY seminar, March 1998, available on WWW from: `http://www-zeus.desy.de/exo/ZEUS_PUBLIC/exo_talks.html`.
429. U. Bassler and G. Bernardi, Z. Phys. **C 76** (1997) 223.
430. F. Jacquet and A. Blondel, Detection of the charged current event – method II, in *Proc. Study of an ep-Facility for Europe (1979), Hamburg, Germany*, ed. U. Amaldi, DESY, 1979, p. 393, DESY Report 79/48.
431. B. Straub, for the ZEUS and H1 Collaborations, New results on neutral and charged current scattering at high Q^2 from H1 and ZEUS, talk at *18th Lepton–Photon Symp. (LP97), Hamburg*, July 1997, available on WWW from: `http://www-zeus.desy.de/exo/ZEUS_PUBLIC/exo_talks.html`.
432. B. Straub, New results on neutral and charged current scattering at high Q^2 from H1 and ZEUS, in *Proc. 18th Int. Symp. on Lepton–Photon Interactions (LP97), Hamburg*, ed. A. De Roeck and A. Wagner, World Scientific, Singapore, 1998, p. 5.
433. G. Bernardi, for the H1 Collaboration, New H1 results at high Q^2, talk at DESY seminar, March 1998, available on WWW from: `http://www-h1.desy.de/h1/www/publications/hiq2.html`.
434. ZEUS Collaboration, Measurement of high-Q^2 neutral-current DIS cross sections at HERA, Abstract 752, *XXIX International Conference on High-Energy Physics, Vancouver*, July 23–29, 1998.
435. H1 Collaboration, Measurement of neutral and charged current cross-sections at high Q^2, Abstract 533, *XXIX International Conference on High Energy Physics, Vancouver*, July 23–29, 1998.
436. ZEUS Collaboration, Measurement of high-Q^2 charged-current DIS cross sections at HERA, Abstract 751, *XXIX International Conference on High-Energy Physics, Vancouver*, July 23–29, 1998.
437. V. Blobel, Unfolding methods in high-energy physics experiments, in *CERN School of Computing (1984), Aiguablava, Spain*, CERN, 1985, p. 88, also available as Report DESY-84-118.
438. G. D'Agostini, Nucl. Instrum. Methods **A 362** (1995) 487.
439. E. Elsen, for the H1 and ZEUS Collaborations, New high-Q^2 results in *ep* collisions at HERA, talk at *Europhysics Conf. on High-Energy Physics (HEP97), Jerusalem*, August 1997, available on WWW from: `http://www-h1.desy.de/h1/www/publications/hiq2.html`.
440. A. Quadt, private communication.
441. A. Arbuzov et al., Comp. Phys. Commun. **94** (1996) 128.
442. M.A.J. Botje, for the ZEUS Collaboration, QCD fits to ZEUS and fixed target structure function data, in *Proc. 5th Int. Workshop on Deep Inelastic Scattering and QCD (DIS97), Chicago, IL, USA*, ed. J. Repond and D. Krakauer, American Institute of Physics, Woodbury, NY, 1997, p. 375, hep-ph/9707289.

443. M.A.J. Botje, Preprint ZEUS-97-066 (unpublished), DESY (1997), see `http://www.nikhef.nl/user/h24/qcdnum`.
444. C.Y. Prescott et al., Phys. Lett. **B 77** (1978) 347; C.Y. Prescott et al., Phys. Lett. **B 84** (1979) 524.
445. D. Haidt and H. Pietschmann, *Electroweak Interactions,* Landolt–Börnstein: Numerical Data and Functional Relationships in Science and Technology, Group I: Nuclear and Particle Physics, Volume 10 (Springer, Berlin, Heidelberg, 1988).
446. M. Drees, Phys. Lett. **B 403** (1997) 353.
447. M. Botje and G. Wolf, Preprint DESY-98-140, DESY (1998), hep-ex/9809027v3.
448. Particle Data Group, R.M. Barnett et al., Phys. Rev. **D 54** (1996) 1, and 1997 off-year partial update for the 1998 edition available on the PDG WWW pages (`http://pdg.lbl.gov/`).
449. H1 Collaboration, T. Ahmed et al., Preprint DESY-94-248, DESY (1994).
450. J. Meyer, High P_{T} leptons and W production at HERA, to appear in *Proc. Ringberg Workshop: New Trends in HERA Physics 1999, Ringberg Castle, Tegernsee, Germany.*
451. N. Brümmer, for the H1 and ZEUS Collaborations, Searches for SUSY at HERA, to appear in *Proc. International Europhysics Conference on High Energy Physics (EPS HEP99), Tampere, Finland.*
452. H. Fritzsch and D. Holtmannspötter, Phys. Lett. **B 457** (1999) 186.

Index

Springer Tracts in Modern Physics

150 **QCD at HERA**
The Hadronic Final State in Deep Inelastic Scattering
By M. Kuhlen 1999. 99 figs. X, 172 pages

151 **Atomic Simulation of Electrooptic and Magnetooptic Oxide Materials**
By H. Donnerberg 1999. 45 figs. VIII, 205 pages

152 **Thermocapillary Convection in Models of Crystal Growth**
By H. Kuhlmann 1999. 101 figs. XVIII, 224 pages

153 **Neutral Kaons**
By R. Belušević 1999. 67 figs. XII, 183 pages

154 **Applied RHEED**
Reflection High-Energy Electron Diffraction During Crystal Growth
By W. Braun 1999. 150 figs. IX, 222 pages

155 **High-Temperature-Superconductor Thin Films at Microwave Frequencies**
By M. Hein 1999. 134 figs. XIV, 395 pages

156 **Growth Processes and Surface Phase Equilibria in Molecular Beam Epitaxy**
By N.N. Ledentsov 1999. 17 figs. VIII, 84 pages

157 **Deposition of Diamond-Like Superhard Materials**
By W. Kulisch 1999. 60 figs. X, 191 pages

158 **Nonlinear Optics of Random Media**
Fractal Composites and Metal-Dielectric Films
By V.M. Shalaev 2000. 51 figs. XII, 158 pages

159 **Magnetic Dichroism in Core-Level Photoemission**
By K. Starke 2000. 64 figs. X, 136 pages

160 **Physics with Tau Leptons**
By A. Stahl 2000. 236 figs. VIII, 315 pages

161 **Semiclassical Theory of Mesoscopic Quantum Systems**
By K. Richter 2000. 50 figs. IX, 221 pages

162 **Electroweak Precision Tests at LEP**
By W. Hollik and G. Duckeck 2000. 60 figs. VIII, 161 pages

163 **Symmetries in Intermediate and High Energy Physics**
Ed. by A. Faessler, T.S. Kosmas, and G.K. Leontaris 2000. 96 figs. XVI, 316 pages

164 **Pattern Formation in Granular Materials**
By G.H. Ristow 2000. 83 figs. XIII, 161 pages

165 **Path Integral Quantization and Stochastic Quantization**
By M. Masujima 2000. 0 figs. XII, 282 pages

166 **Probing the Quantum Vacuum**
Pertubative Effective Action Approach in Quantum Electrodynamics and its Application
By W. Dittrich and H. Gies 2000. 16 figs. XI, 241 pages

167 **Photoelectric Properties and Applications of Low-Mobility Semiconductors**
By R. Könenkamp 2000. 57 figs. VIII, 100 pages

168 **Deep Inelastic Positron-Proton Scattering in the High-Momentum-Transfer Regime of HERA**
By U.F. Katz 2000. 96 figs. VIII, 237 pages

Location: http://www.springer.de/phys/
You are one click away from a world of physics information!
Come and visit Springer's
Physics Online Library
Books
• Search the Springer website catalogue
• Subscribe to our free alerting service for new books
• Look through the book series profiles
You want to order?
Email to: orders@springer.de
Journals
• Get abstracts, ToC´s free of charge to everyone
• Use our powerful search engine LINK Search
• Subscribe to our free alerting service LINK Alert
• Read full-text articles (available only to subscribers of the paper version of a journal)
You want to subscribe?
Email to: subscriptions@springer.de
Electronic Media
• Get more information on our software and CD-ROMs
You have a question on an electronic product?
Email to: helpdesk-em@springer.de
Bookmark now:
http://www.springer.de/phys/
Springer · Customer Service
Haberstr. 7 · D-69126 Heidelberg, Germany
Tel: +49 6221 345 200 · Fax: +49 6221 300186
d&p · 6437a/MNT/SF · Gha.
Springer

Printing: Mercedes-Druck, Berlin
Binding: Stürtz AG, Würzburg